QUANTITATIVE X-RAY FLUORESCENCE ANALYSIS

QUANTITATIVE X-RAY FLUORESCENCE ANALYSIS

THEORY AND APPLICATION

Gerald R. Lachance
Formerly, Geological Survey of Canada

Fernand Claisse
Corporation Scientifique Claisse, Inc., Québec, Canada
Formerly Professor, Department of Mining and Metallurgy,
Université Laval, Canada

JOHN WILEY & SONS

Chichester · New York · Brisbane · Toronto · Singapore

Other Wiley Editorial Offices

John Wiley & Sons, Inc., 605 Third Avenue,
New York, NY 10158-0012, USA

Jacaranda Wiley Ltd, 33 Park Road, Milton,
Queensland 4064, Australia

John Wiley & Sons (Canada) Ltd, 22 Worcester Road,
Rexdale, Ontario M9W IL1, Canada

John Wiley & Sons (SEA) Pte Ltd, 37 Jalan Pemimpin #05-04,
Block B, Union Industrial Building, Singapore 2057

Library of Congress Cataloging-in-Publicalion Data

Lachance, G. R.
    Quantitative X-ray fluorescence analysis : theory and application
    Gerald R. Lachance, Fernand Claisse
        p.   cm.
    Includes bibliographical references and index.
    ISBN 0-471-95167-6
    1. X-ray spectroscopy.   2. X-ray spectroscopy—Industrial
applications.   I. Claisse, F. (Fernand)
II. Title.
QC482.S6L33   1994
543'.08586—dc20   94-10486
                                                         CIP

British Library Cataloguing in Publication Data

A catalogue record for this book is available from the British Library

ISBN 0 471 95167 6

Typeset in 10/12pt Times by Mathematical Composition Setters Ltd, Salisbury, Wiltshire
Printed and bound in Great Britain by Biddles Ltd, Guildford, Surrey

The authors gratefully acknowledge the contribution of

Professor **Henry Chessin**

in the elaboration of the initial chapters of this book, prior to his untimely death in July 1991. Professor Chessin made an invaluable contribution to the field of X-ray spectrometry as the devoted and untiring director of the SUNY Annual X-Ray Clinics over a period of some thirty years. The authors are therefore very pleased to dedicate this book to his memory.

CONTENTS

Preface   xv

Symbols, Definitions and Terminology   xix

PART I: THEORY   1

Chapter 1. X-RAY PHYSICS   3

  1.1   General Considerations   3
  1.2   Atomic Structure   4
  1.3   Binding Energies of Orbital Electrons   6
  1.4*   Critical Excitation Energy   8
  1.5*   Critical Excitation Wavelength   9
  1.6*   Characteristic X-radiations   9
  1.7*   Excitation Sources   11
       1.7.1 Emanations from X-ray tubes   12
       1.7.2 The continuous spectrum   12
       1.7.3 Characteristic X-radiations from the target element   15
  1.8*   Attenuation of X-rays   18
       1.8.1 Experimental determination of μ 20
  1.9*   Relation between μ, Z and λ; Absorption Edges   21
  1.10 Scattering of X-rays   24
       1.10.1 Coherent scatter   24
       1.10.2 Incoherent scatter   25
  1.11* Line Intensities; Probability of Excitation   26
       1.11.1 Absorption jump ratios   27
       1.11.2 Relative line intensities within the K, L and M series   28
       1.11.3 Fluorescence yield   30
  1.12 Discussion   31
  1.13 Summary   32

Chapter 2. X-RAY FLUORESCENCE EMISSIONS   33

2.1   General Considerations   33
2.2   Spectrometer Configurations   34
2.3*   Primary Fluorescence   37
        2.3.1   Description of primary fluorescence   37
        2.3.2   Intensity expression for primary fluorescence   39
        2.3.3   Experimental verification   44
        2.3.4   Primary fluorescence, polychromatic excitation
                source   45
2.4*   Secondary Fluorescence   45
        2.4.1   Description of secondary fluorescence   45
        2.4.2   Intensity expression for secondary fluorescence   47
        2.4.3   Experimental verification   49
        2.4.4   Secondary fluorescence, polychromatic excitation
                source   53
2.5   Tertiary Fluorescence   54
        2.5.1   Description of tertiary fluorescence   54
        2.5.2   Intensity expression for tertiary fluorescence   55
2.6   Relative Contribution from Primary, Secondary and
        Tertiary Emissions   58
2.7   Discussion   59
2.8   Summary   60

Chapter 3. THE FUNDAMENTAL PARAMETERS APPROACH   63

3.1   General Considerations   63
3.2*   Concept of the Fundamental Parameters Approach   64
        3.2.1   Replacing integrals by finite summations   64
        3.2.2   Spectral distributions   65
        3.2.3   Iteration procedure   66
3.3   Classical Formalism   67
        3.3.1   Primary emission   67
                3.3.1.1 Numerical example of the calculation of
                        $P_{(i)}$ and P_i 68
        3.3.2   Secondary emission   70
                3.3.2.1 Numerical example of the calculation of
                        $(P_i + S_{ij})_\lambda$ 70
                3.3.2.2 Experimental confirmation   74
3.4*   Alternate Formalism   76
        3.4.1   Primary fluorescence emission   77
                3.4.1.1 Numerical example   79
        3.4.2   Secondary fluorescence emission   80
                3.4.2.1 Numerical example   82

3.4.3   General expression   83
        3.4.3.1 Numerical example   83
3.5   Discussion   86
        3.5.1   Monte Carlo model   86
        3.5.2   General expression for quantitative X-ray
                fluorescence analysis   87
                3.5.2.1 Numerical examples   89
3.6*   Summary   90

Chapter 4. FUNDAMENTAL INFLUENCE COEFFICIENTS   93

4.1   General Considerations   93
4.2   The Concept of Influence Coefficients   94
4.3   The Domain of Fundamental Influence Coefficients   95
4.4   Initial Developments   95
        4.4.1   von Hamos   95
        4.4.2   Beattie and Brissey   96
        4.4.3   Lachance and Traill   97
4.5   Introducing Polychromatic Excitation   97
        4.5.1   Claisse and Quintin   97
        4.5.2   Lachance   98
4.6*   Fundamental Influence Coefficient Algorithms   99
        4.6.1   Tertian and Broll–Tertian formalisms   100
        4.6.2   Rousseau formalism   102
        4.6.3   Lachance formalism   104
        4.6.4   De Jongh formalism   105
4.7   Numerical Examples   107
        4.7.1   Effect of target material   109
        4.7.2   Effect of instrumental geometry   111
        4.7.3   Effect of specimen composition   112
        4.7.4   Effect of characteristic tube target lines   112
        4.7.5   Binary m_{ij} arrays   115
        4.7.6   Fundamental influence coefficients, multielement
                specimens   116
4.8   Discussion   120
4.9   Summary   121

Chapter 5. ALPHA COEFFICIENT CONCEPT   125

5.1   General Considerations   125
5.2   Concept and Domain of Alpha Coefficients   126
5.3   Concept of Theoretical Influence Coefficients   129
5.4*   Extension to Polychromatic Excitation   130
        5.4.1   Binary m_{ij} influence coefficient arrays   132

5.5*   Linear Approximation of Binary m_{ij} Arrays   134
       5.5.1   Numerical example   138
5.6*   Hyperbolic Approximations of Binary m_{ij} Arrays   139
       5.6.1   Numerical example   140
5.7*   Third Element Effect   143
       5.7.1   Cross-coefficient formalisms   144
       5.7.2   Tertian formalism   146
       5.7.3   Numerical examples   147
5.8   Discussion   152
       5.8.1   The equivalent wavelength concept   152
       5.8.2   Variation of α_{ij} with specimen composition   153
       5.8.3   Limited theoretical justification of alpha coefficient
               models   154
5.9   Summary   156

Chapter 6. EMPIRICAL INFLUENCE COEFFICIENTS   159

6.1   General Considerations   159
6.2   Concept and Domain of Empirical Influence Coefficients   160
6.3*   Criss–Birks and Lachance–Traill Formalisms   161
6.4*   Method of Rasberry and Heinrich   163
6.5*   Method of Lucas-Tooth and Price   166
6.6*   Method of Lucas-Tooth and Pyne   168
6.7   The Japanese Industrial Standard Method   170
6.8*   Discussion   171
6.9   Summary   173

Chapter 7. GLOBAL MATRIX EFFECT CORRECTION
           APPROACHES   175

7.1   General Considerations   175
       7.1.1   Analytical context   176
7.2*   The Double Dilution Method   177
       7.2.1   Principle of the method   177
               7.2.1.1 Numerical example   179
       7.2.2   Effect of enhancement   181
       7.2.3   Practical considerations   183
7.3*   Compton (Incoherent) Scatter Methods   183
       7.3.1   Theoretical considerations   184
               7.3.1.1 Numerical example   187
7.4   The Internal Standard Method   190
       7.4.1   Numerical example   191
7.5*   The Standard Addition Method   193
       7.5.1   Numerical example   194

7.6*   The High Dilution–Heavy Absorber Method   195
    7.6.1 Numerical example   196
7.7*   Matrix Matching Methods   199
    7.7.1 Numerical example   200
7,8   Discussion   201
7.9   Summary   202

Chapter 8. SPECIAL ANALYTICAL CONTEXTS   203

8.1   General Considerations   203
8.2*   Low Z Analytes and Matrices   204
    8.2.1 Low Z analytes   204
    8.2.2 Low Z matrices   205
8.3   Limited Sample Quantity   208
8.4*   Thin Specimens   209
8.5*   Thin Films   211
    8.5.1 Determination of film thickness   211
        8.5.1.1 Numerical examples   212
    8.5.2 Multi-layer thin films   215
8.6   Particulates   216
8.7   Oxide Systems   216
    8.7.1 Numerical example   217
8.8*   Fused Disc Specimens   219
    8.8.1 Influence coefficients modified for flux   219
        8.8.1.1 Numerical example   221
    8.8.2 Correction for loss on fusion (LOF)   224
        8.8.2.1 Numerical example   225
    8.8.3 Correction for gain on fusion (GOF)   227
        8.8.3.1 Numerical example   229
    8.8.4 General theory of LOF and GOF   231
        8.8.4.1 Numerical example   233
8.9   Liquids   236
8.10   Conclusions   237

Chapter 9. ANALYTICAL STRATEGIES   239

9.1   General Considerations   239
9.2   Analytical Context   241
    9.2.1 Nature of samples submitted   242
    9.2.2 Instrumental resources   242
    9.2.3 Sample composition   243
    9.2.4 Precision and accuracy   243
9.3*   Specimen Preparation   245
    9.3.1 Bulk solids—alloys, glasses, plastics, ceramics   245

9.3.2   Pressed powder pellets   246
9.3.3   Fused discs   247
9.3.4   Liquids   250
9.4* Intensity Measurements   251
9.4.1   Precision, counting statistics   251
9.4.2   Excitation source selection   252
9.4.3   Line selection   253
9.4.4   Collimation   253
9.4.5   Crystal selection   254
9.4.6   Detector selection   254
9.4.7   Correction for background   255
9.4.8   Correction for spectral interference   256
9.4.9   Correction for dead time   258
9.5* Calibration   259
9.5.1   Definition   260
9.5.2   Mathematical models   261
9.6* The Analytical Process   266
9.6.1   Confirmation of the validity of a calibration   266
9.6.2   Counting monitors   267
9.6.2.1 Numerical example   268
9.6.3   Quality control   270
9.6.4   Analytical precision and the detection limit   271
9.6.5   Software   273
9.7   Summary   275

PART II: APPLICATION   279

II.1 X-RAY PHYSICS   281

II.1.4   Critical Excitation (Binding) Energies   281
II.1.5   Critical Excitation (Absorption Edge) Wavelengths   282
II.1.6   Energies and Wavelengths of Characteristic X-radiations   283
II.1.7.1   Experimentally measured spectral distributions   284
II.1.8.1   Experimental determination of μ 285
II.1.8.2   Algorithms for generating μ values   286
II.1.9   Generating μ as a Function of Z and λ 290
II.1.11.1 Absorption jump ratios; absorption jump factors   294
II.1.11.2 Relative line intensities within series   295
II.1.11.3 Fluorescence yields   296
I.1.12   Summary and Conclusions   297

II.2 X-RAY FLUORESCENCE EMISSIONS   299

II.2.1   Two-theta Values for Characteristic Lines   299

II.2.2   Excitation Probability Factor   299
II.2.3   Experimental Verification of Primary Emission   301
II.2.4   Experimental Verification of Secondary Emission   302
II.2.5   Summary   305

II.3 THE FUNDAMENTAL PARAMETERS APPROACH   307

II.3.2   Concept of the Fundamental Parameters Approach   307
II.3.2.3 The iterative process   311
II.3.4   Comprehensive Tabulation of $P_i + S_i$, A_{ij} and E_{ij} Data   313
II.3.6   Summary and Conclusions   322

II.4 FUNDAMENTAL INFLUENCE COEFFICIENTS   323

II.4.6.1 Broll–Tertian formalism   323
II.4.6.2 Rousseau formalism   324
II.4.6.3 Lachance formalism   325
II.4.6.4 The de Jongh formalism   327
II.4.6.5 Typical analytical procedure   330
II.4.6.6 Interrelation between fundamental influence coefficients   333
II.4.6.7 Summary and conclusions   336

II.5 ALPHA INFLUENCE COEFFICIENTS   337

II.5.4   The Lachance–Traill Algorithm   338
II.5.5   The Modified Claisse–Quintin Algorithm   340
II.5.6
II.5.7.1 The COLA Algorithm   344
II.5.6
II.5.7.2 The Tertian Algorithm   348

II.6 EMPIRICAL INFLUENCE COEFFICIENTS   351

II.6.3   The Lachance–Traill Model: Eqn (6.7)   351
II.6.3.1 Numerical example   354
II.6.4   The Rasberry–Heinrich Model: Eqn (6.11)   355
II.6.5   The Lucas-Tooth and Price Model: Eqn (6.14)   357
II.6.6   Lucas-Tooth and Pyne Model: Eqn (6.17)   358
II.6.8   Discussion   358

II.7 GLOBAL MATRIX EFFECT CORRECTION METHODS   361

II.7.2   Double Dilution Method: Eqn (7.20)   361
II.7.3   Compton (Incoherent) Scatter Methods: Eqn (7.30)   362
II.7.5   Standard Addition Method   364

II.7.6   High Dilution–Heavy Absorber Method   365
II.7.7   Matrix Matching Methods   365

II.8 SPECIAL ANALYTICAL CONTEXTS   367

II.8.2   Low Z Analytes and Matrices   367
II.8.4   Thin Specimens   368
II.8.5   Thin Films, Film Thickness   369
II.8.8   Fused Disc Specimens   370

II.9 ANALYTICAL STRATEGIES   371

II.9.1   Steels, Nickel-base, Copper-base, Broad Mixtures   372
II.9.2   Cements and Raw Materials in the Cement Industry   374
II.9.3   Geological Contexts   378
II.9.4   Discussion and Summary   383

References and Bibliography   387

Author Index   395

Subject Index   398

PREFACE

Improvements in the sensitivity and stability of spectrometers, coupled with phenomenal advances in the speed and data storage capabilities of computers, have gradually expanded the field of X-ray fluorescence analysis. Not only are analysts expected to determine lower concentrations for all but a few of the lowest atomic number elements, but also to provide accurate data for ever increasing types of materials at high productivity levels. This is generally achieved by maximizing automation at every step: programmable specimen preparation apparatus, computer controlled setting of all preselected instrumental parameters for intensity measurements while verifying spectrometer stability, and processing of the data on-line by applying corrections that compensate for absorption and enhancement.

In broad terms, quantitative X-ray fluorescence analysis evolved from specificity to universality. Initially, methods were limited to the determination of few elements in well defined concentration ranges by statistical treatment of experimental data from reference materials (linear or second order curves), or by compensation methods (dilution, internal standards, Compton scatter, etc.). The reason for this was not so much that the theory of X-ray fluorescence emission was not known, but that the computational labour was too onerous in practice, given the computational facilities available at that time. To expand concentration ranges, semi-empirical influence coefficient methods were introduced, generally by considering absorption effects only for monochromatic excitation in two or three element systems. Universality came about by the development of fundamental parameters approaches for the correction of total matrix effects (absorption and enhancement) in either classical or fundamental influence coefficient formalisms. In current practice, fundamental parameters and semi-empirical methods are frequently combined for greater versatility.

Due to space constraints, it is not possible to treat in equal detail all aspects of X-ray fluorescence spectrometry. Specimen preparation and intensity measurement procedures, which are quite specific to analytical contexts: material analysed, spectral deconvolution, productivity and accuracy expected, etc., are treated in general terms only in this book, it being considered that these topics are adequately covered in the current literature. From a theoretical point of view, at the core of calibration and analysis is the need to describe and define the absorption, emission and enhancement processes within the irradiated material. This consists in deriving mathematical expressions quantifying absorption along the incident path, emission of characteristic spectra including enhancement if present, and absorption along the emergent path. These processes are a common feature of wavelength and energy dispersion spectrometry, and are therefore treated in greater detail.

The authors opted to divide the text into two parts. The first eight chapters of Part I cover in turn: the basic principles of X-ray physics pertaining to spectral emission; spectrometer configuration and the quantification of X-ray fluorescence emissions, absorption and enhancement; the fundamental parameters approach as the cornerstone for theoretical correction for matrix effects; fundamental, semi-empirical and empirical coefficients methods that quantify matrix effects as a sum of individual effects; global methods in the sense that matrix effects are evaluated as an undivided whole; special situations in the sense that additional steps or considerations have to be taken into account. Simpler analytical contexts are generally chosen to illustrate specific processes along with detailed numerical examples. Comprehensive tabulation of theoretical data enables readers to carry out parallel calculations and verify the results obtained. The last chapter of Part I is a synthesis of the previous eight in that it examines analytical strategies, namely guidelines concerning the interrelation between specimen preparation, intensity measurements, calibration and analysis of samples. Some readers may very well conclude that attempts to produce more, better and faster is somewhat akin to doing a balancing act on a high wire.

If Part I is visualized as dealing with the derivation of mathematical expressions for 'what' processes need quantification in X-ray fluorescence spectrometry, Part II is then concerned with 'how' these expressions are used in practice. For example, algorithms that generate fundamental parameters are used in turn to generate theoretical intensity data, which are used in turn to generate influence coefficients that are then applied to correct experimental data for matrix effects. To this end, many numerical examples are given. Whenever possible, identical data are treated sequentially by various algorithms in current use in order to bring out similarities and differences. Also included is a comprehensive tabulation of data relating to analytical strategies, namely a cross-section of the options exercised by analysts pertaining to specimen preparation, intensity measurements, calibration and analysis when dealing with similar analytical contexts.

We wish to express our thanks to all authors whose works are referenced and our regret that due to space considerations, many valuable contributions to the field of X-ray fluorescence spectrometry could not be acknowledged individually.

Nepean, Ontario, Canada and   Gerald R. Lachance
Sainte-Foy, Quebec, Canada   Fernand Claisse
May 1994

SYMBOLS, DEFINITIONS
AND TERMINOLOGY

SYMBOLS

General

A	ampere, unit of electric current
A	atomic weight
Å	Angstrom, unit of wavelength, $1\ \text{Å} = 10^{-8}\ \text{cm}$
C	concentration, weight fraction; Eqn (1.27)
E	energy; Eqns (1.1) and (1.4)
I	intensity, measured experimentally; Eqn (2.43)
N	total number of counts measured; Eqn (9.6)
V	volt
Z	atomic number
c	speed of light
d	interplanar spacing of crystal; Eqn (2.2)
e	charge on the electron
eV	electron volt, unit of energy (keV = kiloelectron volt)
h	Planck's constant
i	electric current; Eqn (1.9)
n	diffraction order; Eqn (2.2)
r	absorption jump ratio; Eqn (1.43)
t	thickness

Greek letters

θ	Bragg angle; Eqn (2.2)
λ	wavelength, unit Å
λ_{min}	short wavelength limit of the continuum; Eqn (1.8)
λ_{abs}	wavelength of an absorption edge; Section 1.9
λ_{eff}	effective wavelength, all incident wavelengths that can lead to the emission of a given characteristic line i.e. all wavelengths in the range $\lambda_{min} - \lambda_{abs}$

λ_{eqv} equivalent wavelength, single wavelength that is representative of the effective spectral distribution; Eqn (5.11)

μ mass attenuation coefficient, unit $cm^2\,g^{-1}$; Eqn (1.25)

μ_l linear absorption coefficient, unit cm^{-1}; Eqn (1.23)

μ_a atomic absorption coefficient, unit $cm^2/atom$

ν frequency; Eqn (1.1)

ρ density, unit $g\,cm^{-3}$

σ mass scattering coefficient; Eqns (1.29) and (1.30)

τ mass photoelectric absorption coefficient; Eqn (1.38)

ψ' incidence angle of excitation source measured from the specimen surface; Figure 2.2

ψ'' take-off angle of fluorescent radiation measured from the specimen surface; Figure 2.2

ω fluorescence yield; Eqn (1.46)

Subscripts

app   apparent, e.g. $C_{i(app)}$, $C_{i,app}$ refer to the apparent concentration (uncorrected for matrix effects) of element i; Eqns (6.2) and (II.6.7)

b, bkg   background, e.g., I_b, I_{bkg}

d   diluent, e.g., C_d; Eqn (7.6)

f   flux, e.g., C_f; Eqn (8.36)

i   element (compound), usually the analyte, e.g., C_i, I_i

(i)   pure element (compound), e.g., $C_{(i)}$, $I_{(i)}$

$j, k, \ldots, n$ matrix elements (compounds), e.g., C_j, I_j

r   reference material of known composition used for calibration, e.g. C_{ir}, I_{ir} refer to concentration and emitted intensity of element i in a reference material, respectively; Eqn (3.47)

s   refers to any specimen, whether reference or unknown

u   specimen analysed, i.e. composition unknown, e.g. C_{iu} I_{iu}; Eqn (3.46)

v   volatile, element or compound loss on ignition or during fusion

Z   atomic number, e.g., $\mu_{Z\lambda}$ refers to the mass attenuation coefficient of elements of atomic number Z for a given wavelength λ

DEFINITIONS

Fluorescence emission

p_{λ_i} probability that a characteristic line of element i will be emitted; Eqn (2.6)

G_i, g_i proportionality constants, usually cancelled out; Eqns (2.23) and (2.1)

$P_{i\lambda}$ theoretical primary fluorescence emission for monochromatic excitation λ; Eqns (2.24) and (2.27)

P_i theoretical primary fluorescence emission, polychromatic excitation; Eqns (2.29), (3.2) and (3.30)

$P_{(i)}$ theoretical fluorescence emission for pure element i, polychromatic excitation

$S_{ij\lambda}$ theoretical secondary fluorescence emission (enhancement) of element i from element j, monochromatic excitation λ; Eqns (2.35) and (2.40)

S_{ij} theoretical secondary fluorescence emission (enhancement) of element i from element j, polychromatic excitation; Eqn (2.34)

S_i theoretical secondary fluorescence emission of element i, polychromatic excitation; Eqn (2.32)

T_i theoretical tertiary fluorescence emission; Eqn (2.62)

T_{ijk} theoretical tertiary fluorescence emission of element i, wherein element k enhances element j, which in turn enhances element i; Eqns (2.56)–(2.60)

I_i net intensity emitted by a characteristic line of element i, i.e. corrected for background, spectral overlap, dead time, drift etc.

$I_{(i)}$ calculated experimental intensity for pure element i; Eqns (5.10) and (II.4.24)

K_i calibration constant; Eqns (3.48)

$K_{(i)}$ calibration constant (sensitivity) equal to concentration divided by intensity corrected for matrix effects for any given reference material: C is known by definition, I is measured experimentally and the correction term is calculated from fundamental theory

R_i relative intensity of element i, intensity relative to that emitted by pure element i; Eqns (3.4), (3.5) and (4.34)

Mass attenuation coefficient (MAC)

μ_i' effective MAC of element i for incident wavelength λ

μ_j' effective MAC of element j for incident wavelength λ

μ_s' effective MAC of specimen s for incident wavelength λ; Eqn (2.18)

μ_i'' effective MAC of element i for emergent wavelength λ_i

μ_j'' effective MAC of element j for emergent wavelength λ_i

μ_s'' effective MAC of specimen s for emergent wavelength λ_i; Eqn (2.19)

μ_i^* total effective MAC of element i (incident wavelength λ); Eqn (2.26)

μ_j^* total effective MAC of element j (incident wavelength λ); Eqn (3.19)

μ_s^* total effective MAC of specimen s (incident wavelength λ); Eqns (2.20), (3.3) and (3.20)

Effective influence coefficients defined explicitly from fundamental theory, multielement specimens

$A_{ij\lambda}$ absorption coefficient, monochromatic excitation λ; Eqn (3.25)

$E_{ij\lambda}$ enhancement coefficient, monochromatic excitation λ; Eqn (3.37)

A_{ij} absorption coefficient, polychromatic excitation; Eqn (3.32)

E_{ij} enhancement coefficient, polychromatic excitation; Eqn (3.44)

M_{ij} matrix coefficient combining absorption and enhancement, polychromatic excitation; Eqns (4.60) and (II.4.7)

Fundamental influence coefficients (FIC) defined explicitly from fundamental theory, multielement specimens

$a_{ij\lambda}$ FIC, absorption effect, monochromatic excitation λ; Eqn (4.14)

$e_{ij\lambda}$ FIC, enhancement effect, monochromatic excitation λ; Eqns (2.36)–(2.39), (3.35)

a_{ij} FIC, absorption effect, polychromatic excitation; Eqn (4.54)

a_{ijp} FIC, absorption effect, defined as a function of polychromatic primary emission; Eqn (4.30)

e_{ij} FIC, enhancement effect, polychromatic excitation; Eqn (4.55)

e_{ijp} FIC, enhancement effect, defined as a function of polychromatic primary emission; Eqn (4.35)

m_{ij} FIC, matrix effect combining absorption and enhancement, polychromatic excitation; Eqn (4.56)

Semi-empirical influence coefficients

α_{ij} coefficient computed from theory but involving a degree of approximation; Eqns (5.1), (5.16), (5.29) and (5.49)

Empirical influence coefficients

r_{ij} coefficients computed by regression of C versus I data; Eqns (6.7), (6.11), (6.14), (6.17) and (6.20)

PART I
THEORY

Part I aims to examine the theoretical principles that underlie the X-ray fluorescence emission process and, more specifically, the mathematical expressions that have been derived in the search for quantitative relationships between intensity and concentration. Because of absorption and enhancement effects, X-ray fluorescence intensities are very rarely directly proportional to concentration. It therefore follows that effective corrections for matrix effects must be applied so that the inherent high precision of experimental intensities will lead to accurate compositional data. Sections in Part I having a component in Part II are designated by an asterisk (*).

1 X-RAY PHYSICS

This introductory chapter aims at identifying, describing and quantifying the basic concepts and processes of X-ray physics that underlie X-ray fluorescence spectrometry. It sets the stage for the derivation of fundamental expressions relating the emitted intensity to the concentration of the element in the specimen whose characteristic X-radiations are measured. Most topics, e.g. atomic structure, electronic transitions, absorption and scatter of X-rays etc., are covered only to the extent that they relate to conventional X-ray fluorescence analysis. Consider the case of the atom, for example. Spruch and Spruch (1974) devote a 450 page textbook exclusively to the atom, while Moeller (1952) devotes a chapter to atomic nuclei and one to extranuclear structures of the atom. In the following presentation, atomic structure will be limited to one section. For a more comprehensive treatment of any one particular topic, the reader should consult the original publications.

1.1 GENERAL CONSIDERATIONS

X-radiations form part of the electromagnetic wave spectrum which includes radiowaves, microwaves, infrared and visible radiation, gamma rays etc. The waves differ in amplitude and wavelength, but all electromagnetic radiations travel through space at the speed of 3.00×10^8 metres per second (3.00×10^{10} cm s^{-1}). Some of the physical quantities and units that are of prime concern to X-ray fluorescence are:

Planck's constant h, a universal proportionality constant that has the value 6.625×10^{-34} J s^{-1}.

Velocity of light c, which has the value 3.0×10^{10} cm s^{-1}.

Ångstrom Å, a unit of length. 1 ångström $= 10^{-8}$ cm or 10^{-10} m.

Volt V, a measure of electric potential difference. A potential of one volt exists between two points if 1 joule of work is done in moving 1 coulomb of charge. $1 \text{ V} = 1 \text{ J C}^{-1}$.

Joule J, a unit of work or energy. $1 \text{ J} = 1 \text{ A} \times 1 \text{ V} \times 1 \text{ s}$.

Electron charge e, the natural unit of electricity that represents the charge associated with a single electron. It is equal to 1.602×10^{-19} C.

Coulomb, a convenient unit of electric charge that corresponds to 1 ampere of current flowing for 1 second. 1 coulomb is equal to 6.24×10^{18} electron charges.

Electron volt eV, a unit of energy acquired by an electron when it is accelerated by a potential of 1 volt. It is equivalent to 1.602×10^{-19} J.

Photon, a unit of electromagnetic radiation having a definite amount of energy, a quantum, that depends on the wavelength. Radiation is absorbed and emitted in whole quanta; no fractional parts of a quantum are allowed.

Energy, symbol E, generally expressed in eV and keV units.

Intensity I, the detected photon counting rate generally expressed in units of 'counts per second'.

X-ray fluorescence spectrometry is one of many techniques used for the determination of elemental concentrations. It falls under the general category of spectrometric techniques because elements are identified and their concentration is determined by generating spectra from specimens submitted for analysis. More specifically, X-ray fluorescence is classified under atomic spectrochemical techniques because the characteristic X-radiations are generated in atoms, i.e. at the atomic level rather than at the molecular level.

1.2 ATOMIC STRUCTURE

Since the processes leading to the emission of characteristic X-radiations occur at the atomic level, it is necessary for X-ray analysts to have a general knowledge of atomic structure. Although Bohr's original model of the atom, in which orbital electrons revolve in fixed circular orbits around a dense nucleus (Figure 1.1), was subsequently modified and ultimately abandoned, it can still be used as the basis for demonstrating schematically the origin of X-ray spectra. According to the classical theory, the Rutherford model in which negatively charged electrons revolve around a positively charged central nucleus should be unstable, as the electrons would spiral towards the nucleus

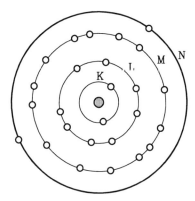

Figure 1.1   Schematic diagram of the iron atom

within a fraction of a second. Based on studies of optical hydrogen spectra, Bohr retained the notion of circular orbits but postulated that:

- from all the possible orbits, only a few highly restricted orbits are permitted;
- electrons circulating along these orbits are prohibited from emitting any electromagnetic waves;
- electrons may jump from a higher orbit to a lower one, in which case electromagnetic waves (photons) are emitted equal to the energy difference between the two orbits.

The visualization of electrons in orbits of limited and well defined shapes can only be regarded as a rough pictorial view. It became evident as a result of in-depth studies of optical and X-ray spectra that Bohr's model could not adequately describe the property of an electron in terms of the size of the orbit only. Modern concepts view electrons as having a high probability of occupying well defined energy regions or levels governed by quantum theory. These levels, sometimes referred to as shells, are designated by the letters K, L, M, N, etc. as they become farther and farther removed from the nucleus (Figure 1.1). Subsequently, the various energy levels were found to be dependent on four quantum numbers, namely:

- n the principal quantum number, a positive integer, where n = 1 designates the K shell electrons, n = 2 designates the L shell electrons, etc.;
- l the angular quantum number introduced by Sommerfeld to accommodate elliptical orbits; it can take all integer values between 0 and n − 1;
- m the magnetic quantum number introduced to account for the behaviour of spectra in magnetic fields; m can take the integer values −1, 0 and +1 only;

Table 1.1   Quantum numbers of K and L energy level electrons

	K level			L level						
n	1	1	2	2	2	2	2	2	2	2
l	0	0	0	0	1	1	1	1	1	1
m	0	0	0	0	-1	-1	0	0	1	1
s	$+\frac{1}{2}$	$-\frac{1}{2}$	$+\frac{1}{2}$	$-\frac{1}{2}$	$+\frac{1}{2}$	$-\frac{1}{2}$	$+\frac{1}{2}$	$-\frac{1}{2}$	$+\frac{1}{2}$	$-\frac{1}{2}$

- s the spin quantum number introduced to explain the close grouping of some spectral lines. Since only two directions of electron spin are possible, the spin quantum number can have only two values, $+\frac{1}{2}$ and $-\frac{1}{2}$, to designate opposite directions of spin.

Given the rule that it is impossible for any two electrons in the same atom to have their four quantum numbers identical, it follows that, in the case of n = 1, the maximum number of electrons is two. Thus, the K level may contain one electron in the case of hydrogen, or two electrons in the case of all the other elements. For the L level, n = 2, and therefore the number of possible combinations, or the maximum number of L level electrons, is eight. The possible combinations of the four quantum numbers for the K and L levels are shown in Table 1.1. Proceeding in a similar manner, it can be deduced that the maximum number of electrons in the next two shells is 18 for the M level and 32 for the N level. Electrons also occupy O and P shells, but X-ray fluorescence is generally not concerned with these because the inner levels are those mainly involved in the generation of the characteristic X-radiations used in quantitative X-ray fluorescence analysis.

1.3   BINDING ENERGIES OF ORBITAL ELECTRONS

To a first approximation, the energy bonding the electrons to the nucleus is determined by the atom's configuration. Thus, the electrons within a level do not all have the same energy. Subsequent more detailed spectroscopic studies indicate that the electrons in the L, M and N levels could be regarded as occupying sublevels. This leads to the concept that the energy levels within a level are dependent not only on the quantum numbers n and l but also on another number j, the vector sum of l and s; i.e. j = l + s, with the restriction that j cannot be negative. Thus, if l = 0, j can only take the value $+\frac{1}{2}$; if l = 1, j can take on the values $+\frac{1}{2}$, $+\frac{3}{2}$, etc. As a result, the K level electrons of an element are bound to the nucleus by energy levels designated by E_K. The L shell is regarded as having three energy sublevels designated by E_{L_1}, E_{L_2}, and E_{L_3}. By the same token, the M and N shells consist of five and seven sublevels designated by E_{M_1}, E_{M_2}, ..., E_{M_5} and E_{N_1}, E_{N_2}, ..., E_{N_7}, respectively. Lastly, the number of electrons that sublevels may accommodate is given by 2j + 1.

The configurations of the energy levels for the first four shells are listed in Table 1.2 along with values of the binding energies for the element tin.

The above notation, namely that of identifying the different energy levels and sublevels as K, L_1, L_2 etc., is referred to as 'X-ray notation'. However, a different notation was developed in parallel, based on studies of optical spectroscopy. In this case the letters s, p, d and f stood for descriptions of spectral lines, i.e. sharp, principal, diffuse and fundamental, and were used to designate the shells and subshells. This terminology is no longer relevant but is still found in the literature. The letters are used as follows:

- the inner shell is designated as $1s^2$;
- the electrons in the next outer shell are designated as belonging to two groups, $2s^2$ and $2p^6$;
- the electrons in the next outer shell are designated as belonging to three groups; $3s^2$, $3p^6$ and $3d^{10}$;
- the electrons in the fourth outer shell are designated as belonging to four groups; $4s^2$, $4p^6$, $4d^{10}$ and $4f^{14}$,

where the numbers in superscript give the maximum number of electrons in the individual groups. Table 1.3 shows the orbital electron configuration for the iron atom in both notations.

Table 1.2 Energy levels and E_{keV} values for Sn ($Z = 50$)

Level	Sublevel	n	l	j	Max. number of electrons	E_{keV}[a]
K		1	0	1/2	2	29.200
L	L_1	2	0	1/2	2	4.465
	L_2	2	1	1/2	2	4.156
	L_3	2	1	3/2	4	3.929
M	M_1	3	0	1/2	2	0.884
	M_2	3	1	1/2	2	0.756
	M_3	3	1	3/2	4	0.714
	M_4	3	2	3/2	4	0.493
	M_5	3	2	5/2	6	0.485
N	N_1	4	0	1/2	2	0.137
	N_2	4	1	1/2	2	0.089
	N_3	4	1	3/2	4	0.089
	N_4	4	2	3/2	4	0.024
	N_5	4	2	5/2	6	0.024
	N_6	4	3	5/2	6	
	N_7	4	3	7/2	8	

[a] E_{keV} data (rounded off) from Bearden and Burr (1967).

Table 1.3 Detailed electron configuration of the K, L, M and N energy levels for iron ($Z = 26$), shown in both electronic and optical configurations[a]

1	1		2				3			4	5
2	K		L				M			N	O
3	2		8				18			32	50
4	K	L_1	L_2	L_3	M_1	M_2	M_3	M_4	M_5	$N_1 \ldots$	
5	2	2	2	4	2	2	4	4	2	2	
6	s	s	p		s	p		d		s	
7	2	2	6		2	6		10		2	
8	$1s^2$	$2s^2$	$2p^6$		$3s^2$	$3p^6$		$3d^6$		$4s^2$	

[a] Row 1: value of n, the principal quantum number; 2: energy level designation; 3: maximum number of electrons permitted; max. = $2n^2$; 4: sublevel designation, X-ray notation; 5: electronic configuration of Fe atom ($Z = 26$); 6: energy level designation, optical notation; 7: maximum number of electrons permitted; 8: optical configuration of Fe atom.

1.4* CRITICAL EXCITATION ENERGY

Recalling Bohr's postulate that radiations are emitted only as a result of transfer of electrons between energy levels, it is therefore necessary to eject an electron from an atom in order to initiate the process for the emission of a radiation that is characteristic of that atom. For example, in X-ray fluorescence spectrometry, the ejection of K level electrons from atoms to generate the K spectral series requires that the incident photons overcome the energy binding those electrons to the nucleus; i.e., the incident photons must have energies equal to or greater than E_K. Similarly, the ejection of L level electrons to generate the L spectral series requires that the energy of the incident photon be equal or greater than E_{L_1}, E_{L_2}, or E_{L_3}. Thus, binding energies are also referred to as critical excitation energies, since they set the minimal energy that must be equalled or exceeded in order to eject electrons

Table 1.4 Critical excitation (binding) energies (keV) and their corresponding wavelengths for the elements silicon, chromium, zirconium and lead[a]

		K	L_1	L_2	L_3	M_1	M_2	M_3
Si	E (keV)	1.839	0.149	0.099	0.099			
	λ (Å)	6.742	83.2	123				
Cr	E (keV)	5.989	0.695	0.584	0.575	0.074	0.043	0.043
	λ (Å)	2.070	17.839	21.230	21.562	168		
Zr	E (keV)	17.998	2.532	2.307	2.222	0.430	0.344	0.331
	λ (Å)	0.689	4.897	5.374	5.580	28.83		
Pb	E (keV)	88.005	15.861	15.200	13.035	3.851	3.554	3.066
	λ (Å)	0.141	0.782	0.816	0.951	3.219	3.489	4.044

[a] Data (rounded off) from Bearden and Burr (1967).

from atoms, leaving the latter in an unstable or ionized state necessary to initiate the process that leads to the emission of characteristic X-rays. Critical excitation energies for the K, L and M energy levels of Si, Cr, Zr and Pb are listed in Table 1.4.

1.5* CRITICAL EXCITATION WAVELENGTH

Consider the two simple but fundamental expressions

$$\nu = \frac{c}{\lambda} \quad \text{and} \quad E = h\nu \tag{1.1}$$

where ν is the frequency of the wave motion in vibrations per second; c is the speed of light, 3.0×10^{10} cm s^{-1}; λ is the wavelength in cm; E is the energy in J; h is Planck's constant, 6.625×10^{-34} J s^{-1}. The relation between energy and wavelength may then be obtained by combining the two above equations, giving

$$E = \frac{hc}{\lambda} \quad \text{or} \quad \lambda = \frac{hc}{E} \tag{1.2}$$

In X-ray fluorescence, it is useful to retain the above expressions with E in eV or keV and wavelength in Ångström units. Substituting the values of h and c and inserting the necessary conversion factors, i.e. 1 cm $= 10^8$ Å and 1 eV $= 1 \cdot 602 \times 10^{-19}$ J, yields

$$\lambda = \frac{6.625 \times 10^{-34} \times 3.0 \times 10^{10} \times 10^8}{\text{eV} \times 1.602 \times 10^{-19}}$$

$$= \frac{12\,398.1}{\text{eV}}$$

and, with E expressed in keV units

$$\lambda = \frac{12.3981}{E} \tag{1.3}$$

$$E = \frac{12.3981}{\lambda} \tag{1.4}$$

Table 1.4 lists the critical excitation energies and critical excitation wavelengths for the elements Si, C, Zr and Pb.

1.6* CHARACTERISTIC X-RADIATIONS

If a specimen is irradiated with a beam of photons of energy E_0, where E_0 exceeds the critical excitation energy of electrons in a given atom, some electrons are ejected from that atom. We refer to such an atom as being in an excited or ionized state. The same process may be studied in terms of wavelength; namely, if a specimen is irradiated with a beam of radiations of wavelength λ shorter than a given critical excitation wavelength, electrons are ejected and the atom is in an ionized state. Atoms in an ionized state are unstable and, quasi-instantly, a process of transition of electrons takes place in order to fill the vacancies left by the ejected electrons. Thus, if a K level electron is ejected, the vacancy may be filled by a L, M, . . . level electron. Similarly, L level vacancies may be filled by M, N, . . . level electrons. Each transition or transfer constitutes an energy loss resulting in the emission of an X-ray photon, its energy being equal to the difference between the two energy levels involved. For example, the transition of an L_3 level electron to the K level results in the emission of a photon labelled $K-L_3$ (K_{α_1} in Siegbahn notation). This is expressed as

$$E_{K-L_3} = E_K - E_{L_3} \tag{1.5}$$

and is referred to as a $K-L_3$ transition. If the transition is from the L_2 to the K level, the emitted photon is labelled $K-L_2$ (K_{α_2} in Siegbahn notation), i.e.

$$E_{K-L_2} = E_K - E_{L_2} \tag{1.6}$$

It is also possible for a K level vacancy to be filled by M level electrons. If the transition is from the M_3 level, the photon is labelled $K-M_3$ (K_{β_1} in Siegbahn notation), thus

$$E_{K-M_3} = E_K - E_{M_3} \tag{1.7}$$

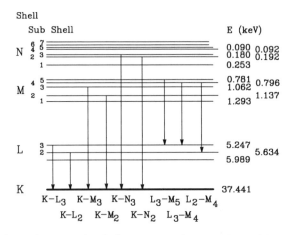

Figure 1.2   Schematic energy level diagram (partial) showing origin of characteristic X-ray spectra. Energy data relate to the element barium ($Z = 56$)

Table 1.5 K and L (partial) spectra of barium

Line	Binding energies	$E_{line,keV}$[a]	λ_{line} (Å)
$K-L_2$	$37.441 - 5.634$	31.807	0.390
$K-L_3$	$37.441 - 5.247$	32.194	0.385
$K-M_2$	$37.441 - 1.137$	36.304	0.342
$K-M_3$	$37.441 - 1.062$	36.379	0.341
$K-N_2$	$37.441 - 0.192$	37.249	0.333
$K-N_3$	$37.441 - 0.180$	37.261	0.333
L_3-M_1	$5.247 - 1.293$	3.954	3.136
L_3-M_4	$5.247 - 0.796$	4.451	2.786
L_3-M_5	$5.247 - 0.781$	4.466	2.776
L_2-M_4	$5.634 - 0.796$	4.838	2.568
L_1-M_2	$5.989 - 1.137$	4.852	2.555
L_1-M_3	$5.989 - 1.062$	4.927	2.516
L_3-N_1	$5.247 - 0.253$	4.994	2.483
L_3-N_4	$5.247 - 0.093$	5.154	2.404
L_2-N_4	$5.634 - 0.093$	5.541	2.242
L_1-N_2	$5.989 - 0.192$	5.797	2.139

[a] E_{keV} data (rounded off) from Bearden and Burr (1967).

The emission of the L series is the result of similar transitions, in this case electrons from M, N, ... levels filling L level vacancies. Figure 1.2 is a schematic representation of the transitions leading to the emission of the main lines of the K and L series, and Table 1.5 lists the binding energies and the energies of the characteristic lines of the element barium, along with their respective wavelengths. For a more complete treatment of nomenclature, symbols and units for X-ray spectroscopy, the reader is referred to *Pure and Applied Chemistry*, Vol. **63**, No. 5 (1991), pp. 736–746.

1.7* EXCITATION SOURCES

It follows from the above considerations that X-ray fluorescence necessitates that specimens be irradiated by highly energetic photons. The energy source for irradiating specimens in most commercial X-ray fluorescence spectrometers is the polychromatic primary beam emanating from X-ray tubes coupled to highly stabilized high voltage generators. Although other sources are available, for example radioactive substances, these have tended to be limited to specialized instrumentation such as portable X-ray spectrometers or to research and development studies. The following treatment is therefore limited to the properties of polychromatic primary beams that concern X-ray fluorescence.

1.7.1   Emanations from X-ray tubes

The processes that lead to the emanation of X-radiations from Coolidge type high vacuum X-ray tubes are:

- a tungsten filament (cathode) is heater to incandescence, giving off electrons;
- the negatively charged electrons are accelerated towards and focused on a metallic target (anode) by means of a high potential applied between the two electrodes;
- X-radiations are emitted due to the braking action (i.e., deceleration of high speed electrons) in matter, and exit through a thin beryllium window.

The two interactions of the incident electrons with the target that are of more concern in X-ray fluorescence are those that give rise to (see Figure 1.3):

(a) a continuous X-ray spectrum (also referred to as continuum, white or Bremsstrahlung radiation;
(b) characteristic X-radiations of the target element.

1.7.2   The continuous spectrum

The continuum results from the loss of energy when the highly accelerated electrons collide with free electrons and undergo deceleration within the target. The three main attributes of the continuum that concern X-ray fluorescence analysts are as follows.

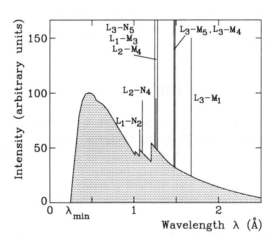

Figure 1.3   Typical spectrum emitted from a tungsten target X-ray tube, 45 kV applied voltage

1.7.2.1 The short wavelength limit

Knowledge of the short wavelength limit is important because it governs which characteristic lines of the elements present in the specimen may or may not be emitted. Although the probability that an electron striking the target transforms all its kinetic energy into a single photon is very low, it is nevertheless possible. When this is the case, the energy of the photon equals the energy of the electron; e.g., an electron accelerated by a potential of 45 kV gives rise to a photon of 45 keV. In most cases, however, the deceleration gives rise to photons of lower energies. The wavelength corresponding to the highest photon energy emanating from X-ray tubes is referred to as the short wavelength limit λ_{min} or minimum wavelength, given by

$$\lambda_{min} = \frac{12.3981}{kV} \tag{1.8}$$

where λ is in Å and kV is the applied voltage in kilovolts. This is also known as the Duane–Hunt (1915) limit. Note the similarity to Eqn (1.3).

1.7.2.2 The total emitted intensity

The total emitted intensity I is proportional to the area under the curve in Figure 1.3. It is a function of applied voltage, current and target element, and on experimental grounds—Beatty (1912)—is found to be given by

$$I = KiZV^2 \tag{1.9}$$

where I is in $W\,m^{-2}$; K is a proportionality constant equal to 1.4×10^{-9}; i is the current in A; Z is the atomic number of the target element; V is the applied voltage in V.

Consider, for example, a wavelength dispersive spectrometer with a tungsten target tube operated at 45 kV and 30 mA; the emitted intensity or useful output energy is equal to

$$\begin{aligned} I &= (1.4 \times 10^{-9}) \times 0.030 \times 74 \times 45\,000^2 \\ &= 6\ W \end{aligned} \tag{1.10}$$

and the input energy is 45 kV × 0.030 A = 1350 W. The difference is large and is dissipated as heat; hence the target must be continuously water cooled during operation. On the other hand, energy dispersive spectrometers are generally operated at lower voltages (in the microampere range); thus, the heat generated by the X-ray tubes under those conditions may be dissipated by air cooling.

1.7.2.3   The spectral distribution of the continuum

Subsequent theoretical treatment to quantify the emitted fluorescence intensity for specimens irradiated by a polychromatic incident beam will require knowledge of the spectral distribution of the continuum. Kulenkampff (1922), as a result of a study of experimentally generated spectra from a number of targets, proposed an empirical relationship for the spectral distribution of the continuum which is generally expressed as

$$I_\lambda = CZ \frac{1}{\lambda^2} \left(\frac{1}{\lambda_{min}} - \frac{1}{\lambda} \right) + BZ^2 \frac{1}{\lambda^2} \tag{1.11}$$

where B and C are constants and Z is the atomic number of the target element. B is so small in comparison to the first term that the second term is generally neglected. Kramers (1923), on the other hand, from purely theoretical concepts, derived an expression for the spectral distribution in the continuum as a function of X-ray tube operating conditions. Although originally expressed in different units, Kramers' expression is equivalent to

$$I_\lambda = CZ \frac{1}{\lambda^2} \left(\frac{1}{\lambda_{min}} - \frac{1}{\lambda} \right) \tag{1.12}$$

which is essentially the same as Kulenkampff's empirical relationship.

In X-ray spectrometry, it is generally preferable to express spectral distribution intensities in number of photons or counts rather than in units of radiant energy, which is the case in Eqns (1.11) and (1.12). This involves multiplication by λ/hc, yielding

$$I_\lambda = CZ \frac{1}{\lambda^2} \left(\frac{\lambda}{\lambda_{min}} - 1 \right) \tag{1.13}$$

However, recent studies have shown that the above equation does not quantify I_λ with sufficient accuracy to meet the more rigorous demands of the fundamental parameters approach (Chapter 3). This has led a number of workers to measure very accurately the spectral distribution emanating from a number of X-ray tubes currently available commercially. For example, a comparison of experimentally measured spectral distributions emanating from side-window and end-window tubes by Arai *et al.* (1986) is shown in Table 1.6. Although the data in this table only relate to operation at 45 kV, additional data relating to operation at 40, 50 and 60 kV are given in the original publication. The availability of more reliable experimentally measured spectral distribution data led other authors to propose modifications to Kramers' law in the form of new equations and algorithms. These provide much more flexibility for the computation of the spectral distribution of the continuum and the intensities of the emitted characteristic lines from the target material.

Table 1.6 Spectral distribution for tungsten target (side- and end-window) X-ray tubes; 45 keV (CP) applied voltage[a]

λ (Å)	Side-window	End-window		λ (Å)	Side-window	End-window
0.28	17.9	5.5		1.4	37.6	50.5
0.32	61.5	43.9		1.5	30.1	45.0
0.36	78.9	64.0		1.6		39.1
0.40	92.9	81.0		2.0	11.8	25.3
0.44	98.9	92.5		2.5	4.0	15.1
0.48	100.0	97.0		3.0		8.8
0.52	98.5	99.0		3.5		4.4
0.56	97.1	99.7		4.0		2.2
0.60	93.3	98.9		4.5		1.3
0.64	89.5	96.5		5.0		0.7
0.70	83.9	92.1		5.5		0.4
1.0	45.8	62.0		6.0		0.2
1.025 −	43.0	59.0		6.5		0.1
1.025 +	46.2	62.1				
1.075 −	42.0	57.5	$W_{L_3-M_5}$	1.476	936.0	1451.0
1.075 +	48.2	63.2	$W_{L_3-M_4}$	1.487	107.0	154.0
1.1	46.0	60.8	$W_{L_2-M_4}$	1.282	570.0	725.0
1.2	36.5	50.9	$W_{L_3-N_5}$	1.247	249.0	306.0
1.216 −	35.3	49.0	$W_{L_1-M_3}$	1.263	90.9	113.0
1.216 +	54.0	64.8	$W_{L_2-N_4}$	1.099	101.2	141.0
1.3	46.5	57.3				

[a] Data from Arai *et al.* (1986). Values normalized to maximum of continuum.

1.7.3 Characteristic X-radiations from the target element

The study of experimental spectral distributions of various targets for X-ray tubes at several operating voltages shows a number of features that are quite important in quantitative X-ray fluorescence analysis. Gilfrich and Birks (1968) noted that the sharp discontinuities at the absorption edges of the target element, although previously recognized but generally disregarded, are quite significant. The discontinuity is particularly significant in the case of Cr target tubes. Even more important was the realization that the characteristic X-ray lines may constitute a major fraction of the total emitted intensity. For example, a W target tube (Machlett OEG-50 type) operated at 45 kV (constant potential) generates tungsten *L* lines that are 24% of the total emissions. In the case of a Cr target tube operated under the same conditions, the chromium *K* lines constitute 75% of the total emitted intensity. The fact that the characteristic lines from the target may constitute the major portion of the incident beam makes it mandatory that they be taken into consideration in any study of the theoretical aspects of quantitative X-ray fluorescence analysis.

A case in point would be the major contribution of the characteristic Cr target lines to titanium fluorescence emissions.

One proposed modification to Kramers' law was made by Tertian and Broll (1984). Based on the premise that an accurate definition of spectral intensity distributions is required for the development of mathematical correction procedures, and that a direct measurement is costly, time consuming and difficult, these authors proposed an expression that takes into account self-absorption by the target, absorption by the tube window, etc. The expression for the continuum distribution for an X-ray tube operated at constant potential is given as

$$I_\lambda \propto \frac{1}{\lambda^2}\left(\frac{\lambda}{\lambda_{min}} - 1\right)^q \exp(-Q'\mu_{i\lambda}^n)\exp(-\mu_{\lambda}\rho t)_{Be} \tag{1.14}$$

where i is the target element and ρ and t are the density and the window thickness, respectively.

The equation is used by adjusting the values of q, Q' and n so as to fit carefully measured and corrected experimental spectra. In the case of the experimental data reported by Loomis and Keith (1976), the values of q, Q' and n for a Cr target tube operated at 45 kV are given as 1.15, 0.0150 and 0.7, respectively; those for a W target tube operated at the same voltage are 0.95, 0.0535 and 0.5, respectively. In the case of the experimental data reported by Brown *et al.* (1975), the equivalent values of q, Q' and n for Cr and W targets operated under the same conditions are 1.15, 0.0106, 0.7 and 0.95, 0.0440, 0.5, respectively. An expression for the intensity of characteristic lines was not within the scope of Tertian and Broll's study.

A similar study was carried out by Pella *et al.* (1985, 1991), who also recognized the need for expressions to calculate spectral distributions for a large variety of target elements over a wide range of operating voltages. Extensive experimental data from electron microprobe spectrometry using a Si(Li) detector and a number of thick metal targets (i.e. Cr, Cu, Se, Zr, Mo, Rh, Ag, Sb, Te, Dy, Tb, W, Pt and Au) were employed in developing the algorithm. The required corrections for detector efficiency, incomplete charge collection tails, sum peaks and escape peaks were made, as were the conversions to express intensity in the commonly used unit, i.e. counts or photons emitted. The equation for the X-ray tube continuum distribution is given as

$$I_\lambda = 2.72 \times 10^{-6} Z \frac{1}{\lambda^2}\left(\frac{\lambda}{\lambda_{min}} - 1\right) f \exp(-0.35\lambda^{2.86} t) \tag{1.15}$$

where t is the thickness of the Be window and f is defined as

$$f = (1 + C\xi)^{-2} \tag{1.16}$$

Through curve-fitting analysis of experimental data the following expressions were obtained for ξ and C

$$\xi - \left(\frac{1}{\lambda_{min}^{1.65}} - \frac{1}{\lambda^{1.65}}\right)\mu_{tg}\csc\psi \tag{1.17}$$

where ψ is the X-ray emergence angle from the target, and μ_{tg} is the mass absorption coefficient of the target element. The following algebraic expression was obtained for the factor C

$$C = \frac{1 + (1 + 2.56 \times 10^{-3}Z^2)^{-1}}{(1 + (2.56 \times 10^3)\lambda_{min}Z^{-2})(0.25\xi + 1 \times 10^4)} \tag{1.18}$$

The derivation of an expression for quantifying the intensities of the characteristic X-ray lines of the target element was based on the work of Green and Cosslett (1968) and Green (1962). The approach involved deriving an expression for the ratio of the intensity of a characteristic line N_{chr} to that of the continuum N_{con} directly under it. The expression obtained can be presented as the product of three factors, i.e.

$$N_{chr}/N_{con} = RST \tag{1.19}$$

where

$$R = \exp\left[-0.5\left(\frac{U_0 - 1}{1.17U_0 + 3.2}\right)^2\right] \tag{1.20}$$

$$S = \frac{a}{b + Z^4} + d \tag{1.21}$$

$$T = \left[\frac{U_0 \ln U_0}{U_0 - 1} - 1\right] \tag{1.22}$$

and U_0 is the overvoltage, i.e. $U_0 = \lambda_{abs}/\lambda_{min}$.

The constants a, b and d in Eqn (1.21) were determined from a least squares fit of experimental data, and their values for the $K-L_{2,3}$, $K-M_3$, $L_3-M_{4,5}$ and L_2-M_4 characteristic lines are listed in Table 1.7. In order to express N_{chr} and

Table 1.7   Values of constants in Eqn $(1.21)^a$

Line	a	b	d
$K-L_{2,3}$	3.22×10^6	9.76×10^4	-0.39
$K-M_3$	5.13×10^5	2.05×10^5	-0.014
$L_3-M_{4,5}$	2.02×10^7	2.65×10^6	0.21
L_2-M_4	1.76×10^7	6.05×10^6	-0.09

^{a} Data from Pella et al. (1985)

N_{con} in similar units, so that their ratio is dimensionless, N_{con} is evaluated by multiplying I_λ in Eqn (1.15) by the wavelength interval $\Delta\lambda$; i.e. N_{con} is equal to the sum of the products $I_\lambda \, \Delta\lambda$.

1.8* ATTENUATION OF X-RAYS

The fact that X-rays undergo attenuation to a different degree in passing through various substances was observed very shortly after their discovery, and found immediate medical application. The observation that bones absorb X-rays 'rather more' than organic tissues soon proved inadequate, and systematic studies were and are still being carried out to quantify the attenuation processes. The decrease in intensity observed when a beam of X-radiation passes through matter will be examined in the following context; see Figure 1.4.

A parallel, monochromatic X-ray beam, wavelength λ, of intensity I_0 is directed onto a thin layer of homogeneous material of uniform thickness. The emergent beam intensity I_x is always less than I_0, owing to absorption and scattering. The loss of intensity through an infinitely thin layer dx is given by

$$dI = -\mu_l I \, dx \tag{1.23}$$

where dx is the thickness of the layer in cm; μ_l is the linear attenuation co-efficient of the material (usually a pure element) at the wavelength under consideration, and has the unit 'per centimetre' (cm^{-1}).

To get the attenuation for a finite thickness x, one integrates Eqn (1.23), resulting in

$$I_x = I_0 \exp(-\mu_l x) \tag{1.24}$$

The linear absorption coefficient (for an irradiated cross-section expressed in cm^2) represents the fraction of the intensity absorbed per centimetre of

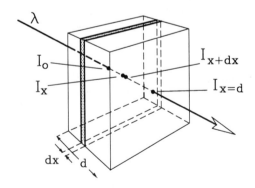

Figure 1.4 Attenuation of X-rays by matter

material traversed. In X-ray fluorescence, it is more convenient to deal with absorption per gram. The attenuation coefficient that is by far the most useful in X-ray fluorescence is therefore defined as

$$\mu = \frac{\mu_1}{\rho} \tag{1.25}$$

where ρ is the density of the material in g cm^{-3}.
Substituting Eqn (1.25) in Eqn (1.24) yields

$$I_x = I_0 \exp(-\mu\rho x) \tag{1.26}$$

where ρx is the mass per cm^2.

The symbol μ is referred to as the mass attenuation coefficient, expressed in cm^2 g^{-1}. The great advantage of mass attenuation coefficients is that they are directly comparable; i.e., the total mass attenuation coefficient μ_s of a multielement specimen is given by the simple relationship

$$\mu_s = C_i\mu_i + C_j\mu_j + \cdots = \sum_i C_i\mu_i \tag{1.27}$$

For example, the total mass attenuation coefficient of an alloy consisting of 18.0% Cr, 74.0% Fe and 8.0% Ni is given by

$$\mu_{\text{alloy}} = \frac{18}{100}\,\mu_{\text{Cr}} + \frac{74}{100}\,\mu_{\text{Fe}} + \frac{8}{100}\,\mu_{\text{Ni}} \tag{1.28}$$

The coefficient μ is an atomic property of each element and is a function of wavelength. It is a measure of the fraction of the intensity that is not transmitted in the same direction as the incident photons, i.e., $I_0 - I_x$. This loss in intensity is actually due to two processes, namely absorption and scatter; hence

$$\mu = \tau + \sigma \tag{1.29}$$

where τ is the mass photoelectric absorption coefficient, and σ is the mass scattering coefficient.

Photoelectric absorption accounts for the number of incident photons that use up their energy either by ejecting orbital electrons, leading to the emission of characteristic radiations, or to the ejection of Auger electrons.

Scatter accounts for the incident photons that are not absorbed per se but are scattered after collision with atoms. The scattered photons having the same wavelength as the original incident photons are referred to as resulting from unmodified, coherent or Rayleigh scattering, symbol σ_{coh}, and those that are scattered at a longer wavelength are referred to as resulting from modified, incoherent or Compton scattering, symbol σ_{inc} (Section 1.10). Thus the mass attenuation coefficient μ is the sum of three coefficients, namely

$$\mu = \tau + \sigma_{\text{coh}} + \sigma_{\text{inc}} \tag{1.30}$$

1.8.1 Experimental determination of μ

The experimental measurement of mass attenuation coefficients for an element as a function of wavelength presents quite a challenge. The following outlines the procedure followed by Dalton (1969). Briefly, the determination of mass attenuation coefficients consists in measuring the intensity of a collimated monochromatic beam of X-rays with and without an absorbing foil in its path. The foil must be a single element of the highest purity, uniform in thickness and such that its area can be accurately measured. The procedure is as follows.

(1) The foil's thickness is reduced by 'pack-rolling'; i.e., the material, with Mylar film on each side, is repeatedly passed between the rollers of a hand rolling mill in order to reduce the thickness to approximately 15 μm.

(2) The Mylar-coated foil is then cut to size, the Mylar films are floated off in carbon tetrachloride, and the foil is rinsed in benzene and ethyl ether and weighed two or three times on a Mettler Micro Gram-atic analytical balance.

(3) The area is measured with a Gaertner toolmaker's microscope. At least three foils of each element are prepared.

(4) Thickness uniformity is assessed by measuring μ in different areas of the foils, the centre and each of four quadrants.

(5) Pure elements or compounds are used to generate $K-L_{2,3}$ or $K-M_3$ wavelengths as the incident radiation. Typical data for the element erbium, for incident radiation $\lambda = 1.436$ Å (Zn $K-L_{2,3}$), are shown in Table 1.8, along with values for μ_{Er} computed from the relationship

$$\mu = \frac{\text{area}}{\text{weight}} \ln \frac{I_0}{I_x} \tag{1.31}$$

derived from Eqn (1.26), where area is in cm^2 and weight is in grams; hence μ is expressed in cm^2 g^{-1}.

In addition to Er, mass attenuation coefficients were determined for Ti, V, Fe, Ni, Cu, Pr and Gd; each reported μ value is the average of 30 measurements, five readings on six different foils. Measurements were carried out for

Table 1.8 Experimental data, determination of mass attenuation coefficients (cm^2 g^{-1}) for Er ($Z = 68$); incident radiation 1.436 Å

Foil	Area (cm^2)	Weight (g)	I_0	I_x	μ_{Er}
1	5.52554	0.064512	12127	397.02	292.86
			12138	396.09	293.14
			12127	397.49	292.76
			12081	393.39	293.32

[a] Data from Dalton (1969).

different monochromatic incident radiations, in most cases from 10 to 14 monochromatic wavelengths in the range 0.472–5.373 Å.

1.9* RELATION BETWEEN μ, Z AND λ; ABSORPTION EDGES

Mass attenuation coefficients vary as a function of the absorbing element and the wavelength of the impinging X-radiation. This variation may be examined in part using values reported for iron and nickel by Dalton and Goldak (1969); see Table 1.9. Consider the value of μ_{Fe}, which rises from 43.1 to 388.0, then drops to 53.0 and rises again to 438.0. It is also observed that the values for μ_{Ni} show a similar pattern. Detailed experiments have shown that the drastic drop in μ values takes place exactly at each critical excitation wavelength. Plots of μ_{Fe} versus λ on a linear and a logarithmic scale are shown in Figures 1.5a and 1.5b, respectively. Both clearly show the sudden drop in μ values that takes place at the critical excitation wavelength (1.743 Å), i.e. at the critical excitation energy (7.112 keV) of iron. In this case, the sudden drop is referred to as the 'K absorption edge of iron', symbols Fe K_{abs} and is hence expressed as λ (Fe K_{abs}) or E(Fe K_{abs}), depending on whether it is expressed in wavelength or energy units.

From plots such as Figures 1.5 and tabulated values of μ, large numbers of $\mu_{i\lambda}$ may be fitted to various algorithms. For example, Heinrich (1966) proposed the expression

$$\mu = C\lambda^n \tag{1.32}$$

In the case of iron, the values of C and n for wavelengths shorter than the K

Table 1.9 Experimentally determined mass attenuation coefficients for iron and nickel[a]

λ (Å)	μ_{Fe} (cm^2 g^{-1})	λ (Å)	μ_{Ni} (cm^2 g^{-1})
0.746	43.1	0.746	53.6
0.876	66.8	0.876	82.6
1.177	149.7	1.106	154.5
1.295	193.5	1.295	237.0
1.542	309.9	1.436	320.0
1.659	388.0	1.500	44.0
1.757	53.0	1.937	89.0
2.085	86.5	2.291	144.0
2.748	189.0	2.504	165.9
3.359	334.2	2.748	243.4
3.742	438.0	3.359	419.8

[a] Data from Dalton and Goldak (1969).

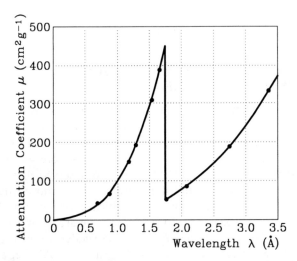

Figure 1.5a   Plot of μ versus λ for the element iron (linear scale); data from Table 1.9

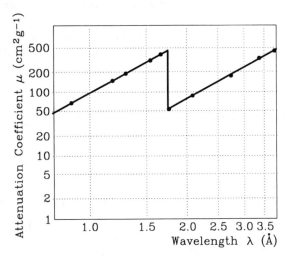

Figure 1.5b   Plot of μ versus λ for the element iron (logarithmic scale); data from Table 1.9

absorption edge of iron as computed by Dalton (1969) are 95.91 and 2.736, respectively. For wavelengths longer than the K absorption edge but shorter than the L_1 absorption edge, the values for C and n are 11.14 and 2.794, respectively. For Ni, the equivalent values are 118.2, 2.718, 14.13 and 2.804, respectively.

Absorption experiments carried out over wider ranges of wavelengths show that there are also three sudden drops in μ values at longer wavelengths, and hence three absorption edges associated with the three critical excitation energies E_{L_1}, E_{L_2} and E_{L_3}. Similarly, five absorption edges associated with

Table 1.10   Mass attenuation coefficients ($\mathrm{cm^2\,g^{-1}}$) for Er over the range 1–100 keV[a]

Abs. edge	E (keV)	μ_{Er}	Abs. edge	E (keV)	μ_{Er}
	100	3.607		5	468.1
	80	6.494		3	1818
	60	13.79	M_1	2.212	4099
K	57.486	15.41		2.211	3836
	57.485	3.291	M_2	2.006	4974
	50	4.598		2.005	4479
	30	17.57		2	4509
	15	114.0	M_2	1.811	5879
	10	345.2		1.810	4409
L_1	9.752	369.8		1.5	7282
	9.751	319.4	M_4	1.455	7900
L_2	9.265	367.4		1.454	5498
	9.264	261.9	M_5	1.410	5969
L_3	8.358	346.2		1.409	1475
	8.357	121.4		1	3678
	8	136.1			

[a] Data from McMaster *et al.* (1969).

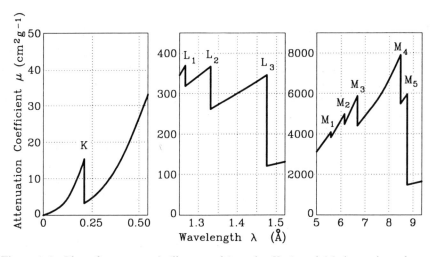

Figure 1.6   Plot of μ_{Er} versus λ (linear scale) at the K, L and M absorption edges

the five M subshells are observed. The fact that the L and M absorption edges occur within a fairly narrow wavelength range makes it somewhat difficult to obtain reliable μ values within those ranges. Table 1.10 lists mass attenuation coefficients within the energy range 100–1.0 keV and, more specifically, values slightly above and below the K, L and M edges for the element erbium. Plots of these data in Figure 1.6 illustrate the general pattern of K, L and M absorption edges as a function of wavelength.

1.10   SCATTERING OF X-RAYS

As mentioned in Section 1.8, the decrease in intensity when X-rays pass through matter is the resultant of two processes, photoelectric absorption and scatter. Just as absorption was described as the sum of the photoelectric and the Auger phenomena, scattering actually involves two processes, namely:

(1) coherent scatter (also referred to as unmodified, elastic or Rayleigh scatter), in which scattering occurs without loss of energy, i.e. without increase in wavelength; and (2) incoherent scatter (also referred to as modified, inelastic or Compton scatter), which involves loss of energy, i.e. increase in wavelength.

The two types of scattering may be observed by carrying out the following experiment on any wavelength dispersive spectrometer. The incident X-ray beam should contain an intense characteristic line of the target element, for example Mo $K-L_3$. In turn, the specimen, i.e. the scattering material, can be a very light element such as C, or elements of slightly higher atomic number, e.g. Al, Ti, etc. In Figure 1.7, the Mo peaks at $20.3°$ 2θ are the result of coherent scatter and the peaks at $21.0°$ 2θ are the result of incoherent scatter.

1.10.1   Coherent scatter

If the incident photons impinging on a specimen cause the orbital electrons to oscillate at the same frequency as the incident beam, the oscillating electrons emit X-radiations at the same frequency as the incident beam. This can be readily observed experimentally; for example, by irradiating a specimen consisting of high Z elements with an incident beam from a Mo target X-ray tube. It is observed that peaks occur, in this case, at the Mo $K-L_{2,3}$ and Mo $K-M_3$ position. Since molybdenum is not present in the specimen being irradiated, the presence of these lines is due to coherent scatter by the specimen. Sproull (1946) described the process as follows: 'If the photon bounces off a firmly bound electron, like the inner electrons of a heavy atom like lead, the recoil of the electron is not vigorous enough to shake it loose from the atom and it is unable to acquire kinetic energy from the photon. Hence the photon emerges with the same energy and hence the same wavelength as it had originally, and the scattering is coherent.'

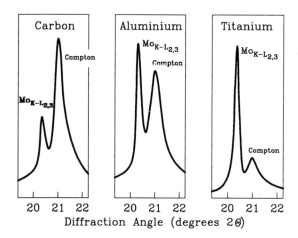

Figure 1.7 Incoherent and coherent scatter of incident Mo $K-L_{2,3}$ radiation by specimens of carbon, aluminium and titanium. The coherent peak is at 20.34 degrees 2θ; the incoherent peak is at 21.03 degrees 2θ

1.10.2 Incoherent scatter

The process of incoherent or Compton scatter may be described as follows.

(1) An incident X-ray photon collides (randomly) with a loosely bound electron (outer or lower binding energy level).

(2) The orbital electron recoils from the impact.

(3) The electron leaves the atom, carrying some of the energy of the incident photon.

(4) The deflection results in loss of energy (or an increase in wavelength of the emergent radiation.

The change in wavelength is given by

$$\Delta\lambda = \frac{h}{m_0 c} (1 - \cos \phi) \tag{1.33}$$

where h is Planck's constant, m_0 is the mass of the electron at rest, c is the speed of light, and ϕ is the scattering angle in degrees. Substitution of the proper values and converting to Å gives

$$\Delta\lambda = 0.0243(1 - \cos \phi) \tag{1.34}$$

Most commercial X-ray spectrometers have a scattering angle of about $90°$ to the primary beam. Given that the cosine of $90°$ is equal to zero, Eqn (1.34) reduces to

$$\Delta\lambda = 0.0243 \text{ Å} \tag{1.35}$$

Table 1.11 Photoelectric, coherent, incoherent and total mass absorption coefficients $(cm^2 g^{-1})$ for iron, wavelength range 0.4–4 Å[a]

keV	λ (Å)	τ	σ_{coh}	σ_{inc}	μ
30.0	0.413	7.665	0.293	0.128	8.087
20.0	0.620	25.10	0.532	0.117	25.75
15.0	0.827	56.78	0.768	0.105	57.66
10.0	1.240	171.5	1.241	0.085	172.8
8.0	1.550	306.8	1.594	0.074	308.5
7.112	1.7433	413.4	1.796	0.068	415.2
7.111	1.7435	50.30	1.796	0.068	52.17
6.0	2.066	82.45	2.217	0.059	84.63
5.0	2.480	138.9	2.486	0.051	141.4
4.0	3.100	259.7	2.917	0.042	262.6
3.0	4.133	570.2	3.419	0.031	573.6

[a] Data (rounded off) from McMaster *et al.* (1969).

Thus the scattering process leads to the emission of two peaks λ_{coh} and λ_{inc}, with

$$\lambda_{inc} = \lambda_{coh} + \Delta\lambda \tag{1.36}$$

The incoherent or Compton scatter is a peak *broader than the characteristic line* on the long wavelength side of each coherently scattered X-ray tube target line, as shown in Figure 1.7. Table 1.11, on the other hand, shows the relative contribution of photoelectric absorption, coherent and incoherent scatter to the total mass attenuation coefficient for iron in the range 30–3 keV (wavelength 0.413–4.133 Å). Details of how Compton scatter is the basis for a matrix effect (absorption) correction method are given in Chapter 7.

1.11[*]   LINE INTENSITIES; PROBABILITY OF EXCITATION

The description of the emission of characteristic X-radiations in Section 1.6 was simplified in that it was given in terms of a single incident photon ejecting a single electron from an atom, leading to the emission of a characteristic X-ray. In practice, the beam of incident photons strikes a multitude of atoms in a given specimen. In that case, the emission of the different characteristic lines and their intensities depend on a number of factors, three of which are now examined, namely:

   (1) the absorption jump ratio r;
   (2) the relative intensity f of a characteristic line within its series;
   (3) the fluorescence yield ω.

1.11.1 Absorption jump ratios

For an element to emit characteristic X-radiations, its atoms must first absorb photons leading to the ejection of K, L, M, ... shell electrons, i.e. the photo-electric absorption process. Subsequently, it will be necessary to quantify each of the components of this process individually, say for example that of the K shell only. The experimental determination of absorption at any one given wavelength may be the resultant of absorption at many energy levels (Figure 1.8). The mass attenuation coefficient is therefore the sum of the attenuation at each level, i.e.

$$\mu = \mu_K + \mu_{L_1} + \mu_{L_2} + \mu_{L_3} + \mu_{M_1} + \cdots \tag{1.37}$$

where μ_K, μ_{L_1}, μ_{L_2}, ... are the partial attenuation coefficients for the K, L_1, L_2, ... energy levels. However, recalling Eqn (1.30)

$$\mu = \tau + \sigma_{\text{coh}} + \sigma_{\text{inc}}$$

it is in terms of photoelectric absorption coefficients that the jump ratio must be defined. Consider for example the element tin ($Z = 50$) as the absorbing material (E_K, E_{L_1}, E_{L_2} and $E_{L_3} = 29.2$, 4.465, 4.156 and 3.929 keV, respectively), and impinging radiation of 3 keV (i.e. $\lambda = 4.13$ Å). Because the incident photons cannot eject K and L level electrons, the photoelectric absorption process can be due only to the ejections of M, N and succeeding level electrons. Thus, in this context

$$\tau = (\tau_M + \tau_N + \tau_O + \cdots)_{E = 3.0 \text{ keV}} \tag{1.38}$$

where τ is the symbol used to designate the photoelectric absorption coefficient. If the incident photon energy is greater than 3.929 keV but less than

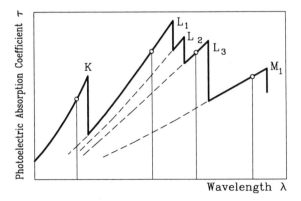

Figure 1.8 Schematic representation of τ_K, τ_{L_1}, τ_{L_3} and τ_{M_1} absorption jumps

4.156 keV, then the L_3 level electrons also contribute to photoelectric absorption; hence

$$\tau = (\tau_{L_3} + \tau_M + \tau_N + \ldots)_{4.156 \geqslant E > 3.929 \text{ keV}} \tag{1.39}$$

Should the incident photon energy be greater than 4.465 but less than 29.2 keV, then all these L sublevel electrons also can be ejected and τ is now given by

$$\tau = (\tau_{L_1} + \tau_{L_2} + \tau_{L_3} + \tau_M \cdots + \tau_N \cdots + \cdots)_{29.2 \geqslant E > 4.465 \text{ keV}} \tag{1.40}$$

Finally, if the energy of the incident photons is greater than 29.2 keV, then all levels contribute and

$$\tau = (\tau_K + \tau_{L_1} + \tau_{L_2} + \tau_{L_3} + \tau_M \cdots + \tau_N \ldots)_{E > 29.2 \text{ keV}} \tag{1.41}$$

The magnitude of the change in absorption at any absorption edge can be interpreted as a measure of the probability that a specific absorption takes place. By convention, the larger value is divided by the smaller; thus the K absorption jump is given by

$$r_K = \frac{\tau_K + \tau_{L_1} + \tau_{L_2} + \tau_{L_3} + \tau_{M_1} + \cdots}{\tau_{L_1} + \tau_{L_2} + \tau_{L_3} + \tau_{M_1} + \cdots} \tag{1.42}$$

In practice, the absorption jump ratio is calculated from the simple relationship

$$r = \frac{\tau_{\lambda_{\text{abs}}} - \Delta\lambda}{\tau_{\lambda_{\text{abs}}} + \Delta\lambda} \tag{1.43}$$

where $\Delta\lambda$ is any small value near zero, i.e. r is the ratio of τ maximum to τ minimum at the absorption edge of the characteristic line being measured. Absorption jump ratios at the L_1, L_2, L_3, M_1 (etc.) absorption edges are computed in a similar manner. Typical values for r_K and r_{L_3} are listed in Table 1.12. Algorithms using a least squares fit of r as a function of atomic number to generate r_K and r_{L_3} are given in Section II.1.11.1.

1.11.2 Relative line intensities within the K, L and M series

The second of the determining factors takes in consideration the likelihood, say for the K series lines, that the $K-L_3$, $K-L_2$, $K-M_3$ etc. lines will be emitted. Experimental results show that the $K-L_3$ line is very nearly twice as strong or intense as the $K-L_2$ line. Sproull (1946) attributes this to the fact that there are four electrons in the L_3 sublevel and only two in the L_2 sublevel (Table 1.3). On the other hand, the relative intensity of the other K lines, say the $K-M_3$ line to the unresolved $K-L_{2,3}$ line, is quite variable with Z. As expected, the relative intensities of L lines within the L series show even wider variations, there being many more energy levels to deal with. The symbol f is

Table 1.12 Typical absorption jump ratio values[a]

Element (Z)	r_K	r_{L_3}
Mg (12)	11.59	
Ti (22)	8.53	
Zn (30)	7.60	5.68
Zr (40)	6.75	3.98
Sn (50)	6.47	3.06
Nd (60)	5.99	2.66
W (74)	5.12	2.62
Pb (82)	4.99	2.47
U (92)		2.28

[a] Data (rounded off) from McMaster et al. (1969).

used to designate the intensity of a given line as a fraction of the total intensity within its series. Thus, $f_{K-L_{2,3}}$ is generally given by

$$f_{K-L_{2,3}} = \frac{I_{K-L_{2,3}}}{I_{K-L_{2,3}} + I_{K-M_3}} \tag{1.44}$$

and is normally determined experimentally. Obviously

$$f_{K-M_3} = 1 - f_{K-L_{2,3}} \tag{1.45}$$

if the other much less intense K lines are disregarded. Typical $f_{K-L_{2,3}}$ and $f_{L_3-M_{4,5}}$ values are listed in Table 1.13. Algorithms using a least squares fit of f as a function of atomic number to generate values of the probability of the $K-L_{2,3}$ or $L_3-M_{4,5}$ lines being emitted are given in Section II.1.11.2.

Table 1.13 Typical f values[a]

Element (Z)	$f_{K-L_{2,3}}$	$f_{L_3-M_{4,5}}$
Mg (12)	1.00	
Ti (22)	0.898	
Zn (30)	0.890	0.948
Zr (40)	0.854	0.945
Sn (50)	0.829	0.858
Nd (60)	0.809	0.823
W (74)		0.804
Pb (82)		0.777
U (92)		0.748

[a] Data from Broll (1986).

1.11.3 Fluorescence yield

The third factor to be considered relates to the following process. The ejection of inner shell electrons leaves the atom in an ionized state. The vacancies are filled by the transfer of electrons from higher energy levels, each releasing its energy of transition. This energy may be released in one of two ways:

(1) as an X-ray photon, the basis of X-ray fluorescence emissions;

(2) as an Auger electron, i.e. the emitted photon is lost by absorption within the atom with ejection of higher level electrons, each being characteristic of the element under consideration.

The fluorescence yield ω is defined as the probability for the photon to leave the atom without being absorbed within the atom itself, and is expressed by

$$\omega = \frac{n_S}{n_P} \tag{1.46}$$

where n_P is the number of primary photons that resulted in the ejection of electrons from a given energy level, and n_S is the number of secondary photons that effectively leave the atom without being absorbed.

It follows that

$$n_P = n_S + n_A \tag{1.47}$$

where n_A is the number of secondary photons that are absorbed within the atom on their way out, i.e. the Auger effect, and that the sum of the fluorescence yield and the Auger electron yield is equal to unity. The K fluorescence yield ω_K is given by the number of photons n_K emitted by all K lines during a given

Table 1.14 Typical fluorescence yield values[a]

Element (Z)	ω_K	ω_{L_3}
Mg (12)	0.022	
Ti (22)	0.215	
Zn (30)	0.490	0.005
Zr (40)	0.735	0.025
Sn (50)	0.855	0.060
Nd (60)	0.922	0.125
W (74)		0.263
Pb (82)		0.367
U (92)		0.525

[a] Data from Bambynek et al. (1972).

time interval divided by the number of K level vacancies N_K created during that same time interval, i.e.

$$\omega_K = \frac{n_{K-L_3} + n_{K-L_2} + n_{K-M_3} + \cdots}{N_K} = \frac{\Sigma(n_K)}{N_K} \tag{1.48}$$

Similarly, it is possible to define and quantify L fluorescence yields but, in this case, there are three different values for ω_L, namely individual values for ω_{L_1}, ω_{L_2} and ω_{L_3}. Typical fluorescence yields are listed in Table 1.14. It should be noted that ω_K values are known with a higher degree of accuracy than ω_L values. Algorithms using a least squares fit of ω as a function of atomic number to generate ω_K and ω_L are given in Section II.1.11.3.

1.12   DISCUSSION

Although there is no doubt that the introductory chapter or chapters in a text on X-ray spectrometry should deal with the principles of X-ray physics, space considerations in the present publication limit the coverage to the basic principles necessary to comprehend and visualize those aspects pertaining to conventional X-ray fluorescence analysis. For example, the determination of the concentrations of low Z elements, i.e. $Z < 10$, would almost have to be dealt with on an individual basis, and probably for each analytical context or application that can be envisaged. The same criteria apply for those cases where analysts wish to use M characteristic radiations. Readers concerned with these applications should preferably consult the recent literature in order to profit from current developments.

Considering that X-ray fluorescence may be classified under the broader heading of atomic spectrometry, atomic structure was dealt with initially in order to show the origin of the characteristic spectra of the elements. Excitation sources were dealt with next in order to examine the components that are available to initiate the process of X-ray fluorescence emission. In fact, Section 1.7, on Excitation Sources, is somewhat independent of the other topics in this chapter and could equally well have been considered in Chapter 2. The reason for this is that the choice of excitation sources is subjective and the operating conditions, for incident sources comprising X-ray tubes, are even more so, because the operator may select the target material, the operating voltage and so on. In other words, the binding energies of $K, L, \ldots$ level electrons are fundamental parameters, common to all aspects of X-ray physics, but the spectral distribution, although subject to physical laws, cannot be regarded in the same light. In a final analysis, no two X-ray tubes can be considered as identical, if for no other reason than that X-ray tube emanations gradually change with usage.

Absorption was covered next, in some detail, because—as will become evident in subsequent chapters—absorption is ever present in X-ray fluorescence emissions and enters into all facets of relating intensities to concentrations. Scattering is briefly discussed because the scattering process is the basis of methods for dealing with absorption effects in limited analytical contexts. The factors quantifying line intensities underlie calculations dealing with enhancement effects, which are covered in the next chapter.

1.13 SUMMARY

Inasmuch as subsequent treatments will involve the following entities directly or indirectly, the reader should be conversant with the following symbols, definitions and terminology. For space considerations, examples refer to K spectra only.

Entity	Symbol	Definition
Binding or critical excitation energy	E_K, E_{L_3}	Fundamental data
Characteristic X-ray photon energy	E_{K-L_3}	$E_K - E_{L_3}$
Wavelength absorption edge Critical excitation wavelength	$\lambda_{abs,K}$	$12.3981/E_K$
Wavelength characteristic line	λ_{K-L_3}	$12.3981/E_{K-L_3}$
Minimum wavelength	λ_{min}	$12.3981/kV$
Spectral distribution continuum	I_λ	see Eqns (1.13)–(1.15)
Mass attenuation coefficient of element i for wavelength λ	$\mu_{i\lambda}$	$C\lambda^n$
Wavelength Compton scatter	λ_{inc}	$\lambda_{coh} + \Delta\lambda$
Absorption jump ratio	r	$\dfrac{\tau(\lambda_{abs} - \Delta\lambda)}{\tau(\lambda_{abs} + \Delta\lambda)}$
Fluorescence yield	ω_K	$\dfrac{\Sigma(n_K)}{N_K}$
Probability of line emission within series	$f_{K-L_{2,3}}$	$\dfrac{I_{K-L_{2,3}}}{I_{K-L_{2,3}} + I_{K-M_3}}$

2 X-RAY FLUORESCENCE EMISSIONS

The aim of this chapter is to examine how the fundamental principles of X-ray physics covered in the previous chapter can be combined in order to derive expressions that will make it practical to calculate emitted intensities from theory, as will be discussed in Chapter 3. These theoretical intensities relate to experimentally measured intensities, and thus provide the means to account for the variation of emitted intensities as a function of specimen composition. The word relate is chosen to indicate clearly that theoretical expressions cannot be used (except in a very few circumstances) to calculate *a priori* the actual intensity that is measured in the laboratory. However, net intensities, i.e. intensities of characteristic lines of elements that are measured experimentally and corrected for background, spectral overlap, detector dead-time etc., are related to theoretical intensities by the expression

$$I_{i,\text{experimental}} = g_i \times I_{i,\text{calculated from theory}} \tag{2.1}$$

where I_i is the intensity of a characteristic line of element i, and g_i is a proportionality constant.

In addition to fundamental parameters such as mass absorption coefficients, fluorescence yields, wavelengths of absorption edges etc., instrumental parameters must also be taken into consideration; it will thus be necessary to define the spectrometer configuration and operating conditions under which I_i is measured in order to match those used for the theoretical computation of I_i.

2.1 GENERAL CONSIDERATIONS

Basically, intensities of characteristic X-radiations emitted by a specimen are the resultant of two processes, namely excitation and absorption. Generally the excitation process involves the following sequential steps:

- the generation of a polychromatic beam of X-radiations, i.e. high energy photons, as the excitation source;

• the irradiation of multielement specimens, resulting in the emission of characteristic X-radiations of the elements present.

Both excitation and emission take place within a certain depth in the specimens; hence the X-radiations are attenuated to a greater or lesser degree depending on the nature of the individual specimen. Some attenuation is due to absorption of the incident beam prior to excitation, and further attenuation is due to absorption of the characteristic lines after excitation. However, in multielement specimens, many combinations of excitation and absorption are possible; hence it is preferable to define and derive, firstly, expressions that quantify emitted intensities due to direct excitation by the incident beam. This is referred to as primary fluorescence P_i. The next process to be defined and quantified deals with the case in which the emitted intensity is due in part to primary fluorescence and in part to excitation by characteristic radiations emitted by other elements within the specimen. This process is referred to as secondary fluorescence S_i, or the enhancement effect. Finally, the more complicated process whereby the primary emission of one element induces an enhancement in a second element, which in turn leads to excitation of atoms in a third element—the analyte—will be defined. This is referred to as tertiary fluorescence T_i.

It should be noted at this stage that, in this and subsequent chapters, the X-ray fluorescence process is described as 'an *incident* beam of X-radiations which impinges on specimens resulting in the *fluorescence* emission of X-radiations characteristic of the elements present', rather than 'a *primary beam* of X-radiations which impinges on specimens resulting in *secondary* X-radiations'. This choice does not imply that the latter is not an acceptable definition of X-ray fluorescence; however, the former is retained in order to avoid possible confusion in using the term primary to describe the incident radiation and the term secondary to describe characteristic radiations, which may be due to primary or to secondary fluorescence processes.

A cursory examination of manufacturers' literature describing the spectrometers commercially available at any one time shows the wide variety in which the basic components that make up an X-ray fluorescence spectrometer can be combined. It is therefore mandatory that some instrumental parameters be known if theoretical intensity calculations are to be made. These are examined next for both wavelength and energy dispersive instruments.

2.2   SPECTROMETER CONFIGURATIONS

A schematic representation of conventional wavelength dispersive and energy dispersive spectrometers is given in Figures 2.1–2.3. Figure 2.1 is typical of a wavelength dispersive spectrometer, the portion enclosed by the broken line representing the components that are common to all X-ray fluorescence

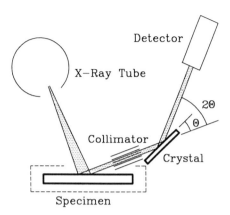

Figure 2.1 Configuration of the flat-crystal wavelength dispersive X-ray spectrometer

spectrometers, namely, an incident beam that impinges on a specimen, which emits characteristic radiations in the direction of a collimator.

In the case of wavelength dispersive spectrometers, polychromatic radiations emanating from the X-ray tube (continuum plus characteristic lines) impinge on the specimen (composition C_i, C_j, C_k, . . .) and eject K, L, . . . level electrons from atoms of the elements present in the specimen. This results in the emission of fluorescence radiations (K, L, . . . series lines) which are then defined by a collimator or slit arrangement (in order to produce a quasi-parallel beam), and are then dispersed by an appropriate crystal according to the well known Bragg equation

$$n\lambda = 2d \sin \theta \tag{2.2}$$

where n is the order of the reflection and is equal to 1, 2, 3, etc.; λ is the wavelength of the diffracted line; d is the interplanar spacing of the crystal; θ is the diffraction angle.

Detectors may be set permanently at 2θ positions which correspond to the diffracted characteristic lines of a number of elements, in which case the spectrometer is referred to as a simultaneous model because many such 'channels' can be accommodated. This permits the measurement of intensities for many elements at the same time, which lends itself to high productivity in a given analytical context. On the other hand, the detector may be coupled to a crystal such that both the crystal and the detector can be rotated while maintaining their relative angular position θ and 2θ. This arrangement, known as the sequential model because intensities are measured each in turn, is more versatile, as the choice of 2θ is unlimited.

In Figure 2.2, the left hand side diagram is typical of end-window X-ray tube spectrometers, and the right hand side diagram is typical of side-window instruments.

Figure 2.3 clearly illustrates the simpler configuration of energy dispersive spectrometers. The emergent fluorescence radiations impinge directly onto a detector that has the property of dispersing radiations on the basis of their energy. For example, Fe $K-L_{2,3}$ radiations are isolated and their intensity is measured based on the fact that their energy is 6.401 keV, whereas in the case of the wavelength dispersive spectrometer described above, the same radiations would be isolated and their intensity measured based on the fact that their wavelength is 1.937 Å. Figure 2.1 also illustrates how a proportionality constant is required in order to account for the loss of intensity due to collimation, absorption along the path length from specimen to detector, and detector efficiency. Figure 2.3, on the other hand, illustrates that the path length from specimen to detector is quite short and there is no loss of intensity due to collimation. The rectangle enclosed by broken lines in Figure 2.1 represents the only components that enter into the derivation of theoretical

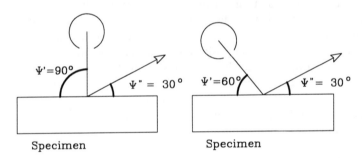

Figure 2.2   Typical geometries encountered in X-ray fluorescence calculations

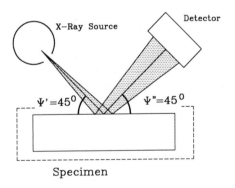

Figure 2.3   Configuration of the energy dispersive spectrometer

Table 2.1 Wavelengths of the K and L_1 absorption edges; $K-M_3$ and $K-L_{2,3}$ characteristic lines of Cr, Fe and Ni

Element	Wavelength, Å			
	K_{abs}	$K-M_3$	$K-L_{2,3}$	$L_{1_{abs}}$
Cr ($Z = 24$)	2.070	2.085	2.291	17.8
Fe ($Z = 26$)	1.743	1.757	1.937	14.7
Ni ($Z = 28$)	1.488	1.500	1.659	12.2

expressions for emitted intensities and are common to both types of spectrometers; i.e., the following treatment applies equally to wavelength dispersion and energy dispersion provided that the necessary λ to E conversions are made.

Although the derivation of theoretical expressions will be carried out for the general case, in which the elements are designated by the subscripts i, j, k, ..., it is desirable to illustrate each process using numerical values relating to specific elements and instrumental operating conditions. The three elements chosen for this purpose are chromium, iron and nickel. The wavelengths relating to these elements that need to be taken into consideration in this Chapter are listed in Table 2.1.

2.3* PRIMARY FLUORESCENCE

Depending on the nature of the specimen being irradiated by a beam of X-radiations, i.e. pure element, binary or multielement system, the emitted characteristic radiations may be the resultant of more than one process. The following treatment examines the processes often encountered in specimens composed of one, two or three elements using nickel, iron and chromium as examples.

2.3.1 Description of primary fluorescence

The simplest example wherein the emitted intensity is due to excitation by the incident beam only is that of a single element specimen. Any one of the three elements could be used; therefore let us consider the situation when a specimen of pure nickel is irradiated by a polychromatic incident beam emanating from an X-ray tube operated at 45 kV. The upper portion of Figure 2.4 represents an atom of nickel in a specimen of pure nickel being excited by incident photons of wavelength λ, resulting in the emission of characteristic nickel $K-L_{2,3}$ X-radiations. The lower portion shows the plot of I_λ versus λ for a

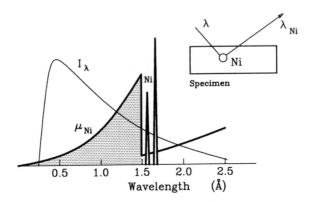

Figure 2.4 Primary fluorescence emission in pure element specimen; Ni K line spectrum

tube operated at this voltage (for clarity, characteristic target lines are not shown) and a plot of μ_{Ni} versus λ for the same λ range. The minimum wavelength λ_{min} is given by Eqn (1.3)

$$\lambda_{min} = \frac{12.3981}{kV} = \frac{12.3981}{45} = 0.276 \text{ Å} \tag{2.3}$$

The longest wavelength that can result in the ejection of K level electrons in nickel atoms is given by substituting the value for the binding (critical excitation) energy of Ni K level electrons, namely 8.333 keV, into Eqn (1.8)

$$\lambda = \frac{12.3981}{8.333} = 1.488 \text{ Å}$$

which is the wavelength of the nickel K absorption edge. The effective wavelength range for the ejection of Ni K level electrons is that between 0.276 and 1.488 Å, represented by the shaded portion in Figure 2.4. It therefore follows that the K series lines of Ni are emitted (only the $K-L_{2,3}$ and $K-M_3$ being shown) and that these emitted X-radiations are due to excitation by the incident beam only. A similar presentation can be made in the case where L characteristic lines are considered except, that three L absorption edges would be involved. Whatever line is chosen for theoretical intensity calculations (generally the $K-L_{2,3}$ or $L_3-M_{4,5}$ lines are selected because they are more sensitive), the symbol $P_{(i)}$, in this case $P_{(Ni)}$, is used to designate the emitted intensity, the parentheses indicating that the intensity is that of the pure element.

Let us now consider the emission of the Ni $K-L_{2,3}$ and Ni $K-M_3$ X-radiations in the system Cr–Fe–Ni under the same conditions as above. From Table 2.1, the critical excitation wavelength of iron is 1.743 Å and that of chromium is 2.070 Å. It therefore follows that the K series lines of Fe and Cr are emitted also as shown in the lower diagram in Figure 2.5. However, the wavelengths of all four lines are longer than the K absorption edge of Ni; thus, the characteristic K lines of Fe and Cr cannot eject K level electrons from Ni atoms and therefore are not part of the process that leads to the emission of Ni $K-L_{2,3}$ and Ni $K-M_3$ radiations. As a result, the emitted Ni $K-L_{2,3}$ and Ni $K-M_3$ intensities are again due to excitation by the incident primary beam only, and the emission of characteristic nickel $K-L_{2,3}$ photons can be represented as shown in the upper portion of Figure 2.5. The theoretically calculated intensity due to primary fluorescence emission for a given element in a multielement specimen is represented by the symbol P_i, in this case P_{Ni}.

2.3.2 Intensity expression for primary fluorescence

Many authors have contributed to the derivation of theoretical expressions for fluorescence emissions. The following treatment is therefore the end-product of many varied approaches. In anticipation of the formalisms in Chapter 3, the derivation will be presented in a formalism that facilitates their linkage. Figure 2.6 (a schematic representation of the portion enclosed by the broken lines in Figures 2.1 and 2.3) illustrates the components that enter into the derivation of an expression for primary fluorescence. Although the entity to quantify will only be in terms of the $K-L_{2,3}$ line from a volume element in a specimen, the expressions apply equally well to other characteristic lines provided the proper substitutions are made. It is assumed that the specimen is homogeneous, flat and infinitely thick with respect to X-ray fluorescence emissions, and that the volume element is a layer labelled dt located at depth t within the specimen. The incident polychromatic beam of known spectral

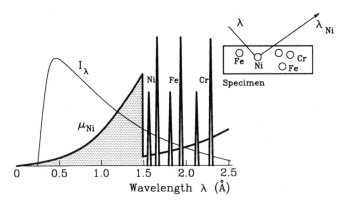

Figure 2.5 Primary fluorescence emission in a multielement specimen, Ni in the presence of Cr and Fe

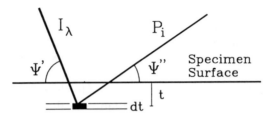

Figure 2.6   Schematic representation of the components that enter into the calculation of primary fluorescence emissions

intensity distribution is assumed to be parallel and to impinge on the specimen at an angle ψ' from the surface, and that the characteristic radiations are emitted at an angle of emergence ψ'' (defined by the collimator). The process is more easily understood if presented firstly for an effective monochromatic wavelength, the polychromaticity of the incident beam being dealt with subsequently.

Following Bertin (1975), the process is considered as involving five factors:

(1) the incident photon intensity at the given incident wavelength;

(2) the attenuation of the incident radiation by the specimen in reaching a layer dt;

(3) the excitation of characteristic X-radiations within this layer;

(4) the fraction of the characteristic photons emitted in the direction of the collimator;

(5) the attenuation of the characteristic radiation on its way out of the specimen by layer t.

(1) The incident photon intensity at wavelength λ is represented by I_λ.

(2) The intensity that reaches the layer dt is the fraction of the incident intensity I_λ transmitted through the effective path length from the specimen surface to the layer. The path length may be expressed as $t/\sin \psi'$ or $t \csc\psi'$, the latter being retained for use. Thus the intensity that reaches the volume element in layer dt, represented by I_v, is given by Eqn (1.26)

$$I_v = I_\lambda \exp(-\mu_{s\lambda}\rho t \csc \psi') \tag{2.4}$$

where I_λ is the intensity of the incident source at wavelength λ; $\mu_{s\lambda}$ is the mass attenuation coefficient of the specimen at wavelength λ; ρ is the specimen density; t is the depth of layer dt in the specimen; ψ' is the angle of incidence.

(3) The quantification of the intensity emitted by a given characteristic line due to the presence of the analyte in layer dt is equal to the product of: (a) the fraction of I_v absorbed by the analyte in the layer dt which is given by Eqns (1.23) and (1.26)

$$\text{fraction absorbed} = C_i\mu_{i\lambda}\rho \, \text{d}t \csc \psi' \tag{2.5}$$

and (b), the fraction of the absorbed incident X-radiations that leads to the fluorescence emission of the selected characteristic line λ_i of element i. That fraction is the probability p_{λ_i} (also referred to as the excitation factor), which is equal to the product of three probabilities, namely

$$p_{\lambda_i} = p_{\text{level}} \times p_{\text{line}} \times p_{\text{fluorescence}} \qquad (2.6)$$

where p_{level} is the *probability* that the absorbed incident radiations ejects electrons from a given level; p_{line} is the *probability* that a given *line* is emitted within its series; $p_{\text{fluorescence}}$ is the *probability* that the energy of the characteristic photon results in a *fluorescence* emission rather than an Auger electron.

The probability that the absorbed electron ejects electrons from a given level rather than electrons from any other level (also known as the absorption jump factor, symbol J) relates to the fraction absorbed by that level. It is given by

$$p_{\text{level}} = \frac{\text{absorption at the level}}{\text{absorption at all levels}} \qquad (2.7)$$

For example, the probability p_K that a K level electron is ejected is given by

$$p_K = \frac{\text{abs. at all levels} - \text{abs. at levels other than } K}{\text{abs. at all levels}} \qquad (2.8)$$

and, from the definition of the absorption jump ratio, Eqn (1.43) and Figure 1.8

$$p_{\text{level}} = \frac{r-1}{r} \qquad (2.9)$$

Thus the probability p_K that a K level electron will be ejected is given by

$$p_K = \frac{r_K - 1}{r_K} \qquad (2.10)$$

By the same token, the probability that an L_1 level electron is ejected is given by

$$p_{L_1} = \frac{r_{L_1} - 1}{r_{L_1}} \qquad (2.11)$$

The probability that a particular line within a series is emitted, p_{line}, is given by the relative intensity of that line within its series. Thus, the probability that $K-L_{2,3}$ photons are emitted rather than $K-M_3$ photons is given by the factor

$$f_{K-L_{2,3}} = \frac{I_{K-L_{2,3}}}{\Sigma I_K \text{ lines}} \qquad (2.12)$$

Similarly, the probability that the $L_3-M_{4,5}$ line will be emitted rather than the other L lines is given by the factor

$$f_{L_3-M_{4,5}} = \frac{I_{L_3-M_{4,5}}}{\Sigma I_L \text{ lines}} \tag{2.13}$$

The probability that the emitted photons are characteristic photons rather than Auger electrons is given by the fluorescence yield ω. Thus, ω_K quantifies the probability that $K-L_{2,3}$ or $K-M_3$ photons produces within the atom will result in a fluorescence emission. In the case of transitions to L levels, there are three distinct fluorescence yields, namely ω_{L_1}, ω_{L_2} and ω_{L_3}.

Thus, in the present case, if only the $K-L_{2,3}$ line of a given element is being considered, the excitation factor or the overall probability that a $K-L_{2,3}$ photon is emitted in a specimen is obtained by combining Eqns (2.10) and (2.12) with ω_K:

$$p_{\lambda_i} = p_{\text{level } K} \times p_{\text{line } K-L_{2,3}} \times p_K \text{ fluorescence yield} \tag{2.14}$$

$$p_{K-L_{2,3}} = \frac{r_K - 1}{r_K} \times f_{K-L_{2,3}} \times \omega_K \tag{2.15}$$

Therefore the intensity of a characteristic line generated in layer dt, represented by I_c, is given by multiplying the intensity I_v of the radiation that reaches the layer dt by the fraction absorbed (Eqn (2.5)) and the probability p_{λ_i} that the characteristic line under consideration is emitted:

$$I_c = I_v C_i \mu_{i\lambda} p_{\lambda_i} \rho \, \csc \psi' \, dt \tag{2.16}$$

(4) The characteristic photons are emitted in all directions; the fraction emitted towards the collimator and the detector is given by $\Omega/4\pi$, where Ω is the solid angle defined by the collimator.

(5) Finally, one must account for the absorption of the emergent characteristic X-radiations of element i by the specimen through the path length $t \csc \psi''$. Thus the emitted intensity I_e of a characteristic line at the specimen surface is given by

$$I_e = I_c \exp(-\mu_{s\lambda_i} \rho t \csc \psi'') \tag{2.17}$$

where $\mu_{s\lambda_i}$ is the mass attenuation coefficient of the specimen at the characteristic wavelength of element i, and ρ, t and ψ'' have been defined previously.

Defining

$$\mu_s' = \mu_{s\lambda} \csc \psi' \tag{2.18}$$

$$\mu_s'' = \mu_{s\lambda_i} \csc \psi'' \tag{2.19}$$

$$\mu_s^* = \mu_s' + \mu_s'' \tag{2.20}$$

and combining the five factors leads to an expression for the primary fluorescence $dP_{i\lambda}$ for a given line excited by an incident beam of wavelength λ in layer dt

$$dP_{i\lambda} = I_\lambda C_i \mu_{i\lambda} \rho p_{\lambda_i} \csc \psi' \frac{\Omega}{4\pi} \exp(-\mu_s^* \rho t) \, dt \tag{2.21}$$

Integrating from $t = 0$ to $t =$ infinite thickness, and rearranging terms, the primary fluorescence $P_{i\lambda}$ emitted by an element in a specimen irradiated by a monochromatic excitation source wavelength λ is given by

$$P_{i\lambda} = I_\lambda p_{\lambda_i} \frac{\Omega}{4\pi} \csc \psi' \frac{C_i \mu_{i\lambda}}{\mu_s^*} \tag{2.22}$$

where μ_s^* is the effective mass attenuation coefficient of the specimen. A similar equation for primary fluorescence derived in te---- of energies instead of wavelengths is given in Jenkins *et al.* (1981).

Given that p_{λ_i} is constant for any characteristic that $\Omega/4\pi \csc \psi'$ is constant for any one instrume proportionality constant

J.E. Willis (1988) p.215

$$G_i = p_{\lambda_i} \frac{\Omega}{4\pi} \csc \psi'$$

Substitution in Eqn (2.22) and regrouping term

$$P_{i\lambda} = G_i C_i \frac{I_\lambda \mu_{i\lambda}}{\mu_s^*} \tag{2.24}$$

For the intensity emitted by a pure element, the last expression reduces to

$$P_{(i)\lambda} = G_i \frac{I_\lambda \mu_{i\lambda}}{\mu_i^*} \tag{2.25}$$

where by analogy with Eqn (2.20), i.e., $\mu_s^* = \mu_s' + \mu_s''$

$$\mu_i^* = \mu_{i\lambda} \csc \psi' + \mu_{i\lambda_i} \csc \psi'' = \mu_i' + \mu_i'' \tag{2.26}$$

The fact that nickel was used to describe the process of primary fluorescence emission served simply to make it possible to introduce actual numerical values for illustrative purposes. In the case of nickel K series emissions, any element with atomic number Z less than 27 (Co) could be a constituent of the sample, as none of the characteristic lines of these elements have emissions shorter than the K absorption edge of nickel that could excite Ni. Equations (2.21) and (2.22) therefore quantify the fluorescence emission process for any element in a specimen, provided that none of the other elements present generate characteristic radiations that are shorter than the absorption edge related to the line of element i selected for measurement. In a multicomponent specimen, only one element is likely to meet this condition, namely the element having the shortest absorption edge wavelength.

2.3.3 Experimental verification

The experimental data given in Section II.2.3 were obtained on an energy dispersive spectrometer using the characteristic $K-L_{2,3}$ line of zinc ($\lambda = 1.437$ Å) to irradiate a specimen containing 3.29% of Ni and 95.49% of Fe and a specimen of pure nickel. The incidence and emergence angles ψ' and ψ'' are equal to 45 degrees, hence csc $\gtreqqless' = \csc \psi w = 1.4142$. The irradiation time was 1000 s. The emitted Ni $K-L_{2,3}$ intensity for the Fe$-$Ni specimen was 19.7 c/s, whereas that for pure Ni was 1017.8 c/s.

The first step involves the combination of Eqns (2.24) and (2.25) so as to define $P_{i\lambda}$ in terms of $P_{(i)\lambda}$. Dividing Eqn (2.24) by Eqn (2.25) results in the cancellation of G_i, $\mu_{i\lambda}$ and I_λ, leading to

$$P_{i\lambda} = P_{(i)\lambda} C_i \frac{\mu_i^*}{\mu_s^*} \tag{2.27}$$

The link between experimentally measured and theoretically calculated intensities, given by Eqn (2.1), leads to (the g_i proportionality constant cancels out)

$$I_{i\lambda} = I_{(i)\lambda} C_i \frac{\mu_i^*}{\mu_s^*} \tag{2.28}$$

Recalling the C and n values for nickel and iron determined experimentally in Section 1.9, the four mass attenuation coefficients required can be computed

$$\mu_i' = \mu_{Ni}' = 118.2 \times 1.437^{2.718} \times 1.4142 = 447.81 \text{ cm}^2 \text{g}^{-1}$$

$$\mu_i'' = \mu_{Ni}'' = 14.13 \times 1.659^{2.804} \times 1.4142 = 82.62 \text{ cm}^2 \text{g}^{-1}$$

$$\mu_j' = \mu_{Fe}' = 95.91 \times 1.437^{2.736} \times 1.4142 = 365.74 \text{ cm}^2 \text{g}^{-1}$$

$$\mu_j'' = \mu_{Fe}'' = 95.91 \times 1.659^{2.736} \times 1.4142 = 541.85 \text{ cm}^2 \text{g}^{-1}$$

$$\mu_s' = (0.0329 \times 447.81) + (0.9549 \times 365.74) = 363.98 \text{ cm}^2 \text{g}^{-1}$$

$$\mu_s'' = (0.0329 \times 82.62) + (0.9549 \times 541.85) = 520.13 \text{ cm}^2 \text{g}^{-1}$$

$$\mu_i^* = 447.81 + 82.62 = 530.43 \text{ cm}^2 \text{g}^{-1}$$

$$\mu_s^* = 363.98 + 520.13 = 884.11 \text{ cm}^2 \text{g}^{-1}$$

Substituting in Eqn (2.28)

$$I_{Ni\lambda} = 1017.8 \times 0.0329 \times \frac{530.43}{884.11}$$

$$= 20.1 \text{ c/s}$$

which compares reasonably well with the measured value of 19.7 c/s. By direct proportion, the measured intensity would be

$$I_{(Ni)\lambda} C_i = 1017.8 \times 0.0329 = 33.5 \text{ c/s}$$

Thus, the intensity emitted by nickel is reduced from 33.5 c/s (the intensity expected if iron and nickel had identical mass absorption coefficients) to 20.1 c/s (measured value 19.7 c/s) because of the higher absorption of the nickel X-radiations by iron.

Experimental data relating to three other binary specimens and pure iron and nickel for incident $\lambda = 0.492$, 0.711, 1.437 and 1.659 Å are listed in Table II.2.2.

2.3.4 Primary fluorescence, polychromatic excitation source

To get the total emitted intensity generated by a polychromatic incident beam, Eqns (2.24) and (2.25) must be integrated over the effective wavelength range, i.e. over the range λ_{min} to λ_{abs}, resulting in

$$P_i = G_1 C_1 \int_{\lambda_{min}}^{\lambda_{abs,i}} \frac{I_\lambda \mu_{i\lambda}}{\mu_s^*} \, d\lambda \qquad (2.29)$$

and

$$P_{(i)} = G_i \int_{\lambda_{min}}^{\lambda_{abs,i}} \frac{I_\lambda \mu_{i\lambda}}{\mu_i^*} \, d\lambda \qquad (2.30)$$

where I_λ is expressed by equations of the type of Eqn (1.13)

$$I_\lambda = CZ \frac{1}{\lambda^2} \left(\frac{\lambda}{\lambda_{min}} - 1 \right) \qquad (2.31)$$

or obtained from experimental data.

2.4* SECONDARY FLUORESCENCE

As shown schematically in Figure 2.5, the presence of chromium and iron does not contribute to the emission process that leads to the fluorescence emission of Ni $K-L_{2,3}$ and Ni $K-M_3$ radiations. If we now examine the fluorescence emission of the characteristic K lines of iron in the same Cr–Fe–Ni system, a new phenomenon or process is involved, that of secondary fluorescence or, as it is more familiarly expressed, the process of enhancement.

2.4.1 Description of secondary fluorescence

The difference between the emission processes related to the emission of characteristic Ni radiations and characteristic Fe radiations may be observed by comparing Figure 2.5 relating to Ni and its Fe counterpart, Figure 2.7. It is considered that both samples are irradiated under the same condition, i.e. 45 kV; hence λ_{min} is equal to 0.276 Å in both cases. However, the effective excitation wavelength range is somewhat wider because the K absorption edge

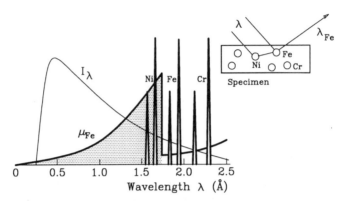

Figure 2.7 Secondary fluorescence emission, Fe in the presence of Ni

or critical excitation wavelength for Fe is equal to 1.743 Å rather than 1.488 Å (Table 2.1). The major difference, however, is the fact that the two nickel K lines are both shorter than the K absorption edge of iron and contribute to the emission of Fe radiation. The two chromium lines, on the other hand, are longer than the K absorption edge of iron and, thus, the effect of chromium in the process of the fluorescence emission of characteristic iron K lines is analogous to that of iron towards nickel emissions, namely, chromium is a strong absorber of iron radiations.

As shown schematically in Figure 2.7, the characteristic nickel photons generated within the specimen may encounter iron atoms, so that iron K level electrons are ejected, leading to additional emission of Fe $K-L_{2,3}$ and Fe $K-M_3$ photons. There is a fairly high probability of this happening because iron is a strong absorber of nickel K radiations; hence the corollary: nickel is a strong enhancer of iron. Therefore, the emission of Fe $K-L_{2,3}$ and Fe $K-M_3$ X-radiations in the presence of nickel is the resultant of two processes:

- some K level electrons are ejected directly by reacting with incident beam photons;
- some K level electrons are ejected indirectly; i.e., some incident beam photons eject nickel K level electrons, leading to the emission of Ni $K-L_{2,3}$ and Ni $K-M_3$ photons, which in turn have enough energy to eject iron K level electrons.

Both directly and indirectly produced characteristic iron radiations undergo absorption by chromium and nickel in addition to self absorption within the specimen.

2.4.2 Intensity expression for secondary fluorescence

By analogy with the symbol P_i used to designate theoretically calculated *Primary* fluorescence emissions, the symbol S_i is used to designate theoretically calculated *Secondary* fluorescence emissions. In multielement specimens, some emitted X-ray lines are likely to be the resultant of enhancement by more than one element. Thus the symbol S_i represents the total enhancement or intensity due to secondary fluorescence emissions, and is the sum of each individual enhancement effect, i.e.

$$S_i = S_{ij} + S_{ik} + \cdots \tag{2.32}$$

where S_{ij} is the total enhancement of element i by element j, and S_{ik} is the total enhancement of element i by element k.

In the case of secondary fluorescence emissions of iron in the presence of nickel, both Ni $K-L_{2,3}$ and Ni $K-M_3$ radiations can eject K level iron electrons; thus, in such a case, S_{ij} is the sum of two enhancement terms (omitting other minor K lines), namely

$$S_{ij} = S_{ij,K-L_{2,3}} + S_{ij,K-M_3} \tag{2.33}$$

Should both K and L lines of element j induce enhancement, then

$$S_{ij} = S_{ij,K-L_{2,3}} + S_{ij,K-M_3} + S_{ij,L_3-M_{4,5}} + \cdots \tag{2.34}$$

For space considerations, the following development will be limited to enhancement by a single characteristic line; e.g., $S_{ij,K-L}$. Figure 2.8 is a simplified schematic representation of the components that enter into the quantification of secondary fluorescence emissions. Rather than presenting a detailed derivation of each mathematical expression involved, the following is limited to describing what has to be quantified. Readers should refer to

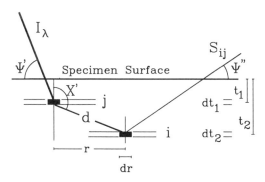

Figure 2.8 Schematic representation of the components that enter into the calculation of secondary fluorescence emissions

Shiraiwa and Fujino (1966) or Tertian and Claisse (1982) for a more detailed mathematical treatment.

As shown in Figure 2.8, there are now two layers involved. Layer dt_1 is the layer in which an element j emits radiations that leads to enhancement, in the present case nickel. The second layer dt_2 is the layer in which the analyte i is enhanced, in the example under consideration iron. Thus, primary emissions in layer dt_1 lead to secondary emissions in layer dt_2. The components to be quantified are as follows:

- attenuation of the incident monochromatic parallel beam wavelength λ by the specimen in reaching layer dt_1;
- the intensity P_j of the X-radiations of wavelength λ_j emitted in layer dt_1. So far, the process is analogous to that described for the emission of nickel in the previous section;
- the intensity that reaches layer dt_2. The complication, as it were, is that now the emanations from layer dt_1 cannot be described by a narrow parallel beam but form a family of cones that subtends annular volume elements of radius r, since the characteristic radiations of element j are emitted in all directions;
- the intensity relation for P_i at wavelength λ is applied in a similar way to a narrow segment of width dr of layer dt_2. After absorption by the specimen along the path length d between the volume elements in layers dt_1 and dt_2 and emission of λ_i in the annular element dr, the result is $S_{ij\lambda}$ for the volume elements selected. One must then integrate for the whole layer dt_2, i.e. for angle $\chi' = 0$ to $\chi' = \pi$;
- the intensity of the characteristic X-radiations of element i generated in layer dt_2 is attenuated as it emerges from the specimen at angle ψ'';
- layers dt_1 and dt_2 are considered to be at depths of t_1 and t_2 within the specimen. Therefore one must integrate over all possible combinations of t_1 and t_2, i.e. from t_1 and $t_2 = 0$ to t_1 and $t_2 = $ infinity.

Thus, compared with the primary fluorescence emission which involves integration over two variables t and λ, the secondary fluorescence emission involves four integrations over the variables t_1, t_2, r and λ when the polychromaticity of the incident beam is taken into consideration. For a given monochromatic incident wavelength λ, the emitted intensity of element i due to secondary excitation by any of the characteristic X-radiations of element j, in a specimen considered infinitely thick with respect to X-ray fluorescence emissions, is

$$S_{ij\lambda} = G_i C_i \frac{I_\lambda \mu_{i\lambda}}{\mu_s^*} e_{ij\lambda} C_j \qquad (2.35)$$

where

$$e_{ij\lambda} = [e(e' + e'')]_{\lambda_j} \qquad (2.36)$$

with

$$e = 0.5 p_{\lambda_j} \mu_{i\lambda_j}(\mu_{j\lambda}/\mu_{i\lambda}) \qquad (2.37)$$

$$e' = \frac{1}{\mu_s'} \ln\left(1 + \frac{\mu_s'}{\mu_{s\lambda_j}}\right) \qquad (2.38)$$

$$e'' = \frac{1}{\mu_s''} \ln\left(1 + \frac{\mu_s''}{\mu_{s\lambda_j}}\right) \qquad (2.39)$$

Combining with Eqn (2.35) leads, for a given monochromatic incident wavelength λ, to

$$S_{ij\lambda} = P_{i\lambda} e_{ij\lambda} C_j \qquad (2.40)$$

As mentioned previously, if more than one line of element j can lead to enhancement, then the total enhancement is given by the sum over each individual line; thus, in the common case where both characteristic K lines contribute to the enhancement effect

$$S_{ij\lambda} = P_{i\lambda} e_{ij\lambda_{K-L_{2,3}}} C_j + P_{i\lambda} e_{ij\lambda_{K-M_3}} C_j \qquad (2.41)$$

In the present case, where iron is the analyte and nickel the enhancing element, the secondary fluorescence of iron for an effective incident wavelength λ short enough to excite both iron and nickel atoms is given by

$$S_{\text{FeNi}\lambda} = S_{\text{FeNi}\lambda_{K-L_{2,3}}} + S_{\text{FeNi}\lambda_{K-M_3}} \qquad (2.42)$$

where the last two terms are expressed by Eqn (2.40).

2.4.3 Experimental verification

The data relating to the incident source, system, specimen, fundamental parameters and experimentally measured intensities are given in Table 2.2. The exercise consists in calculating the emitted intensity of the $K-L_{2,3}$ line of iron in a specimen of known Fe–Ni composition given the emitted intensity of the $K-L_{2,3}$ line of iron in a specimen of pure iron, both irradiated by a monochromatic incident beam. The result can then be compared with the intensity actually measured. Recalling that experimentally measured intensities can be linked to theoretically calculated intensities by a proportionality constant, Eqn (2.1)

$$I_{i\lambda} = g_i(P_i + S_{ij})_\lambda \qquad (2.43)$$

Table 2.2 Analytical context and fundamental parameters used for the calculation of I_{Fe} in the presence of Ni; monochromatic excitation source

Analytical context
Excitation source: Mo $K-L_{2,3}$ ($\lambda = 0.711$ Å), 17.44 keV
Instrument geometry: $\psi' = 45°$; $\psi'' = 45°$; csc $45° = 1.4142$
System: Fe–Ni; specimen: $C_{Fe} = 0.2263$, $C_{Ni} = 0.7711$
Experimental data: $I_{(Fe)} = 626.2$ c/s, $I_{Fe} = 209.9$ c/s

Element	$\lambda_{K-L_{2,3}}$ (Å)	λ_{K-M_3} (Å)	λ_K (Å)	λ_{L_1} (Å)
Fe	1.937	1.757	1.743	14.7
Ni	1.659	1.500	1.488	12.2

	$\lambda < \lambda_K$		$\lambda_K < \lambda < \lambda_{L_1}$	
	C_K	n_K	C_{KL}	n_{KL}
Fe	95.91	2.736	11.14	2.794
Ni	118.2	2.718	14.13	2.804

Data from Dalton and Goldak (1969)

	r_K	$f_{K-L_{2,3}}$	ω_K
Fe	8.221	0.892	0.350
Ni	7.855	0.892	0.420

r_K data from McMaster *et al.* (1969)
$f_{K-L_{2,3}}$ and ω_K data from Broll (1986)

and

$$I_{(i)\lambda} = g_i P_{(i)\lambda} \tag{2.44}$$

it is possible in the present case to write

$$I_{Fe\lambda} = g_{Fe}(P_{Fe} + S_{FeNi})_\lambda \tag{2.45}$$

and

$$I_{(Fe)\lambda} = g_{Fe} P_{(Fe)\lambda} \tag{2.46}$$

where $P_{Fe\lambda}$ is the intensity of the Fe $K-L_{2,3}$ line due to primary fluorescence emission only, and $S_{FeNi\lambda}$ is the intensity due to secondary fluorescence emission, i.e. enhancement of iron by nickel. The iron intensity due to primary fluorescence emission is computed as in Section 2.3.3, and enhancement is computed using the additional data in Table 2.2, i.e.

$$I_{Fe\lambda} = I_{FE\lambda,pri} + I_{Fe\lambda,pri}e_{FeNi\lambda_{K-L_{2,3}}}C_{Ni} + I_{Fe\lambda,pri}e_{FeNi\lambda_{K-M_3}}C_{Ni} \qquad (2.47)$$

$$I_{Fe\lambda} = I_{Fe\lambda,pri} + I_{Fe\lambda,pri}(e_{FeNi\lambda_{K-L_{2,3}}} + e_{FeNi\lambda_{K-M_3}})C_{Ni} \qquad (2.48)$$

The computations of the mass attenuation coefficients using Eqn (1.32) are as follows:

$$\mu_{Fe(0.711\ \text{Å})} = 95.91 \times 0.711^{2.736} = 37.72\ \text{cm}^2\,\text{g}^{-1}$$

$$\mu_{Ni(0.711\ \text{Å})} = 118.2 \times 0.711^{2.718} = 46.77\ \text{cm}^2\,\text{g}^{-1}$$

$$\mu_{Fe(1.937\ \text{Å})} = 11.14 \times 1.937^{2.794} = 70.65\ \text{cm}^2\,\text{g}^{-1}$$

$$\mu_{Ni(1.937\ \text{Å})} = 14.13 \times 1.937^{2.806} = 90.21\ \text{cm}^2\,\text{g}^{-1}$$

$$\mu_{Fe(1.659\ \text{Å})} = 95.91 \times 1.659^{2.736} = 383.15\ \text{cm}^2\,\text{g}^{-1}$$

$$\mu_{Ni(1.659\ \text{Å})} = 14.13 \times 1.659^{2.806} = 58.42\ \text{cm}^2\,\text{g}^{-1}$$

$$\mu_{Fe(1.500\ \text{Å})} = 95.91 \times 1.500^{2.736} = 290.84\ \text{cm}^2\,\text{g}^{-1}$$

$$\mu_{Ni(1.500\ \text{Å})} = 14.13 \times 1.500^{2.806} = 44.05\ \text{cm}^2\,\text{g}^{-1}$$

Instrumental geometry is then taken into consideration to compute effective mass attenuation coefficients

$$\mu'_{Fe} = 37.72 \times 1.4142 = 53.34\ \text{cm}^2\,\text{g}^{-1}$$

$$\mu'_{Ni} = 46.77 \times 1.4142 = 66.14\ \text{cm}^2\,\text{g}^{-1}$$

$$\mu'_s = \mu'_{Fe}C_{Fe} + \mu'_{Ni}C_{Ni} = (53.34 \times 0.2263) + (66.14 \times 0.7711)$$
$$= 63.07\ \text{cm}^2\,\text{g}^{-1}$$

$$\mu''_{Fe} = 70.65 \times 1.4142 = 99.91\ \text{cm}^2\,\text{g}^{-1}$$

$$\mu''_{Ni} = 90.21 \times 1.4142 = 127.57\ \text{cm}^2\,\text{g}^{-1}$$

$$\mu''_s = \mu''_{Fe}C_{Fe} + \mu''_{Ni}C_{Ni} = (99.91 \times 0.2263) + (127.57 \times 0.7711)$$
$$= 120.98\ \text{cm}^2\,\text{g}^{-1}$$

$$\mu_{s(1.659\ \text{Å})} = (383.15 \times 0.2263) + (58.42 \times 0.7711) = 131.76\ \text{cm}^2\,\text{g}^{-1}$$

$$\mu_{s(1.500\ \text{Å})} = (290.84 \times 0.2263) + (44.05 \times 0.7711) = 99.79\ \text{cm}^2\,\text{g}^{-1}$$

$$\mu^*_{Fe} = 53.34 + 99.91 = 153.25\ \text{cm}^2\,\text{g}^{-1}$$

$$\mu^*_{Ni} = 66.14 + 127.57 = 193.71\ \text{cm}^2\,\text{g}^{-1}$$

$$\mu^*_s = 63.07 + 120.98 = 184.05\ \text{cm}^2\,\text{g}^{-1}$$

Substitution into Eqn (2.28) quantifies the emitted iron intensity due to primary fluorescence emission only, i.e. $I_{Fe\lambda,pri}$

$$I_{Fe\lambda,pri} = 626.2 \times 0.2263 \times \frac{153.25}{184.05}$$

$$= 117.8\ \text{c/s}$$

The computation of enhancement of iron by nickel is as follows. The total enhancement is given by Eqns (2.41) and (2.42), indicating that enhancement by both the Ni $K-L_{2,3}$ and Ni $K-M_3$ lines must be taken into consideration. The term $e_{NiK-L_{2,3}}$ is calculated from Eqn (2.37) after p_{Ni} has been calculated from Eqn (2.15)

$$p_{NiK-L_{2,3}} = \frac{r_{K,Ni} - 1}{r_{K,Ni}} \times f_{K-L_{2,3}} \times \omega_{Ni,K}$$

$$= \frac{7.885 - 1}{7.885} \times 0.892 \times 0.420$$

$$= 0.327 \tag{2.49}$$

$$e_{NiK-L_{2,3}} = 0.5 p_{\lambda_{NiK-L_{2,3}}} \mu_{Fe\lambda_{NiK-L_{2,3}}} \mu_{Ni\lambda} / \mu_{Fe\lambda}$$
$$= 0.5 \times 0.327 \times 383.15 \times 46.77/37.72$$
$$= 77.68 \tag{2.50}$$

The terms $e'_{NiK-L_{2,3}}$ and $e''_{NiK-L_{2,3}}$ are calculated from Eqns (2.38) and (2.39)

$$e'_{NiK-L_{2,3}} = \frac{1}{\mu'_s} \ln\left(1 + \frac{\mu'_s}{\mu_{s\ NiK-L_{2,3}}}\right)$$

$$= \frac{1}{63.07} \ln\left(1 + \frac{63.07}{131.76}\right) = 0.00620 \tag{2.51}$$

$$e''_{Ni\ K-L_{2,3}} = \frac{1}{\mu''_s} \ln\left(1 + \frac{\mu''_s}{\mu_{s\ Ni\ K-L_{2,3}}}\right)$$

$$= \frac{1}{120.98} \ln\left(1 - \frac{120.98}{131.76}\right) = 0.00538 \tag{2.52}$$

The term $e_{FeNi\ K-L_{2,3}}$ is given by Eqn (2.36)

$$e_{FeNi\ K-L_{2,3}} = e_{NiK-L_{2,3}}(e'_{Ni\ K-L_{2,3}} + e''_{NiK-L_{2,3}})$$
$$= 77.68 \times (0.00620 + 0.00538) = 0.8995 \tag{2.53}$$

A similar process is followed to calculate $e_{FeNiK-M_3}$, namely

$$e_{Ni\ K-M_3} = 0.5 \times 0.0396 \times 290.84 \times 46.77/37.72 = 7.140$$

$$e'_{Ni\ K-M_3} = \frac{1}{63.07} \ln\left(1 + \frac{63.07}{99.79}\right) = 0.00777$$

$$e''_{NiK-M_3} = \frac{1}{120.98} \ln\left(1 + \frac{120.98}{99.79}\right) = 0.00656$$

$$e_{FeNiK-M_3} = 7.140 \times (0.00777 + 0.00656) = 0.1023$$

Finally, the total emitted intensity is computed from Eqn (2.48)

$$I_{Fe\lambda} = 117.8 + [117.8 \times (0.8995 + 0.1023) \times 0.7711]$$
$$= 117.8 + (117.8 \times 1.0018 \times 0.7711)$$
$$= 208.8, \text{ as compared with } 209.9 \text{ c/s measured experimentally.}$$

In order to simplify the above numerical calculations, some approximations can be made. Subsequent computations involving systems typical of Cr–Fe–Ni are carried out assuming that the same wavelength, that of the $K-L_{2,3}$ line, can be used for the enhancement due to both the $K-L_{2,3}$ and the $K-M_3$ line. In this particular instance the error so introduced is negligible. The probability that the $K-L_{2,3}$ line of nickel is emitted becomes

$$p_{NiK-L_{2,3}} = \frac{r_{K,Ni} - 1}{r_{K,Ni}} (f_{NiK-L_{2,3}} + f_{NiK-M_3}) \omega_{Ni,K}$$

$$= \frac{7.855 - 1}{7.855} \times 1.0 \times 0.420$$

$$= 0.3665 \tag{2.54}$$

Substitution in Eqns (2.37), (2.36) and (2.48) leads to, respectively

$$e_{FeNiK-L_{2,3}} = 0.5 \times 0.3665 \times 383.15 \times 46.77/37.72$$
$$= 87.06$$

$$e_{FeNi} = 87.06 \times (0.00620 + 0.00538) = 1.008$$

and

$$I_{Fe\lambda} = 117.8 + (117.8 \times 1.008 \times 0.7711) = 209.4 \text{ c/s}$$

as compared with 208.8 c/s obtained previously when the enhancement effects of the $K-L_{2,3}$ and $K-M_3$ lines were calculated separately.

2.4.4 Secondary fluorescence, polychromatic excitation source

So far, secondary fluorescence has been calculated by Eqn (2.35), which expresses the emitted intensity of element i for a given *monochromatic incident source*. Considering that the incident beam may be polychromatic, Eqn (2.35) must then be integrated over the effective incident wavelength range that results in element j enhancing element i, i.e.

$$S_{ij} = G_1 C_i \int_{\lambda_{min}}^{\lambda_{abs,j}} \frac{I_\lambda \mu_{i\lambda}}{\mu_s^*} (e_{ij\lambda} C_j) \, d\lambda \tag{2.55}$$

Thus the overall conditions which must be met in order to quantify the enhancement of element i by element j for any given incident wavelength are:

- the incident λ must be shorter than a given absorption edge of element j, i.e. characteristic lines of element j are emitted, and
- the wavelengths of the emitted lines of j must be shorter than the absorption edge of element i, therefore element i is enhanced by element j and $S_{ij\lambda_j}$ is computed.

Thus, in the context of the analyte iron in the Fe–Ni system, all incident wavelengths shorter than 1.743 Å result in primary fluorescence emissions of iron, and all incident wavelengths shorter than 1.488 Å result in secondary fluorescence emissions of iron due to the enhancement effect by nickel.

As far as absorption and enhancement are concerned, the binary system Cr–Fe presents a case analogous to that of the binary system Fe–Ni; i.e. if iron is the analyte, only primary fluorescence emissions are involved; if chromium is the analyte, then the $K-L_{2,3}$ and $K-M_3$ lines of iron give rise to secondary fluorescence emissions of chromium. If chromium is the analyte in the presence of iron and nickel, the characteristic lines of Ni also give rise to enhancement of chromium, but an additional process is involved, which is examined next.

2.5 TERTIARY FLUORESCENCE

The realization that an element enhancing another may already be enhanced by a third element results in an additional component of fluorescence emission. It should be noted that this process will be referred to as *tertiary fluorescence* T_{ijk} rather than third element effect, which on occasion is also used to refer to the process about to be described. The term *third element effect* is retained in subsequent chapters in a different context.

2.5.1 Description of tertiary fluorescence

The upper portion of Figure 2.9 illustrates the four excitation sources involved in tertiary fluorescence, giving rise to the generation of characteristic chromium X-radiations in the presence of iron and nickel.

- One excitation source is the incident beam giving rise to $P_{Cr\lambda}$, i.e. $P_{i\lambda}$;
- another excitation source is the characteristic lines of iron excited by the incident beam giving rise to $S_{CrFe\lambda}$, i.e. $S_{ij\lambda}$;
- another excitation source is the characteristic lines of nickel excited by the incident beam giving rise to $S_{CrNI\lambda}$, i.e. $S_{ik\lambda}$;

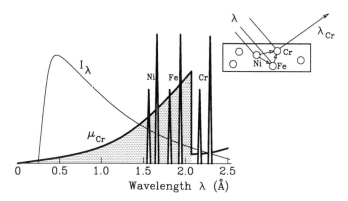

Figure 2.9 Tertiary fluorescence emission, Cr in the presence of Fe and Ni

- another excitation source is the characteristic lines of iron excited by the characteristic lines of nickel which were excited in the first place by the incident beam T_{CrFeNi}, i.e. T_{ijk}.

The lower diagram in Figure 2.9 is similar to those in Figures 2.5 and 2.7, except that chromium is now the analyte. The diagram illustrates that the $K-L_{2,3}$ and $K-M_3$ lines of nickel and iron are all shorter than the K absorption edge of chromium and hence cause secondary fluorescence emissions. However, as shown in Figure 2.8, some nickel photons lead to ejection of K level iron electrons, resulting in secondary iron fluorescence emissions. It is this process that leads in turn to the emission of tertiary fluorescence from chromium atoms. It will be shown subsequently that enhancement due to secondary emission drops off fairly rapidly as the difference between the wavelength of the K edge of the analyte and the wavelength of the lines causing the enhancement increases. In other words, nickel ($Z = 28$) is less efficient than iron ($Z = 26$) in enhancing Cr, and zinc ($Z = 30$) is even less efficient than Ni, etc. Thus, for example, molybdenum ($Z = 42$) is a very weak enhancer of chromium.

2.5.2 Intensity expression for tertiary fluorescence

Based on the experience of the previous section regarding the derivation of an intensity expression for secondary emission, which is somewhat more complex than that for primary emission, the reader may very well surmise that the intensity expression for tertiary fluorescence is even more complex. Figure 2.10 provides an illustration, greatly simplified, of the somewhat complex processes

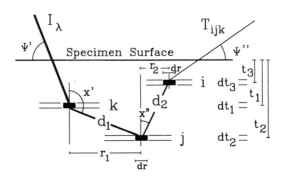

Figure 2.10 Schematic representation of the components that enter into the calculation of tertiary fluorescence emissions

involved in quantifying T_{ijk} in the example under consideration, T_{CrFeNi}. Now we must consider three layers or slabs:

- the layer dt_1 at a distance t_1 from the surface, which relates to element k, i.e. nickel, the source of primary X-radiations;
- the layer dt_2 at a distance t_2 from the surface, which relates to element j, i.e. iron, which constitutes the source of the secondary X-radiations;
- the layer dt_3 at a distance t_3 from the surface, which relates to the analyte i, i.e. chromium, which is the element whose emitted intensity involves tertiary fluorescence.

Up to the emission of X-radiations of wavelength λ_j and intensity I_j from a theoretically infinitesimally thin annular volume in layer dt_2, the processes are similar to those described in the previous section. The tertiary process is further complicated in that we have λ_j radiations emanating from points in an annular volume in layer dt_2 that are absorbed by the specimen along path lengths represented by d_2, which then eject K level electrons of element i in an annular volume of radius r_2 in layer dt_3. This results in the emission of additional intensities, which are labelled T_{ijk}. For a more detailed treatment, the reader should refer to Shiraiwa and Fujino (1966), who showed that, in the case of a system consisting of three elements in which tertiary fluorescence is involved in the generation of characteristic radiations of element i, T_{ijk} is given by

$$T_{ijk} = 0.25 C_i \int_{\lambda_{min}}^{\lambda_{abs,k}} P_{i\lambda} F_1 F_2 \tag{2.56}$$

where

$$F_1 = p_{\lambda_j} C_j p_{\lambda_k} C_k \mu_{k\lambda} \mu_{i\lambda} \mu_{i\lambda_j} \mu_{j\lambda_k} \tag{2.57}$$

$$F_2 = \frac{1}{\mu_s'} \ln\left(1 + \frac{\mu_s'}{\mu_{s\lambda_j}}\right) \frac{1}{\mu_s'} \ln\left(1 + \frac{\mu_s'}{\mu_{s\lambda_k}}\right)$$

$$+ \frac{1}{\mu_s''} \ln\left(1 + \frac{\mu_s''}{\mu_{s\lambda_j}}\right) \frac{1}{\mu_s'} \ln\left(1 + \frac{\mu_s'}{\mu_{s\lambda_k}}\right)$$

$$+ \frac{1}{\mu_s''} \ln\left(1 + \frac{\mu_s''}{\mu_{s\lambda_j}}\right) \frac{1}{\mu_s''} \ln\left(1 + \frac{\mu_s''}{\mu_{s\lambda_k}}\right)$$

$$+ \left(\frac{1}{\mu_s'} + \frac{1}{\mu_s''}\right)\left[\frac{1}{\mu_{s\lambda_j}} \ln(1 + t_{jk}) + \frac{1}{\mu_{s\lambda_k}} \ln(1 + t_{kj})\right]$$

$$- \frac{1}{\mu_s'} \int_{t=0}^{t=t_{jk}} \frac{1}{\mu_s' t + \mu_{s\lambda_k}} \ln\left(\frac{1+t}{t}\right) dt$$

$$- \frac{1}{\mu_s''} \int_{t=0}^{t=t_{kj}} \frac{1}{\mu_s'' t + \mu_{s\lambda_j}} \ln\left(\frac{1+t}{t}\right) dt \qquad (2.58)$$

where

$$t_{jk} = \frac{\mu_{s\lambda_k}}{\mu_{s\lambda_j}} \qquad (2.59)$$

$$t_{kj} = \frac{\mu_{s\lambda_j}}{\mu_{s\lambda_k}} \qquad (2.60)$$

As stated by Tertian and Claisse (1982), the last two terms in the F sum cannot be integrated analytically; hence they must be calculated graphically or numerically. Recalling Eqn (2.58), it is observed that the first three terms in F_2 could be expressed simply as

$$e_j' e_k' + e_j'' e_k' + e_j'' e_k''$$

It must be realized at this point that the quantification of T_{ijk} involves taking into consideration all the combinations between the characteristic lines of elements j and k that can produce tertiary fluorescence emissions. In the case of chromium in the Cr–Fe–Ni system, this results in the intensity of the Cr $K–L_{2,3}$ line due to tertiary fluorescence being equal to the sum of four terms; i.e.

$$T_{CrFeNi} = T_{CrFeK-L_{2,3}NiK-L_{2,3}} + T_{CrFeK-L_{2,3}NiK-M_3} + T_{CrFeK-M_3NiK-L_{2,3}}$$

$$+ T_{CrFeK-M_3NiK-M_3} \qquad (2.61)$$

By the same token, each combination of three elements that leads to tertiary fluorescence must also be taken into consideration; thus the total tertiary fluorescence for element i is given by

$$T_i = T_{ijk} + T_{ijl} + T_{ikl} + \cdots + T_{imn} \qquad (2.62)$$

Given the number of possible permutations and combinations, it is evident that if all or most tertiary emissions were significant, quantitative X-ray fluorescence analysis would be, if not seriously compromised, at least quite difficult, unless the unknown specimen and the reference standard were essentially identical.

2.6 RELATIVE CONTRIBUTION FROM PRIMARY, SECONDARY AND TERTIARY EMISSIONS

Very little data has appeared in the literature regarding the relative contribution of tertiary fluorescence to the total emitted intensity. Table 2.3 is a partial list of data from Shiraiwa and Fujino (1967) for the Cr–Fe–Ni system.

Table 2.3 Relative contribution of primary, secondary and tertiary fluorescence of Cr in the presence of Fe and Ni[a]

Analytical context
Incident source: W target tube, 30 kV
System: Cr–Fe–Ni
Incidence angle $\psi' = 60°$; emergence angle $\psi'' = 30°$

				Fluorescence (%)		
C_{Cr}	C_{Fe}	C_{Ni}	R_{Cr}	primary	secondary	tertiary
0.05	0.90	0.05	0.080	67.71	31.79	0.50
0.05	0.80	0.15	0.077	68.96	29.74	1.30
0.05	0.70	0.25	0.074	70.16	27.83	2.02
0.10	0.80	0.10	0.148	71.31	27.88	0.81
0.10	0.70	0.20	0.143	72.43	26.10	1.47
0.10	0.50	0.40	0.135	74.33	23.23	2.44
0.10	0.40	0.50	0.132	75.09	22.26	2.65
0.10	0.20	0.70	0.129	75.88	21.87	2.26
0.10	0.10	0.80	0.128	75.70	22.82	1.48
0.20	0.70	0.10	0.269	76.25	23.12	0.63
0.20	0.60	0.20	0.262	77.15	21.67	1.18
0.20	0.40	0.40	0.247	79.72	18.42	1.87
0.20	0.30	0.50	0.246	78.99	19.14	1.87
0.20	0.10	0.70	0.242	79.11	19.82	1.07
0.50	0.40	0.10	0.565	87.14	12.53	0.34
0.50	0.25	0.25	0.555	87.60	11.84	0.56
0.50	0.10	0.40	0.543	87.61	11.98	0.41

[a] Data from Shiraiwa and Fujino (1967).

The instrumental parameters were given as: tungsten target tube operated at 30 kV, full-wave rectified; incidence angle = 60°, emergence angle = 30°.

It is observed that the contribution of tertiary fluorescence to the total fluorescence emissions may reach a maximum of $\approx 3\%$ in some instances for this specific analytical context. If not taken into consideration when applying theoretical corrections for matrix effects, this error would only be compensated in part by the calibration process.

2.7 DISCUSSION

At this stage in the present treatment of intensity–concentration expressions in X-ray fluorescence emissions, a point has been reached where the end result of the contributions of von Hamos (1945), Gillam and Heal (1952), Sherman (1955) and Shiraiwa and Fujino (1966) have led to an expression for primary emitted intensity that is fairly elaborate if every physical process is expressed in its most elementary form. Visualizing the intensity–concentration relationship from an analytical point of view, other workers have derived expressions that strive to condense these fundamental relationships in a very few terms. For example, in the context of a specimen irradiated by a monochromatic source of wavelength λ, if Eqn (2.24)

$$P_{i\lambda} = G_i C_i \frac{I_\lambda \mu_{i\lambda}}{\mu_s^*} \tag{2.63}$$

and Eqn (2.25)

$$P_{(i)\lambda} = G_i \frac{I_\lambda \mu_{i\lambda}}{\mu_i^*} \tag{2.64}$$

are ratioed, the proportionality constant G_i and the terms $\mu_{i\lambda}$ and I_λ cancel out, yielding

$$P_{i\lambda} = P_{(i)\lambda} C_i \frac{\mu_i^*}{\mu_s^*} \tag{2.65}$$

a very simple expression for the emitted primary fluorescence intensity. Thus, in the absence of enhancement effects, the emitted intensity is a function of:

● the intensity $P_{(i)\lambda}$ emitted by the pure element;
● the concentration C_i of element i in the specimen;
● the total mass absorption coefficient μ_i^*;
● the total mass absorption coefficient μ_s^*.

An equally simple expression for $P_{i\lambda}$ is obtained if Eqn (2.27) is ratioed to the equivalent equation for any reference material of known composition. The emitted intensity is then given by

$$P_{i\lambda} = P_{ir\lambda} \frac{C_i}{C_{ir}} \frac{\mu_r^*}{\mu_s^*} \qquad (2.66)$$

where the subscript r designates a reference material of known composition. An equally simple expression may be visualized for the emitted intensity in the presence of enhancement. From Eqns (2.32) and (2.40) it follows that

$$(P_i + S_i)_\lambda = P_{i\lambda} + P_{i\lambda} e_{ij\lambda} C_j + P_{i\lambda} e_{ik\lambda} C_k + \cdots \qquad (2.67)$$

However, one major drawback remains to be overcome if the above theoretical expressions quantifying emitted intensity for a monochromatic incident excitation source are to be applicable for polychromatic incident radiation sources, which is by far the more commonly used form of excitation in X-ray fluorescence analysis. The major drawback is that the expressions now involve integrals; i.e., one must integrate over the effective wavelength range λ_{min} to $\lambda_{abs.edge}$ of the analyte, Eqns (2.29) and (2.30) and, in the case of secondary fluorescence, over the range λ_{min} to $\lambda_{abs.edge}$ of the enhancing element, Eqn (2.55). How this drawback is overcome in practical applications is the topic of the next chapter.

2.8 SUMMARY

The following symbols and definitions were introduced in this chapter.

I_i the experimentally measured net intensity of a characteristic line emitted by the analyte element i

P_i the theoretically calculated intensity of the same characteristic line of element i resulting from primary fluorescence emissions only:

$$P_i = G_i C_i \int_{\lambda_{min}}^{\lambda_{i,abs}} \frac{\mu_{i\lambda} I_\lambda}{\mu_s' + \mu_s''} \, d\lambda$$

G_i a proportionality constant:

$$G_i = P_{\lambda_i} \frac{\Omega}{4\pi} \csc \psi'$$

S_i the theoretically calculated intensity of the same characteristic line of element i resulting from secondary fluorescence emissions of elements j, k, . . .:

$$S_i = S_{ij} + S_{ik} + \cdots$$

S_{ij} the sum of the enhancement effect of each line of element j that excites analyte i:

$$S_{ij} = S_{i\lambda_{j1}} + S_{i\lambda_{j2}} + S_{i\lambda_{j3}} + \cdots$$

$S_{i\lambda_j}$ the intensity of element i resulting from secondary fluorescence emissions by characteristic line λ_j

$$S_{i\lambda_j} = G_i C_i \int_{\lambda_{min}}^{\lambda_{j,abs}} \frac{\mu_{i\lambda} I_\lambda}{\mu'_s + \mu''_s} \, (e(e'_j + e''_j))_{\lambda_j} C_j$$

where

$$e = 0.5 p_{\lambda_j} \mu_{i\lambda_j} \mu_{j\lambda} \div \mu_{i\lambda}$$

$$e'_j = \frac{1}{\mu'_s} \ln\left(1 + \frac{\mu'_s}{\mu_{s\lambda_j}}\right)$$

$$e''_j = \frac{1}{\mu''_s} \ln\left(1 + \frac{\mu''_s}{\mu_{s\lambda_j}}\right)$$

p_{λ_i} the probability that a characteristic line wavelength λ_i is emitted:

$$p_{\lambda_i} = \frac{r-1}{r} \, f_{\lambda_i} \omega$$

where r is the absorption jump ratio, f is the probability that line λ_i will be emitted, and ω is the fluorescence yield

T_i the theoretically calculated intensity of the characteristic line of element i resulting from tertiary fluorescence emissions

μ'_i the effective mass attenuation coefficient of element i for incident wavelength λ:

$$\mu'_i = \mu_{i\lambda} \csc \psi'$$

μ''_i the effective mass attenuation coefficient of element i for the characteristic line wavelength λ_i:

$$\mu''_i = \mu_{i\lambda_i} \csc \psi''$$

μ'_s the effective mass absorption coefficient of the specimen for incident wavelength λ:

$$\mu'_s = \sum_i C_i \mu'_i$$

μ''_s the effective mass attenuation coefficient of the specimen for the characteristic line wavelength λ_i:

$$\mu''_s = \sum_i C_i \mu''_i$$

μ_i^* the total effective mass attenuation coefficient of element i for incident wavelength λ:

$$\mu_i^* = \mu_i' + \mu_i''$$

μ_s^* the total effective mass absorption coefficient of the specimen for incident wavelength λ:

$$\mu_s^* = \mu_s' + \mu_s''$$

3 THE FUNDAMENTAL PARAMETERS APPROACH

The derivation of mathematical expressions quantifying fluorescence emissions in terms of fundamental physical parameters and instrumental parameters is an essential first step for applying theoretically computed corrections, although some problems remain. A major problem is that the expressions involve multiple integrals, and thus cannot be applied easily in practice. A second problem is the realization that the spectral distribution of the incident beam may not be adequately known. A giant step to resolve these problems was initiated by Criss and Birks (1968), who proposed a 'workable model' that overcame both problems. Referred to as the fundamental parameters approach in this presentation, it provided a breakthrough for the correction of matrix effects in X-ray fluorescence analysis wherein the correction is based on modifications of equations derived by Sherman (1955) and by Shiraiwa and Fujino (1966). It also provided the impetus for a better understanding of other methods previously used. This chapter examines the premises and concepts underlying the fundamental parameters approach and subsequent developments that have led to a clearer understanding of the sometimes apparently capricious matrix effects.

3.1 GENERAL CONSIDERATIONS

The derivation of expressions quantifying emitted X-ray fluorescence intensities as a function of fundamental parameters, instrumental parameters and specimen composition can be regarded primarily as the domain of the physicist. In other words, given the composition of two specimens, instrumental geometry, incident spectral distribution and the fundamental parameters μ and ω, expressions have been derived that make it possible to calculate *a priori* the intensity that is emitted by one specimen, knowing the emitted intensity of the other. The analyst, on the other hand, is primarily concerned with the inverse task of converting intensities to concentrations. The fact that the intensity–concentration relationships derived in the previous chapter cannot be inverted, i.e., algebraic manipulations cannot produce explicit formulae for the mass fractions of the elements as a function of intensities, implies that the concentrations must be found by iteration procedures. There is no doubt that the advent of increasingly powerful computer facilities in recent years has made this and a host of other tasks, such as automated spectrometer operation, accessible to most analysts.

3.2* CONCEPT OF THE FUNDAMENTAL PARAMETERS APPROACH

The concept underlying the fundamental parameters approach can be stated fairly simply: 'modify the time-honoured equations that quantify the physical processes forming the basis of X-ray fluorescence emission in order to provide a theoretical method for the correction of matrix effects in practice'. The problems to be overcome in order to provide an algorithm that can be used in practice require:

(a) dealing with the integrals in the theoretically derived expressions for emitted fluorescence intensities, i.e. Eqns (2.29), (2.30) and (2.55);

(b) reliable X-ray tube spectral distributions;

(c) development of an iteration equation to determine specimen composition.

3.2.1 Replacing integrals by finite summations

That the integral over the effective wavelength range, λ_{min} to λ_{edge}, presented a major barrier is beyond doubt.

● Sherman (1954):

> The relations become much more involved if the incident beam is polychromatic . . . labor involved makes the relations too complicated for general use.

● Beattie and Brissey (1954):

> . . . its precise form is not usable owing to the continuous spectrum of the incident beam.

● Sherman (1955):

> Even a cursory examination . . . will show that extension to a polychromatic beam introduces great difficulties for a rigorous solution.

To overcome having to deal with integrals, Criss and Birks (1968) proposed that the primary spectrum be divided into a finite number of $\Delta\lambda$ elements. If the $\Delta\lambda$ element is small compared with the effective wavelength range λ_{min} to

λ_{edge}, then the integrals in Eqns (2.29) and (2.55) may for practical purposes be replaced by summations over a number of discrete wavelength intervals. e.g. for primary fluorescence emission

$$P_i = G_i C_i \int_{\lambda_{min}}^{\lambda_{abs}} \frac{I_\lambda \mu_{i\lambda}}{\mu_s^*} \, d\lambda \tag{3.1}$$

is replaced by

$$P_i = G_i C_i \sum_{\lambda_{min}}^{\lambda_{abs}} \frac{I_\lambda \mu_{i\lambda}}{\mu_s^*} \Delta\lambda \tag{3.2}$$

where P_i is the primary fluorescence emission of element i; G_i is the product of geometric factors and the probability that the characteristic photons of the analyte are emitted. Generally, intensities are expressed relative to a reference material, in which case G_i cancels out: C_i is the weight fraction of element i, the analyte; $\mu_{i\lambda}$ is the mass attenuation coefficient of the analyte for incident wavelength λ; I_λ is the incident intensity for the $\Delta\lambda$ element at incident λ; μ_s^* is the effective mass attenuation coefficient of the specimen defined as

$$\mu_s^* = \sum_i C_i(\mu_{i\lambda} \csc \psi' + \mu_{i\lambda_i} \csc \psi'') \tag{3.3}$$

over all the elements of the specimen.

3.2.2 Spectral distributions

Coupled with the concept of replacing integrals by summations was the need for reliable data on the spectral distribution of the polychromatic emanations from X-ray tubes. Criss and Birks (1968) concluded that the theoretical expressions proposed by Kramers, although most useful for showing the general shape of I versus λ curves (functions of target element Z, applied voltage and current), were not adequate for the stringent requirements imposed by quantitative analysis. As part of the same project, it was therefore decided to proceed with the experimental measurement of spectral distributions for a number of commercially available X-ray tubes, for various targets and applied potentials (see Part II.1.7–II.1.9 for an example of experimentally measured spectral distribution of W, Cr and Rh target tubes operated at 45 kV). The continuum was divided into wavelength elements of $\Delta\lambda = 0.02$ Å (Figure 3.1) and integrated intensities (in arbitrary units) were measured for the continuum at wavelengths of 0.29, 0.31, ..., 2.59 Å, i.e. in the middle of the $\Delta\lambda$ elements. Characteristic line intensities were also measured and assumed to occur at the exact wavelength for that line.

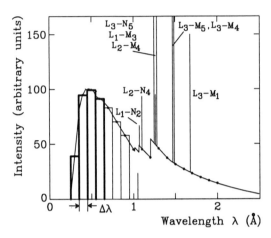

Figure 3.1 Schematic of continuum emanating from X-ray tube divided into $\Delta\lambda$ elements. Each $I_{\lambda+\Delta\lambda}$ represents the area of a rectangle and their sum approximates the area under the curve

3.2.3 Iteration procedure

Although the iteration equation used by Criss and Birks in the fundamental parameters calculations was originally developed to deal with measured intensities relative to the intensity of a pure element, it was shown that any multicomponent specimen of known composition may be used as reference. In other words

$$R_i = \frac{I_{iu}}{I_{(i)}} \tag{3.4}$$

$$= \left(\frac{I_{iu}}{I_{ir}}\right) \times \left(\frac{P_i + S_i}{P_{(i)}}\right)_r \tag{3.5}$$

where R_i is the intensity relative to that of the pure element; I_{iu} is the measured intensity from a specimen of unknown composition; $I_{(i)}$ is the measured or calculated intensity emitted by pure analyte i; I_{ir} is the measured intensity from a reference material of known composition; $(P_i + S_i)_r$ is the theoretically calculated intensity for the reference material; $P_{(i)}$ is the theoretically calculated intensity for the pure analyte.

The iterative procedure consists in:

- making some initial estimate of composition scaled to unity;
- calculating relative intensities for that composition;
- comparing with the measured relative intensities;

- making a better estimate of composition, scaled to unity;
- repeating the process until convergence is achieved.

That procedure may be expressed as

$$C_{i,\text{next estimate}} = \frac{R_{i,\text{measured}}}{R_{i,\text{calculated}}} C_{i,\text{present estimate}} \qquad (3.6)$$

Two examples of application of the iteration process are given in Section II.3.2.

3.3 CLASSICAL FORMALISM

The term classical is used to emphasize the fact that the equations proposed by Criss and Birks retained the form that had been developed from what might be called *a physicist's point of view*. Subsequent developments led to formalisms that are more suited to *an analyst's point of view*.

3.3.1 Primary emission

Dealing firstly with absorption effects, Criss and Birks proposed the following expression for the intensity of emergent characteristic radiation of element i (caused to fluoresce by only the incident spectrum)

$$P_i = G_i C_i \sum_k \frac{D_i(\lambda_k)\mu_i(\lambda_k)I(\lambda_k)\,\Delta\lambda_k}{\mu_M(\lambda_k)\text{cosec }\varepsilon + \mu_M(\lambda_i)\text{cosec }\psi} \qquad (3.7)$$

$D_i(\lambda_k)$ is defined as equal to unity for all values of λ short enough to excite element i, and is equal to zero for all other values of λ. Recalling the definitions of μ_i', μ_i'', μ_s' and μ_s'' from the previous chapter, the Criss and Birks expression for primary fluorescence is the same as Eqn (3.2), it being implied that the summation is over the range to $\lambda_{\min}$ to λ_{edge}, namely for those wavelengths that can eject electrons from the level to which the transitions take place. Thus for computational work, one first verifies whether $\lambda_{\min}$ is less than λ_{edge}; if so one proceeds with the calculation of $P_{i\lambda}$ for the related $\Delta\lambda$ element. Wavelength $\lambda_{\min}$ is then incremented by $\Delta\lambda$ and the operation is repeated until the condition $\lambda > \lambda_{\text{edge}}$ is reached. The incident wavelength can then no longer excite analyte atoms and cannot lead to the emission of the characteristic lines of the analyte. As stated by Criss and Birks and shown in Section 2.3.3, the constant G_i cancels out when one considers relative X-ray intensities $P_i/P_{(i)}$. Thus the term G_i may be disregarded in the calculation of P_i under those conditions.

3.3.1.1   Numerical example of the calculation of $P_{(i)}$ and P_i

In order to show detailed numerical values for all effective wavelengths and bearing in mind space considerations, the system Y–Mo is selected. If molybdenum is the analyte i and yttrium is the matrix element j, one is then dealing with a case where enhancement is not involved, since both characteristic lines of yttrium are longer than the K absorption edge of molybdenum; i.e. λ (Mo$_K$ edge) = 0.620 Å, but λ (Y $K-L_{2,3}$) = 0.830 Å and λ (Y $K-M_3$) = 0.741 Å. The expressions to be calculated are

$$P_{(\mathrm{Mo})} = \sum_\lambda \frac{\mu_{\mathrm{Mo}\lambda} I_\lambda}{\mu_{\mathrm{Mo}}^*} \Delta\lambda \tag{3.8}$$

$$P_{\mathrm{Mo}} = C_{\mathrm{Mo}} \sum_\lambda \frac{\mu_{\mathrm{Mo}\lambda} I_\lambda}{\mu_s^*} \Delta\lambda \tag{3.9}$$

The analytical context chosen and the fundamental parameters used are given in Table 3.1, along with pertinent data for each of the 17 effective $\Delta\lambda$ elements.

- The minimum wavelength in the present context is given by $\lambda_{\min} = 12.3981 \div \mathrm{kV} = 12.3981 \div 45 = 0.276$ Å.
- The values C and n used for the computation of the mass attenuation coefficients in the expression $\mu = C\lambda^n$ are computed from the tables of Leroux and Thinh (1977).

The following terms are independent of the incident wavelength and thus remain constant; they can be taken either from tables or calculated as follows:

$$\mu_{\mathrm{Mo}}'' = 47.3 \times 0.710^{2.73} \times 1.8361 = 34.09 \text{ cm}^2 \text{g}^{-1}$$
$$\mu_Y'' = 273.2 \times 0.710^{2.85} \times 1.8361 = 189.0 \text{ cm}^2 \text{g}^{-1}$$
$$\mu_s'' = C_{\mathrm{Mo}}\mu_{\mathrm{Mo}}'' + C_Y\mu_Y'' = (0.2 \times 34.09) + (0.8 \times 189.0)$$
$$= 158.02 \text{ cm}^2 \text{g}^{-1}$$

The data in the body of Table 3.1 are obtained by calculating the values that are functions of the incident wavelength and introducing the above constants when required. Detailed computation of data for incident wavelength $\lambda = 0.49$ Å are as follows:

$$\mu_{\mathrm{Mo}\lambda} = 330.1 \times 0.49^{2.85} = 43.22 \text{ cm}^2 \text{g}^{-1}$$

$I_\lambda = 99.1$; see Table II.1.7

$$\mu_{\mathrm{Mo}}' = \mu_{\mathrm{Mo}\lambda} \csc \psi' = 43.22 \times 1.1223 = 48.51 \text{ cm}^2 \text{g}^{-1}$$

$$P_{(\mathrm{Mo})\lambda} = \frac{\mu_{\mathrm{Mo}\lambda} I_\lambda}{\mu_{\mathrm{Mo}}^*} = \frac{43.22 \times 99.1}{48.51 + 34.09} = 51.85$$

Table 3.1 Theoretical calculation of the $K-L_{2,3}$ line intensity in a context that does not involve enhancement. See text for details of data relating to incident $\lambda = 0.49$ Å

Analytical context
Excitation source: W target tube; 45 kV (CP); see Table II.1.7
Instrument geometry: $\psi' = 63°$, $\psi'' = 33°$
System: Y–Mo; analyte is Mo; $C_{Mo} = 0.2$, $C_Y = 0.8$

	$\lambda_{K-L_{2,3}}$ (Å)	C_K	n_K	$\lambda_{abs}\ K$ (Å)	C_{KL}	n_{KL}	p_K	f_λ	ω_K
Y	0.831	273.20	2.85	0.728	38.0	2.73	0.854	1.0	0.715
Mo	0.710	330.10	2.85	0.620	47.3	2.73	0.856	1.0	0.765

λ (Å)	μ_{Mo} cm²g⁻¹	I_λ	μ'_{Mo} cm²g⁻¹	μ'_Y cm²g⁻¹	$P_{(Mo)}$	μ'_s cm²g⁻¹	P_{Mo}	R_{Mo}
0.29	9.69	15.5	10.88	9.00	3.34	9.38	0.180	0.0537
0.31	11.72	36.6	13.16	10.89	9.08	11.34	0.507	0.0558
0.33	14.01	56.8	15.72	13.01	15.97	13.55	0.928	0.0581
0.35	16.57	76.6	18.59	15.39	24.09	16.03	1.458	0.0605
0.37	19.41	96.2	21.78	18.03	33.42	18.78	2.112	0.0632
0.39	22.55	111.1	25.31	20.95	42.18	21.82	2.786	0.0661
0.41	26.01	116.4	29.19	24.16	47.84	25.16	3.305	0.0691
0.43	29.79	114.6	33.43	27.67	50.55	28.82	3.654	0.0723
0.45	33.91	109.9	38.06	31.50	51.65	32.81	3.906	0.0756
0.47	38.38	104.5	43.08	35.65	51.97	37.14	4.111	0.0791
0.49	43.22	99.1	48.51	40.15	51.85	41.82	4.287	0.0827
0.51	48.44	93.9	54.37	45.00	51.42	46.87	4.440	0.0864
0.53	54.05	89.0	60.67	50.21	50.76	52.30	4.575	0.0901
0.55	60.07	84.4	67.42	55.80	49.94	58.12	4.692	0.0939
0.57	66.51	80.3	74.65	61.78	49.12	64.35	4.804	0.0978
0.59	73.38	76.6	82.36	68.16	48.27	71.00	4.909	0.1017
0.61	80.69	73.3	90.56	74.95	47.45	78.08	5.011	0.1056
Σ					678.90		55.66	

$$\mu'_{Y\lambda} = 273.2 \times 0.49^{2.85} \times 1.1223 = 40.15 \text{ cm}^2\,\text{g}^{-1}$$

$$\mu'_s = (0.2 \times 48.51) + (0.8 \times 40.15) = 41.82 \text{ cm}^2\,\text{g}^{-1}$$

Substitution in

$$P_{Mo\lambda} = C_{Mo}\ \frac{\mu_{Mo\lambda} I_\lambda}{\mu_s^*}$$

$$= 0.2 \times \frac{43.22 \times 99.1}{41.82 + 158.02} = 4.287$$

The calculated emitted intensities for pure Mo and for the specimen under consideration are obtained by summing the data in the columns headed $P_{(Mo)}$

and P_{Mo}, respectively. Thus, the values 678.90 and 55.66 incorporate the polychromatic nature of the incident beam, and the ratio $55.66/678.90 = 0.0820$ quantifies the absorption effect of yttrium on molybdenum; i.e. a specimen containing 20% of Mo and 80% of Y emits only 8.2% compared with a specimen of pure Mo in this particular analytical context. Similar calculations are carried out for the system Cr–Fe–Ni in Part II, Table II.3.3 for cases in which the theoretical results may be compared with published experimental data.

3.3.2 Secondary emission

If the elements in the above example are reversed, i.e. yttrium becomes the analyte and molybdenum the companion element, the wavelength of the $K-L_{2,3}$ line of molybdenum, which is 0.710 Å, is shorter than the K absorption edge of yttrium. Thus, the emitted yttrium intensity consists of both a primary component, Eqn (2.29), and a secondary fluorescence component, Eqn (2.55). From Criss and Birks (1968), the expression for emitted intensity due to both primary and secondary fluorescence emissions is

$$
P_i + S_i = G_i C_i \sum_\lambda \left(\frac{\mu_{i\lambda} I_\lambda \, \Delta\lambda}{\mu_s' + \mu_s''} \left\{ 1 + \frac{1}{2\mu_{i\lambda}} \sum_j p_{\lambda j} \mu_{i\lambda j} \mu_{j\lambda} C_j \right. \right.
$$

$$
\left. \left. \times \left[\frac{1}{\mu_s'} \ln\left(1 + \frac{\mu_s'}{\mu_{s\lambda_j}}\right) + \frac{1}{\mu_s''} \ln\left(1 + \frac{\mu_s''}{\mu_{s\lambda_j}}\right) \right] \right\} \right)
\tag{3.10}
$$

Recalling Eqn (3.2)

$$
P_i = G_i C_i \sum_{\lambda_{min}}^{\lambda_{abs\ j}} \frac{\mu_{i\lambda} I_\lambda}{\mu_s^*} \, \Delta\lambda
\tag{3.11}
$$

and Eqn (2.55) expressed in summation formalism for enhancement by more than one element

$$
S_i = G_i C_i \sum_{\lambda_{min}}^{\lambda_{abs\ j}} \frac{\mu_{i\lambda} I_\lambda}{\mu_s^*} \sum_j e_{ij\lambda} C_j \, \Delta\lambda
\tag{3.12}
$$

then

$$
P_i + S_i = G_i C_i \sum_\lambda \left[\frac{\mu_{i\lambda} I_\lambda}{\mu_s^*} \left(1 + \sum_j e_{ij\lambda} C_j\right) \right] \Delta\lambda
\tag{3.13}
$$

given the proviso that the summation is over the effective wavelength range, i.e. when $D_i(\lambda_k)$ is equal to unity in Eqn (3.7)

3.3.2.1 *Numerical example of the calculation of* $(P_i + S_{ij})_\lambda$

Considering that we are dealing with relative intensities, G_i cancels out, therefore the expressions to be calculated in the present case (i = yttrium, j = molybdenum) are

$$P_{(Y)} = \sum_\lambda \frac{\mu_{Y\lambda}I_\lambda}{\mu_Y^*} \Delta\lambda \qquad (3.14)$$

and

$$P_Y + S_{YMo} = C_Y \sum_\lambda \left[\frac{\mu_{Y\lambda}I_\lambda}{\mu_s^*} (1 + e_{YMo\lambda}C_{Mo}) \right] \Delta\lambda \qquad (3.15)$$

The analytical context is identical to that in Table 3.1 except that yttrium is the analyte and molybdenum is the companion or matrix element, i.e. $C_Y = 0.2$ and $C_{Mo} = 0.8$. For space considerations, the numerical example is simplified by carrying out the computations assuming that enhancement is due to the $K-L_{2,3}$ line of molybdenum only, i.e. $f_{Mo\,K-L_{2,3}}$ is equated to 1.0. In normal practice, the computation would be carried out using $f_{Mo\,K-L_{2,3}} = 0.847$ and $f_{Mo\,K-L_3} = 0.153$, the two results being added. The following terms are independent of the incident wavelength and thus remain constant:

$$\mu_Y'' = 38.0 \times 0.831^{2.73} \times 1.8361 = 42.09 \text{ cm}^2\text{g}^{-1}$$

$$\mu_{Mo}'' = 47.30 \times 0.831^{2.73} \times 1.8361 = 52.39 \text{ cm}^2\text{g}^{-1}$$

$$\mu_s'' = (0.2 \times 42.09) + (0.8 \times 52.39) = 50.33 \text{ cm}^2\text{g}^{-1}$$

$$p_{MoK-L_{2,3}} = p_{K,Mo}f_{K-L_{2,3}}\omega_{K,Mo}$$

$$= 0.856 \times 1.0 \times 0.765$$

$$= 0.655$$

$$\mu_{sMoK-L_{2,3}} = C_Y\mu_{YMoK-L_{2,3}} + C_{Mo}\mu_{MoMoK-L_{2,3}}$$

$$= (0.2 \times 273.2 \times 0.710^{2.85}) + (0.8 \times 47.3 \times 0.710^{2.73})$$

$$= 35.44 \text{ cm}^2\text{g}^{-1}$$

By substitution in

$$e_{Mo}'' = \frac{1}{\mu_s''} \ln\left(1 + \frac{\mu_s''}{\mu_{sMoK-L_{2,3}}}\right)$$

$$= \frac{1}{50.33} \ln\left(1 + \frac{50.33}{35.44}\right)$$

$$= 0.01756$$

The data in Table 3.2 are obtained by calculating the values that are functions of the incident wavelengths and incorporating the above values when required. For the element yttrium, there are twenty-two discrete effective values of $\Delta\lambda$, seventeen of which involve enhancement.

Table 3.2   Theoretical calculation of the $K-L_{2,3}$ line intensity in a context involving enhancement. See text for details relating to incident $\lambda = 0.49$ Å

Analytical context
Identical to Table 3.1 except that the analyte is yttrium and $C_Y = 0.2$, $C_{Mo} = 0.8$

λ(Å)	$P_{(Y)}$	P_Y	e_{Mo}	e'_{Mo}	e_{YMo}	S_{YMo}	$P_Y + S_Y$	R_Y
0.29	2.43	0.409	40.72	0.0247	1.721	0.563	0.972	0.3993
0.31	6.70	1.127		0.0241	1.697	1.530	2.66	0.3969
0.33	11.95	2.011		0.0235	1.671	2.688	4.70	0.3933
0.35	18.27	3.076		0.0228	1.645	4.048	7.12	0.3897
0.37	25.71	4.331		0.0222	1.617	5.603	9.93	0.3863
0.39	32.90	5.547		0.0215	1.589	7.051	12.60	0.3830
0.41	37.82	6.382		0.0208	1.561	7.970	14.35	0.3794
0.43	40.50	6.840		0.0201	1.532	8.383	15.22	0.3758
0.45	41.91	7.084		0.0194	1.503	8.518	15.60	0.3722
0.47	42.70	7.222		0.0187	1.475	8.522	15.75	0.3689
0.49	43.11	7.297		0.0180	1.447	8.447	15.75	0.3654
0.51	43.23	7.322		0.0173	1.420	8.318	15.64	0.3618
0.53	43.14	7.312		0.0167	1.393	8.148	15.46	0.3584
0.55	42.87	7.271		0.0160	1.367	7.951	15.22	0.3551
0.57	42.56	7.222		0.0154	1.342	7.754	14.98	0.3520
0.59	42.20	7.165		0.0148	1.318	7.555	14.72	0.3489
0.61	41.82	7.106	40.72	0.0142	1.294	7.356	14.46	0.3457
0.63	41.30	13.027	0.0	0.0	0.0	0.0	13.03	0.3155
0.65	40.65	13.176	0.0	0.0	0.0	0.0	13.18	0.3242
0.67	39.95	13.293	0.0	0.0	0.0	0.0	13.29	0.3327
0.69	39.08	13.342	0.0	0.0	0.0	0.0	13.34	0.3413
0.71	38.27	13.388	0.0	0.0	0.0	0.0	13.39	0.3499
$\sum$	759.05	160.95				110.41	271.36	

Detailed calculations of data for incident wavelength $\lambda = 0.49$ Å are as follows: substitution in

$$P_{(Y)\lambda} = \frac{\mu_{Y\lambda} I_\lambda}{\mu_Y^*}$$

$$= \frac{35.77 \times 99.1}{40.15 + 42.09}$$

$$= 43.11$$

$\mu_s' = C_Y \mu_{Y\lambda}\ \csc\ \psi' + C_{Mo}\mu_{Mo\lambda}\ \csc\ \psi'$
$= (0.2 \times 273.2 \times 0.49^{2.85} \times 1.1223) + (0.8 \times 330.1 \times 0.49^{2.85} \times 1.1223)$
$= 46.84\ \text{cm}^2\,\text{g}^{-1}$

$$P_{Y\lambda} = C_Y \frac{\mu_{Y\lambda} I_\lambda}{\mu_s^*}$$

$$= 0.2 \times \frac{35.77 \times 99.1}{46.84 + 50.33}$$

$$= 7.297$$

$$e_{Mo} = 0.5 p_{MoK-L_{2,3}} \mu_{YMoK-L_{2,3}} (\mu_{Mo\lambda}/\mu_{Y\lambda})$$

$$\mu_{YMoK-L_{2,3}} = 273.2 \times 0.710^{2.85} = 102.94 \text{ cm}^2 \text{g}^{-1}$$

$$\mu_{Mo\lambda} = 330.1 \times 0.49^{2.85} = 43.22 \text{ cm}^2 \text{g}^{-1}$$

$$\mu_{Y\lambda} = 273.2 \times 0.49^{2.85} = 35.77 \text{ cm}^2 \text{g}^{-1}$$

$$e_{Mo} = 0.5 \times 0.6552 \times 102.94 \times (43.22/35.77)$$

$$= 40.75$$

$$e'_{Mo} = \frac{1}{\mu_s'} \ln\left(1 + \frac{\mu_s'}{\mu_{sMoK-L_{2,3}}}\right)$$

$$= \frac{1}{46.84} \ln\left(1 + \frac{46.84}{35.44}\right)$$

$$= 0.01798$$

All the terms that are functions of λ having been computed, substitution in Eqn (3.10) yields

$$P_Y + S_{YMo} = C_Y \frac{\mu_{Y\lambda} I_\lambda}{\mu_s^*} \{1 + [e(e' + e'')]_{\lambda_{Mo}} C_{Mo}\}$$

$$= 0.2 \frac{35.77 \times 99.1}{46.84 + 50.33} (1 + 40.75 \times (0.01798 + 0.01756) \times 0.8)$$

$$= 7.297(1 + 1.448 \times 0.8)$$

$$= 15.75$$

or

$$P_Y + S_{YMo} = 7.297 + 8.453$$

$$= 15.75$$

Thus, enhancement accounts for 53.6% of the total intensity emitted by yttrium for incident $\lambda = 0.49$ Å. It is also observed in Table 3.2 that the wide range in values for $P_{(Y)}$, P_Y, S_{YMo} and $P_Y + S_{YMo}$ are not reflected in the values of R_Y, i.e. from a minimum value of 0.3155 to a maximum of 0.3993.

The intensity of yttrium relative to pure yttrium in the analytical context under consideration, i.e. monochromatic incident source, $\lambda = 0.49$ Å, is

$$R_Y = \frac{15.75}{43.11} = 0.3654$$

whereas R_Y for the polychromatic incident beam under consideration is given by the sum of column $P_Y + S_{YMo}$ divided by the sum of column $P_{(Y)}$, namely

$$R_Y = \frac{271.36}{759.05}$$

$$= 0.3575$$

3.3.2.2 Experimental confirmation

At this stage, it is useful to recapitulate how the contributions from numerous workers has made it possible to quantify matrix effects from theoretical principles. The two examples used to compare theoretical and experimental data in Chapter 2 relate to the analytes Fe and Ni in the context of a mon~ ` `matic incident source. Comparison of theoretical and experimental
g to those same two analytes but in the context of a polychromatic
ᴿce are now examined. The procedural steps are as follows.

analyst must establish what elements are under consideration. It
ᴴat six fundamental parameters can be quantified (see Table
ᴠ:

gth or energy of the characteristic lines used to measure the
perimentally; in the present case, Fe $K-L_{2,3} = 1.937$ Å and
Å;

or energy of the absorption edges related to those lines,
dges of Fe and Ni, 1.743 Å and 1.488 Å, respectively;
ᴸenuation coefficients at any given wavelength for each
ᴸne present case, μ_{Fe} and μ_{Ni} values were generated following
ᴸn (1966);

(4) $p_{K\,Fe} = 0.878$ and $p_{K\,Ni} = 0.873$ are taken from Broll (1986);

(5) the transition probabilities which relates to the relative intensity of a given line within its series; in the present case, both $f_{FeK-L_{2,3}}$ and $f_{NiK-L_{2,3}}$ are equal to 0.892. For the actual computations the value 1.0 was used, i.e. $f_{K-L_{2,3}} + f_{K-M_3} = 1.0$.

(6) the fluorescence yield $\omega_{K\,Ni} = 0.420$ is taken from Broll (1986).

Secondly, as shown in Chapter 2, knowledge of three instrumental parameters is required, namely:

(1) the spectral distribution of the incident X-radiation source, in this case a tungsten target tube operated at 45 kV (CP); see Table II.1.7.

(2) an average value for the incidence angle of the excitation source; in this case from Rasberry and Heinrich (1974), an angle of $63°$.

(3) an average value for the emergence angle of the characteristic X-rays; in this case an angle of $33°$.

Thirdly, it is necessary to know the composition of the specimens under consideration, in this case nine binary specimens; see Table 3.3. The column headed $R_{Ni,calc}$ lists R_{Ni} values computed by summation over the 60 effective wavelength elements. All the calculated relative intensities are lower by roughly the same factor, as shown in the column headed R_{calc}/R_{expr}, i.e. an overcorrection for the absorption effect of Fe. Some of the factors which might account for this slight bias are discussed in Chapter 9 and, as indicated by Criss and Birks (1968), bias in results can be adjusted out by comparison with one or two non-pure-element standards. It is also noted that the bias is practically nil if $35°$ is used as the emergence angle.

Corresponding data relating to the same nine specimens when Fe is the analyte are listed in Table 3.4. In this case the calculated R_{Fe} values are all higher, and the ratio R_{calc}/R_{expr} increases as the concentration of Ni increases, indicating an overcorrection for the enhancement effect of Ni. In this case it

Table 3.3 Comparison of theoretical and experimental R_{Ni} values for Fe–Ni binaries for a polychromatic excitation source

Analytical context
Excitation source: W target tube; 45 kV (CP); see Table II.1.7
Instrument geometry: $\psi' = 63°$, $\psi'' = 33°$
System: Fe–Ni; analyte: Ni

	$\lambda_{K-L_{2,3}}$ (Å)	C_K	n_K	$\lambda_{abs\ K}$ (Å)	C_{KL}	n_{KL}	p_K	f	ω_K
Fe	1.937	95.8	2.72	1.743	11.75	2.73	0.878	1.0	0.350
Ni	1.659	115.9	2.71	1.488	14.80	2.73	0.873	1.0	0.420

Specimen	C_{Ni}	C_{Fe}	$R_{Ni,calc}$	$R_{Ni,expr}$	$\dfrac{R_{calc}}{R_{expr}}$
971	0.9516	0.0462	0.8727	0.8782	0.9937
972	0.9322	0.0659	0.8257	0.8321	0.9923
974	0.8964	0.1018	0.7485	0.7595	0.9855
983	0.7711	0.2263	0.5409	0.5483	0.9865
986	0.6931	0.3067	0.4411	0.4515	0.9770
987	0.6552	0.3431	0.4010	0.4073	0.9845
1159	0.4820	0.5100	0.2514	0.2553	0.9847
126B	0.3599	0.6315	0.1694	0.1720	0.9849
809B	0.0329	0.9549	0.0125	0.0125	1.00

Table 3.4 Comparison of theoretical and experimental R_{Fe} values for Fe–Ni binaries for a polychromatic excitation source

Analytical context
Identical to Table 3.3 except that the analyte is iron and the matrix element is nickel

Specimen	C_{Fe}	C_{Ni}	$R_{Fe,calc}$	$R_{Fe,expr}$	$\dfrac{R_{calc}}{R_{expr}}$
809B	0.9549	0.0329	0.9732	0.9659	1.0076
126B	0.6315	0.3599	0.7075	0.6958	1.0168
1159	0.5100	0.4820	0.6037	0.5907	1.0220
987	0.3431	0.6552	0.4476	0.4373	1.0236
986	0.3067	0.6931	0.4108	0.4007	1.0252
983	0.2263	0.7711	0.3254	0.3172	1.0259
974	0.1018	0.8964	0.1683	0.1621	1.0382
972	0.0659	0.9322	0.1147	0.1104	1.0389
971	0.0462	0.9516	0.0830	0.0789	1.0520

is noted that the bias is practically nil if the value 0.892 is used for $f_{Ni\ K-L_{2,3}}$; i.e. if enhancement by Ni $K-M_3$ is not taken into consideration.

3.4* ALTERNATE FORMALISM

Although the fundamental parameters method proposed by Criss and Birks is an analytical method in its own right, it has also been a major factor in the study of matrix effects since its very inception. In a critical survey of mathematical matrix correction procedures for X-ray fluorescence analysis, Tertian noted (1986) that, from 1968 on, the fundamental parameters method revived interest in the theoretical equations of X-ray fluorescence emission, prompted work on the re-evaluation of older methods, and progressively gave influence coefficient methods a sound theoretical status. The following development therefore examines an alternate formalism relating intensity to concentration that can be derived from the equations of Criss and Birks (1968) and thus ultimately from the work of Sherman (1955) and Shiraiwa and Fujino (1966). In essence, Lachance (1988) proposed that the characteristic line intensity (primary and secondary fluorescence) emitted by a specimen may be expressed mathematically as being

(a) directly proportional to the product of the intensity emitted by a specimen of the pure analyte times the concentration of the analyte in the specimen
(b) minus the sum of each individual absorption effect
(c) plus the sum of each individual enhancement effect.

3.4.1 Primary fluorescence emission

Recalling the fundamental parameters expressions, Eqns (2.24) and (2.25), for emitted intensity due to primary fluorescence in multielement and pure element specimens in the context of a monochromatic incident source, wavelength λ

$$P_{i\lambda} = G_i C_i \frac{\mu_{i\lambda} I_\lambda}{\mu_s^*} \tag{3.16}$$

$$P_{(i)\lambda} = G_i \frac{\mu_{i\lambda} I_\lambda}{\mu_i^*} \tag{3.17}$$

it was shown (Section 2.7) that ratioing the two above equations leads to

$$P_{i\lambda} = \frac{P_{(i)\lambda} C_i \mu_i^*}{\mu_s^*} \tag{3.18}$$

Since the term μ_i^* relates to the total absorption of element i (in that it takes into consideration both the incident and the characteristic wavelength and its respective path length) it may be applied to any other element, and equivalent terms for total absorption may be defined similarly, i.e.

$$\mu_j^* = \mu_j' + \mu_j'' = \mu_{j\lambda} \csc \psi' + \mu_{j\lambda_i} \csc \psi'' \tag{3.19}$$

It is therefore possible to define μ_s^* in the denominator of Eqn (3.18) as

$$\mu_s^* = \sum_i C_i \mu_i^* \tag{3.20}$$

where the summation is over every element present, i.e. $\sum C_i = 1.0$. For a specimen comprising three elements, Eqn (3.18) becomes

$$P_{i\lambda} = \frac{P_{(i)\lambda} C_i \mu_i^*}{C_i \mu_i^* + C_j \mu_j^* + C_k \mu_k^*} \tag{3.21}$$

The next step involves visualizing the absorption effect in terms of a specimen of the pure analyte in which portions of the analyte have been replaced by equivalent portions of other elements. In the context of Eqn (3.20) this entails replacing portions of analyte i by equivalent portions of elements j and k. Consider the case in which a certain concentration of the analyte, let us say C_{i1}, is replaced by an equal concentration C_j of element j, and a similar substitution, let us say C_{i2}, is replaced by an equal concentration C_k of element k. The denominator of Eqn (3.21) may then be expressed as

$$1 \, \mu_i^* - C_{i1}\mu_i^* + C_j\mu_j^* - C_{i2}\mu_i^* + C_k\mu_k^* \tag{3.22}$$

Consequently, the effect of a given concentration C_j in a specimen may be envisaged as having an effective absorption effect equal to $C_j(\mu_j^* - \mu_i^*)$. Thus Eqn (3.21) may be expressed as

$$P_{i\lambda} = \frac{P_{(i)\lambda}C_i\mu_i^*}{\mu_i^* + C_j(\mu_j^* - \mu_i^*) + C_k(\mu_k^* - \mu_i^*)} \tag{3.23}$$

Division of both numerator and denominator by μ_i^* and cross multiplication leads to

$$P_{i\lambda} + P_{i\lambda}\left(\frac{\mu_j^* - \mu_i^*}{\mu_i^*}\right)C_j + P_{i\lambda}\left(\frac{\mu_k^* - \mu_i^*}{\mu_i^*}\right)C_k = P_{(i)\lambda}C_i \tag{3.24}$$

Defining

$$A_{ij\lambda} = P_{i\lambda}\left(\frac{\mu_j^* - \mu_i^*}{\mu_i^*}\right)_\lambda \tag{3.25}$$

$$A_{ik\lambda} = P_{i\lambda}\left(\frac{\mu_k^* - \mu_i^*}{\mu_i^*}\right)_\lambda \tag{3.26}$$

and rearranging terms results in

$$P_{i\lambda} = P_{(i)\lambda}C_i - A_{ij\lambda}C_j - A_{ik\lambda}C_k \tag{3.27}$$

which can be interpreted, in the context of a monochromatic excitation source wavelength λ, as: 'the intensity emitted by analyte i due only to primary fluorescence emission from a specimen is equal to the product $P_{(i)\lambda}C_i$, i.e. the intensity that would be emitted if absorption remained constant, minus the net absorption effect of each individual matrix element'. Expressed in generalized form

$$P_{i\lambda} = P_{(i)\lambda}C_i - \sum_j A_{ij\lambda}C_j \tag{3.28}$$

The relationship is obviously applicable to a polychromatic incident beam, namely

$$P_i = P_{(i)}C_i - \sum_j A_{ij}C_j \tag{3.29}$$

where

$$P_i = \sum_\lambda P_{i\lambda}\, \Delta\lambda \tag{3.30}$$

$$P_{(i)} = \sum_\lambda P_{(i)\lambda}\, \Delta\lambda \tag{3.31}$$

$$A_{ij} = \sum_{\lambda} A_{ij\lambda} \, \Delta\lambda \tag{3.32}$$

i.e. the three entities P_i, $P_{(i)}$ and A_{ij} are equal to the sum of their respective values at each $\Delta\lambda$ element.

3.4.1.1 Numerical example

For space considerations, Ag is chosen as the analyte so that the effective wavelength range is not too extensive. The absorption effects of Y and Mo will be evaluated for two incident sources, namely W and Rh targets operated at the same 45 kV potential, in order to bring out the fact that, although the data are quite different depending on the incident source, the intensities relative to that of the pure element are quite similar. This is to be expected from Eqn (1.13). The analytical context and the fundamental parameters utilized are given in Table 3.5 and the pertinent data for the absorption effects of Mo and Y on Ag for each of the ten effective wavelength intervals are listed in Table 3.6, along with their respective summations.

It is observed that $R_{Ag} = 45.89/450.03 = 0.1020$ for the W target tube incident radiation and $19.56/193.83 = 0.1009$ in the case of the Rh target. P_{Ag}, the intensity due to primary fluorescence emission, may be computed using Eqn (3.29)

$$
\begin{aligned}
P_{Ag} &= P_{(Ag)}C_{Ag} - A_{AgMo}C_{Mo} - A_{AgY}C_Y \\
&\doteq (450.03 \times 0.2) - (67.34 \times 0.3) - (47.82 \times 0.5) \\
&= 90.01 - 20.20 - 23.91 \\
&= 45.90
\end{aligned}
$$

Table 3.5 Analytical context and fundamental parameters used for the theoretical calculations in Tables 3.6–3.9. Note that the factors $f_{K-L_{2,3}}$ are equated to unity

Analytical context
Excitation sources:

    (a) W target tube, 45 kV (CP), see Table II.1.7
    (b) Rh target tube, 45 kV (CP), see Table II.1.9

Instrument geometry: $\psi' = 90°$, $\psi'' = 30°$

	$\lambda_{K-L_{2,3}}$ (Å)	C_K	n_K	$\lambda_{abs\ K}$ (Å)	C_{KL}	n_{KL}	p_K	$f_{K-L_{2,3}}$	ω_K
Y	0.831	273.2	2.85	0.728	38.0	2.73	0.854	1.0	0.715
Mo	0.710	330.1	2.85	0.620	47.3	2.73	0.856	1.0	0.765
Ag	0.561	440.0	2.85	0.486	65.6	2.71	0.848	1.0	0.827
Sn	0.492	515.3	2.85	0.425	78.8	2.69	0.845	1.0	0.855

Table 3.6 Comparison of absorption effects for two polychromatic excitation sources in a context that does not involve enhancement

Analytical context
See Table 3.5
System: Ag–Mo–Y $i = Ag$ $j = Mo$ $k = Y$
$C_{Ag} = 0.2$ $C_{Mo} = 0.3$ $C_Y = 0.5$

Context	λ (Å)	I_λ	P_{Ag}	$P_{(Ag)}$	A_{AgMo}	A_{AgY}
(a) W target tube	0.29	15.5	0.379	4.97	0.91	0.69
	0.31	36.6	1.062	13.31	2.37	1.78
	0.33	56.8	1.929	23.05	3.99	2.97
	0.35	76.6	3.008	34.23	5.74	4.23
	0.37	96.2	4.32	46.77	7.58	5.52
	0.39	111.1	5.845	58.18	9.07	6.54
	0.41	116.4	6.631	65.07	9.74	6.92
	0.43	114.6	7.256	67.87	9.73	6.80
	0.45	109.9	7.674	68.48	9.36	6.43
	0.47	104.5	7.989	68.10	8.85	5.95
	Σ		45.89	450.03	67.34	47.82
(b) Rh target tube	0.29	8.0	0.196	2.57	0.47	0.35
	0.31	25.0	0.725	9.09	1.62	1.21
	0.33	33.6	1.141	13.64	2.36	1.76
	0.35	38.2	1.500	17.07	2.86	2.11
	0.37	41.2	1.850	20.03	3.24	2.37
	0.39	43.4	2.205	22.73	3.54	2.55
	0.41	44.4	2.529	24.82	3.72	2.64
	0.43	44.8	2.837	26.53	3.80	2.66
	0.45	45.2	3.156	28.16	3.85	2.64
	0.47	44.8	3.425	29.20	3.80	2.55
	Σ		19.56	193.83	29.26	20.85

This value may also be obtained from the classical formalism, namely summing the column headed P_{Ag}

$$P_{Ag} = \sum_\lambda P_{Ag\lambda} \, \Delta\lambda = 45.89$$

in which case the individual contributions of yttrium and molybdenum to the total absorption effect are not computed separately.

3.4.2 Secondary fluorescence emission

Enhancement effects can also be expressed in a formalism similar to Eqn (3.29). Given that Eqn (3.13) is expressed for a monochromatic incident wavelength λ, the consideration that intensities will ultimately be expressed relative to the pure analyte leads to

$$(P_i + S_i)_\lambda = C_i \frac{\mu_{i\lambda} I_\lambda}{\mu_s^*} \left(1 + \sum_j e_{ij\lambda} C_j\right) \qquad (3.33)$$

The realization that $C_i \mu_{i\lambda} I_\lambda / \mu_s^*$ is none other than $P_{i\lambda}$, as shown by Broll and Tertian (1983), leads to

$$(P_i + S_i)_\lambda = P_{i\lambda}\left(1 + \sum_j e_{ij\lambda} C_j\right) \qquad (3.34)$$

where

$$e_{ij\lambda} = e_{i\lambda_{j1}} + e_{i\lambda_{j2}} + \cdots \qquad (3.35)$$

Expansion for a three element specimen yields

$$(P_i + S_i)_\lambda = P_{i\lambda} + P_{i\lambda} e_{ij\lambda} C_j + P_{i\lambda} e_{ik\lambda} C_k \qquad (3.36)$$

Defining

$$E_{ij\lambda} = P_{i\lambda} e_{ij\lambda} \qquad (3.37)$$

$$E_{ik\lambda} = P_{i\lambda} e_{ik\lambda} \qquad (3.38)$$

and substituting in the above expression gives

$$(P_i + S_i)_\lambda = P_{i\lambda} + E_{ij\lambda} C_j + E_{ik\lambda} C_k \qquad (3.39)$$

and for the generalized form

$$(P_i + S_i)_\lambda = P_{i\lambda} + \sum_j E_{ij\lambda} C_j \qquad (3.40)$$

The relationship is also applicable to a polychromatic incident beam, namely

$$P_i + S_i = P_i + \sum_j E_{ij} C_j \qquad (3.41)$$

where

$$P_i + S_i = \sum_\lambda (P_i + S_i)_\lambda \, \Delta\lambda \qquad (3.42)$$

$$P_i = \sum_\lambda P_{i\lambda} \, \Delta\lambda \qquad (3.43)$$

$$E_{ij} = \sum_\lambda E_{ij\lambda} \, \Delta\lambda \qquad (3.44)$$

3.4.2.1 Numerical example

For space considerations Mo is chosen as the analyte, and therefore data for 17 effective $\Delta\lambda$ elements will be summed; Ag and Sn are chosen as the matrix elements, both contributing a secondary fluorescence component to the emitted Mo intensity. The same two incident sources as in Section 3.4.1 are retained but, in this case, the characteristic lines of Rh must be taken into consideration. The pertinent data are listed in Table 3.7.

It is observed in Table 3.7 that:

- enhancement due to Sn is involved in seven of the 17 effective $\Delta\lambda$, elements;
- enhancement due to Ag is involved in 10 of the 17 effective $\Delta\lambda$ elements;
- in the case of the Rh target tube, the $\Delta\lambda$ element $(K-M_3)$ $\lambda = 0.55$ Å contributes 12.5% of the total emitted intensity, whereas $(K-L_{2,3})$ at $\lambda = 0.61$ Å contributes 48.3% of the total;

Table 3.7 Comparison of absorption and enhancement effects for two polychromatic excitation sources

Analytical context
See Table 3.5
System: Mo–Ag–Sn   i = Mo   j = Ag   k = Sn
                  $C_{Mo} = 0.2$ $C_{Ag} = 0.3$ $C_{Sn} = 0.5$

	(a) W target tube				(b) Rh target tube			
λ (Å)	$P_{Mo} + S_{Mo}$	P_{Mo}	E_{MoAg}	E_{MoSn}	$P_{Mo} + S_{Mo}$	P_{Mo}	E_{MoAg}	E_{MoSn}
0.29	1.03	0.44	0.70	0.75	0.53	0.23	0.36	0.39
0.31	2.78	1.22	1.88	2.00	1.90	0.83	1.28	1.36
0.33	4.86	2.16	3.25	3.44	2.87	1.28	1.92	2.03
0.35	7.28	3.29	4.83	5.08	3.63	1.64	2.41	2.53
0.37	10.03	4.60	6.60	6.89	4.29	1.97	2.83	2.95
0.39	12.58	5.86	8.20	8.51	4.91	2.29	3.20	3.33
0.41	14.18	6.71	9.15	9.45	5.41	2.56	3.49	3.61
0.43	12.92	8.95	13.24	0.0	5.05	3.50	5.18	0.0
0.45	13.49	9.40	13.64	0.0	5.55	3.87	5.61	0.0
0.47	13.87	9.72	13.82	0.0	5.95	4.17	5.93	0.0
0.49	11.96	11.96	0.0	0.0	5.36	5.36	0.0	0.0
0.51	12.36	12.36	0.0	0.0	5.74	5.74	0.0	0.0
0.53	12.69	12.69	0.0	0.0	6.56	6.56	0.0	0.0
0.55	12.98	12.98	0.0	0.0	24.18	24.18	0.0	0.0
0.57	13.25	13.25	0.0	0.0	8.58	8.58	0.0	0.0
0.59	13.50	13.50	0.0	0.0	9.13	9.13	0.0	0.0
0.61	13.75	13.75	0.0	0.0	93.14	93.14	0.0	0.0
$\sum$	183.51	142.85	75.31	36.12	192.78	175.02	32.21	16.20

- the $P_{(Mo)\lambda}$ data for the identical analytical contexts (W and Rh targets) are listed in Tables 3.8 and 3.9.

In the case of the W target tube, the enhancement components computed from Eqn (3.41) are

$$P_{Mo} + S_{Mo} = P_{Mo} + F_{MoAg}C_{Ag} + E_{MoSn}C_{Sn}$$
$$= 142.85 + (75.31 \times 0.3) + (36.12 \times 0.5)$$
$$= 142.85 + 22.59 + 18.06$$
$$= 183.50$$

whereas in the case of the Rh target tube

$$P_{Mo} + S_{Mo} = 175.02 + (32.21 \times 0.3) + (16.2 \times 0.5)$$
$$= 175.02 + 9.66 + 8.10$$
$$= 192.78$$

The values 183.5 and 192.78 can also be obtained directly by the classical formalism, i.e. by summing the columns headed $P_{Mo} + S_{Mo}$, in which case the individual contributions of Ag and Sn to the total enhancement effect are not quantified separately.

3.4.3 General expression

Although expressions have been derived quantifying absorption and enhancement effects separately, it is their combination that underlies matrix effects in practice. Combining Eqns (3.29) and (3.41) leads to

$$P_i + S_i = P_{(i)}C_i - \sum_j A_{ij}C_j + \sum_j E_{ij}C_j \tag{3.45}$$

and enables one to visualize what, under normal circumstances, one can feel intuitively and observe experimentally, namely, that absorption leads to a decrease in intensity and that enhancement results in an increase in emitted intensity. An exception occurs in situations where the total absorption effect of a matrix element is less than that of the analyte itself; i.e. where $\mu_{j\lambda}^{\star}$ is less than $\mu_{i\lambda}^{\star}$. This leads to a negative value for A_{ij}; hence the net result is one of an increase in intensity.

3.4.3.1 Numerical example

In keeping with previous examples, incident radiations from W and Rh target tubes are considered, the analyte is Mo and the two matrix elements are Y (strong absorber) and Ag (strong enhancer). Table 3.8 lists data relating to the

Table 3.8 Computation of fundamental entities relating to Eqn
(3.45), W target tube

Analytical context
Excitation source: W target tube, 45 kV (CP); see Table II.1.7
Instrument geometry: $\psi' = 90°$, $\psi'' = 30°$
System: Y–Mo–Ag   i = Mo   j = Y   k = Ag
                 $C_{Mo} = 0.2$ $C_Y = 0.3$ $C_{Ag} = 0.5$

λ (Å)	$P_{(Mo)}$	$P_{Mo} + S_{Mo}$	A_{MoY}	A_{MoAg}	E_{MoAg}
0.29	3.21	0.45	1.01	0.11	0.32
0.31	8.78	1.24	2.71	0.30	0.89
0.33	15.56	2.23	4.67	0.54	1.59
0.35	23.63	3.45	6.90	0.84	2.42
0.37	33.02	4.90	9.36	1.20	3.40
0.39	41.97	6.34	11.51	1.55	4.35
0.41	47.94	7.37	12.69	1.80	4.99
0.43	51.01	7.98	13.01	1.96	5.33
0.45	52.45	8.36	12.86	2.05	5.49
0.47	53.11	8.61	12.49	2.11	5.58
0.49	53.30	7.19	14.42	−1.71	0.0
0.51	53.15	7.45	13.96	−2.02	0.0
0.53	52.75	7.69	13.44	−2.33	0.0
0.55	52.16	7.89	12.86	−2.63	0.0
0.57	51.53	8.09	12.27	−2.93	0.0
0.59	50.86	8.28	11.69	−3.22	0.0
0.61	50.20	8.46	11.11	−3.51	0.0
$\sum$	694.62	105.96	176.96	−5.89	34.36

W target tube, and Table 3.9 refers to the Rh target tube. Thus, for the
analytical context of Table 3.8, the emitted Mo intensity is given by

$$P_{Mo} + S_{Mo} = P_{(Mo)}C_{Mo} - A_{MoY}C_Y - A_{MoAg}C_{Ag} + E_{MoAg}C_{Ag}$$
$$= 694.62 \times 0.2 = 138.92$$
$$-176.96 \times 0.3 = -53.09$$
$$-(-5.89) \times 0.5 = 2.95$$
$$+ 34.4 \times 0.5 = 17.20$$
$$= 105.98$$

or directly by summing $(P_{Mo} + S_{Mo})_\lambda \, \Delta\lambda$. The absorption effects can be isolated
by neglecting the last term in the above expression; hence $P_{Mo} = 88.78$.
Absorption and absorption/enhancement effects are shown schematically in
Figures 3.2 and 3.3, respectively.

Table 3.9 Computation of fundamental entities relating to Eqn (3.45), Rh target tube

Analytical context

Excitation source: Rh target tube, 45 kV (CP); see Table II.1.9

Instrument geometry: $\psi' = 90°$, $\psi'' = 30°$

System: Y–Mo–Ag $i = Mo$ $j = Y$ $k = Ag$

$C_{Mo} = 0.2$ $C_Y = 0.3$ $C_{Ag} = 0.5$

λ (Å)	$P_{(Mo)}$	$P_{Mo} + S_{Mo}$	A_{MoY}	A_{MoAg}	E_{MoAg}
0.29	1.66	0.23	0.52	0.06	0.17
0.31	6.00	0.85	1.85	0.21	0.61
0.33	9.20	1.32	2.76	0.32	0.94
0.35	11.78	1.72	3.44	0.42	1.21
0.37	14.14	2.10	4.01	0.51	1.46
0.39	16.40	2.48	4.50	0.61	1.70
0.41	18.29	2.81	4.84	0.69	1.90
0.43	19.94	3.12	5.09	0.76	2.08
0.45	21.57	3.44	5.29	0.84	2.26
0.47	22.77	3.69	5.35	0.91	2.39
0.49	23.88	3.22	6.46	−0.76	0.0
0.51	24.68	3.46	6.48	−0.94	0.0
0.53	27.27	3.97	6.94	−1.21	0.0
0.55	97.14	14.70	23.95	−4.90	0.0
0.57	33.37	5.24	7.95	−1.90	0.0
0.59	34.39	5.60	7.90	−2.18	0.0
0.61	340.80	57.31	75.30	−23.76	0.0
$\sum$	722.56	115.24	172.64	−30.33	14.72

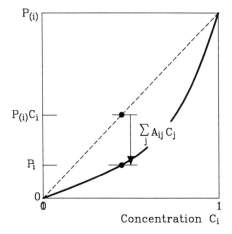

Figure 3.2 Schematic illustration of absorption effects

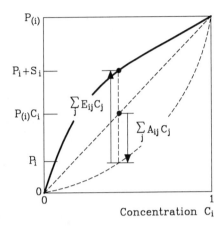

Figure 3.3 Schematic illustration of combined absorption and enhancement effects

3.5 DISCUSSION

It is possible to draw only general conclusions as to the success of the fundamental parameters approach for the correction of matrix effects in X-ray fluorescence analysis. The term approach is used in this instance to indicate that it can be visualized as being open-ended, i.e. amenable to incorporating more advanced principles of X-ray fluorescence spectrometry explicitly, on the one hand, or amenable to simplification for expediency, on the other. These two aspects are examined next.

3.5.1 Monte Carlo model

In a review of the Monte Carlo method to extend the application of the fundamental parameters approach in practical situations, Gardner and Doster (1979) pointed out the necessary assumptions made by Sherman (1955) to obtain analytical closed form solutions for fluorescence X-ray intensities:

(a) the assumption of narrow beam incidence and emergence angles, and the rather drastic simplification that one average angle can be used to represent the range of angles involved;

(b) that specimens are homogeneous and infinitely thick with respect to X-ray fluorescence emissions;

(c) that the excitation source consists of monoenergetic photons;

(d) that scattering within the sample can be neglected; and

(e) that the models are normalized to pure element responses.

The Monte Carlo method is described as a probabilistic technique for simulating the random processes in X-ray fluorescence spectrometer configurations by sampling from probability distribution functions that describe the frequencies of the random events. For example, a large number of photons are followed from birth to death by a series of random walks, their fate along each path being determined by randomly choosing from the appropriate probability distribution function. The procedure is repeated until a sufficient number of histories has been observed to allow conclusions to be drawn about the process of interest. The ultimate goal is the development of models to simulate the entire spectral response to be used with suitable non-linear least squares methods wherein all of the spectral information is used to obtain specimen composition. The authors felt that instrumental geometries, specimen thickness, source distribution, matrix and scattering effects may by modelled precisely without adding significantly to the computational time.

Doster and Gardner (1982) combined a detector response function with a Monte Carlo model to predict the entire pulse height spectra for a specific energy dispersive spectrometer configuration, and Verghese et al. (1988) described the 'Monte Carlo–library least-squares analysis principle' and its application to a Cu–Ni alloy and a stainless steel specimen. Monte Carlo simulations were examined in the context of non-homogeneous matrices by Helsen and Vrebos (1984), in the context of specimens with rough surfaces by Ebel and Poehn (1989), and in the context of thin-film X-ray micro analysis by Rosa and Armigliato (1989).

3.5.2 General expression for quantitative X-ray fluorescence analysis

It is evident from the excitation and emission processes considered so far that experimentally measured intensities of characteristic lines are not linearly related to concentration. Following Jenkins et al. (1981), concentration may be visualized as involving four factors. namely, C is a function of K, I, M and S, where:

(a) K is a function of the incident excitation source, the angles of incidence and emergence, mass attenuation coefficients and instrumental parameters, i.e. a number of factors depending on the collimating, dispersing and detecting components selected by the analyst. For a given element i and fixed operating conditions, K remains constant, and is generally termed the calibration factor for converting intensity to concentration.

(b) I is the net experimentally measured intensity of a characteristic line. A measured intensity must be corrected for background, peak overlap, dead-time etc., so that I_i can be ascribed with confidence as emanating from the analyte only.

(c) M quantifies matrix effects. Owing to absorption and enhancement, every element present has a bearing on the intensity of all the characteristic lines emitted. This need not be interpreted as leading to errors in the determination of the 'true' concentrations, but rather as a generally inherent need to apply a correction to compensate for matrix effects given experimentally valid intensities.

(d) S relates to physical factors such as specimen heterogeneity, thickness, surface texture etc., factors that are not easily amenable to mathematical computation nor experimentally controllable. It is generally preferable to minimize the effect of these factors by developing specimen preparation procedures that make it possible, for practical purposes, to consider all specimens as homogeneous, flat and infinitely thick in respect of X-ray fluorescence emissions. This topic is addressed in Chapter 9.

Given that the physical factors relating to specimens may be reduced to such a degree as to be considered negligible, quantitative analysis then requires that analysts deal with the three remaining factors that relate intensities to concentrations. The general formulation for quantitative X-ray fluorescence analysis of thick, flat and homogeneous specimens of unknown composition may then be expressed as

$$C_{iu} = K_i I_{iu} M_{iu} \qquad (3.46)$$

where C_{iu} is the concentration (weight fraction) of element i in a specimen of unknown composition; K_i is the calibration factor, and is instrument dependent, but remains constant provided that all intensity measurements are made under constant operating conditions (voltage, current, collimator, crystal, detector etc.); I_{iu} is the net intensity of a characteristic line of element i emitted by a specimen of unknown composition; and M_{iu} is an expression or term that corrects for matrix effects on element i.

The constant K_i may be computed from experimental intensity data for specimens of known composition, i.e. reference materials or standards. In this case Eqn (3.46) is expressed as

$$C_{ir} = K_i I_{ir} M_{ir} \qquad (3.47)$$

where the subscript r indicates a reference material of known composition. Solving the above equation for K_i leads to

$$K_i = \left(\frac{C_i}{I_i M_i}\right)_r \qquad (3.48)$$

where C_{ir} is known (by definition), I_{ir} is obtained experimentally, and M_{ir} can be calculated from fundamental parameters, i.e.

$$M_{ir} = \frac{C_{ir} P_{(i)}}{P_{ir} + S_{ir}} = \left(\frac{C_i}{R_i}\right)_r \qquad (3.49)$$

Conversely, K_i may be computed from the following relationship derived from Eqn (3.5)

$$K_i = \frac{(P_i + S_i)_r}{I_{ir}P_{(i)}} = \left(\frac{P_i + S_i}{I_i P_{(i)}}\right)_r \tag{3.50}$$

i.e. K_i is simply the reciprocal of $I_{(i)}$.

It is preferable to use more than one reference material to compute K_i, in which case K_i is simply the average value

$$K_i = \frac{\Sigma K_i}{N} \tag{3.51}$$

3.5.2.1 Numerical examples

The data in Table 3.10, taken from Tables 3.7 and 3.9, relate to the analyte Mo in the systems Mo–Ag–Sn and Mo–Y–Ag. For illustrative purposes, nominal intensities I_{Mo} are computed assuming a value of 3.0 for the proportionality constant in Eqn (2.1), i.e.

$$I_{MO, \text{ nominal c/s}} = 3.0 \times (P_{Mo} + S_{Mo})$$

Substitution of the appropriate data in Eqn (3.49) yields

Mo–Y–Ag: $\quad M_{ir} = \dfrac{722.56 \times 0.2}{115.24} = 1.25401$

Mo–Ag–Sn: $\quad M_{ir} = \dfrac{722.56 \times 0.2}{192.78} = 0.74962$

Substitution in Eqn (3.48) yields

Mo–Y–Ag: $\quad K_i = \dfrac{0.2}{345.72 \times 1.25401} = 4.6132 \times 10^{-4}$

Mo–Ag–Sn: $\quad K_i = \dfrac{0.2}{578.34 \times 0.74962} = 4.6132 \times 10^{-4}$

Conversely, substitution in Eqn (3.50) yields

Mo–Y–Ag: $\quad K_i = \dfrac{115.24}{345.72 \times 722.56} = 4.6132 \times 10^{-4}$

Mo–Ag–Sn: $\quad K_i = \dfrac{192.78}{578.34 \times 722.56} = 4.6132 \times 10^{-4}$

Table 3.10 Data taken from Tables 3.7 and 3.9 to illustrate the computation of M_{ir} and K_i using Eqns (3.49), (3.48) and (3.50), respectively

		I_{Mo} (nominal c/s)[a]
Table 3.9 Mo–Y–Ag:	$P_{(Mo)} = 722.56$	(2167.7)
Table 3.7 Mo–Ag–Sn:	$P_{Mo} + S_{Mo} = 192.78$	578.34
Table 3.9 Mo–Y–Ag:	$P_{Mo} + S_{Mo} = 115.24$	345.72

[a] Assuming a proportionality constant of 3.0.

Hence

$$I_{(Mo)} = 1/(4.6132 \times 10^{-4}) = 2167.7 \text{ nominal c/s}$$
$$= 3.0 \times 722.56 = 2167.7 \text{ nominal c/s}$$

Combination of Eqns (3.46) and (3.48) leads to

$$C_{iu} = \left(\frac{C_i}{I_i M_i}\right)_r I_{iu} M_{iu} \qquad (3.52)$$

which is an expression that will be used widely in the subsequent treatment of calibration and correction for matrix effects in this text. Thus, if Mo–Ag–Sn is treated as the reference and Mo–Y–Ag is treated as the unknown, substitution in Eqn (3.52) yields

$$C_{iu} = \frac{0.2}{578.34 \times 0.74962} \times 345.74 \times 1.254$$

$$= 0.20$$

in which case the value for $M_{iu} = 1.254$ would be calculated by iteration.

3.6 SUMMARY

The reader should be conversant with the following fundamental relationships between emitted intensities and specimen composition, and with how each term is defined and generated numerically.

Emitted intensity due to primary and secondary fluorescence:

(a) Classical formalism:

$$P_i + S_i = G_i C_i \sum_{\lambda} \left(\frac{\mu_{i\lambda} I_\lambda \, \Delta\lambda}{\mu_s' + \mu_s''} \times \left\{1 + \sum_j (0.5 C_j p_{\lambda j} \mu_{i\lambda_j}(\mu_{j\lambda}/\mu_{i\lambda})) \right.\right.$$
$$\left.\left. \times \left[\frac{1}{\mu_s'} \ln\left(1 + \frac{\mu_s'}{\mu_{s\lambda_j}}\right) + \frac{1}{\mu_s''} \ln\left(1 + \frac{\mu_s''}{\mu_{s\lambda_j}}\right)\right]\right\}\right)$$

$$R_{i,\text{measured}} = \frac{I_i}{I_{ir}} \frac{(P_i + S_i)_r}{P_{(i)}}$$

$$= \frac{I_i}{I_{(i)}}$$

Concentrations are determined by iteration using

$$C_{i,\text{next estimate}} = \frac{R_{i,\text{measured}}}{R_{i,\text{calculated}}} C_{i,\text{present estimate}}$$

(b) Alternate formalisms:

$$P_i = P_{(i)} C_i - \sum_j A_{ij} C_j$$

$$P_i + S_i = P_i + \sum_j E_{ij} C_j$$

$$= P_{(i)} C_i - \sum_j A_{ij} C_j + \sum_j E_{ij} C_j$$

where

$$A_{ij} = \sum_\lambda A_{ij\lambda} \, \Delta\lambda$$

$$= \sum_\lambda C_i \frac{\mu_{i\lambda} I_\lambda}{\mu_s^*} \left(\frac{\mu_j^* - \mu_i^*}{\mu_i^*} \right) \Delta\lambda$$

$$E_{ij} = \sum_\lambda E_{ij\lambda} \, \Delta\lambda$$

$$= \sum_\lambda C_i \frac{\mu_{i\lambda} I_\lambda}{\mu_s^*} \left(e_{i\lambda_{j1}} + e_{i\lambda_{j2}} + \cdots \right) \Delta\lambda$$

(c) Algorithm for quantitative XRF analysis:

$$C_{iu} = \left(\frac{C_i}{I_i M_i} \right)_r I_{iu} M_{iu}$$

4 FUNDAMENTAL INFLUENCE COEFFICIENTS

In addition to introducing the concept of influence coefficients used to quantify matrix effects individually (i.e. the matrix effect of element j, the matrix effect of element k, ... on analyte i) in quantitative X-ray fluorescence analysis, the main thrust of this chapter is directed towards defining and delineating the domain of a subset of influence coefficients, namely fundamental influence coefficients. In order to demarcate clearly the distinctive characteristics of these coefficients, the following symbols will be used in this book to designate exclusively influence coefficients that are deemed fundamental:

a_{ij}, a_{ijp} absorption effect of element j on analyte i;
e_{ij}, e_{ijp} enhancement effect of element j on analyte i;
m_{ij} total matrix effect of element j on analyte i;
m_{ijn} matrix effect resulting from substitution of element n by element j on analyte i.

4.1 GENERAL CONSIDERATIONS

The first three chapters may be envisaged as having led, in simplified terms, to the conclusion that emitted intensities may be quantified as a function of specimen composition (C_i, C_j, ... C_n) instrumental parameters (I_λ, ψ' and ψ'') and fundamental parameters ($\mu, r, \omega, f, \lambda_{min}, \lambda_{abs}, \lambda_i$ ...) in expressions that initially precluded their application in practice because of the integrals involved. As shown in the previous chapter, Criss and Birks (1968) proposed a workable model that compensates for matrix effects by means of mass attenuation coefficients, fluorescence yields, instrument geometry, etc., i.e. a fundamental parameters approach. The procedure generates intensities relative to that emitted by the pure element based on measured intensities for any given multielement reference material of known composition, namely

$$R_i = \frac{I_i}{I_{(i)}} = \left(\frac{I_{iu}}{I_{ir}}\right)_{measured} \times \left(\frac{P_i + S_i}{P_{(i)}}\right)_{r,calculated} \tag{4.1}$$

The calculation of concentrations in unknowns is based on making successively better estimates of composition by an iteration procedure. The process is repeated until the differences between the last estimate and the newly computed C_{iu}, C_{ju} etc. are less than some predetermined value.

However, Criss and Birks proposed their fundamental parameters approach as an alternative to another approach, then and still in current use, which is generally referred to as 'influence coefficient methods or models'. The study of influence coefficient models and their interrelation is not straightforward for a number of reasons:

- workers have used the same symbols in identical models to represent entirely different entities;
- workers have used different symbols in identical models when referring to the same entities;
- models are expressed in different formalisms and hence are not easily recognized.

Thus, in this and subsequent chapters treating influence coefficient methods, every effort will be made to bear in mind the judicious recommendation of Tertian (1988), namely, that 'the significance and properties of the influence coefficients in any given model should be fully understood'.

4.2 THE CONCEPT OF INFLUENCE COEFFICIENTS

Consider a multielement specimen of composition C_i, C_j, C_k, ... C_n and measured intensities I_i, I_j, I_k, ... I_n. The term 'influence coefficient' is used to designate any factor, say x_{ij} or y_{ij}, which when multiplied by either C_j or I_j is deemed to quantify the *total* matrix effect of element j on analyte i. In the context of Eqn (3.52)

$$C_{iu} = \left(\frac{C_i}{I_i M_i}\right)_r I_{iu} M_{iu} \tag{4.2}$$

where the term inside the brackets is a calibration constant K_i independent of the reference material used to define it. The correction term M_i is then simply the summation of each individual matrix effect correction in expressions of the type

$$C_{iu} = K_i I_{iu}\left[\ldots \sum_j (\text{influence coefficient})_{ij} C_j \ldots\right]_u \tag{4.3}$$

or

$$C_{iu} = K_i I_{iu} \left[\ldots \sum_j (\text{influence coefficient})_{ij} I_j \ldots \right]_u \qquad (4.4)$$

There is no doubt, from a theoretical point of view, that one of the two terms in the quantification of each elemental matrix effect should involve the concentration of the element as in Eqn (4.3). It is for practical considerations that some workers fully aware of the theoretical aspects of matrix effects have proposed intensity models. These are treated in Chapter 6.

It often happens when seeking the solution to a problem that developments proceed simultaneously along parallel lines, each in turn influencing the course of the other. The quest for a mathematical solution that compensates for the effect of absorption and enhancement is such a case. Over the years a number of workers concentrated their efforts on elucidating the theoretical aspects of influence coefficients, whereas others pursued the study of influence coefficients of a semi-empirical nature. The latter are examined in Chapter 5.

4.3 THE DOMAIN OF FUNDAMENTAL INFLUENCE COEFFICIENTS

Briefly stated, the domain of fundamental influence coefficients encompasses any mathematical expression relating emitted intensities and concentrations in which the influence coefficients are defined and derived explicitly in terms of fundamental parameters. Most of the theoretical developments regarding influence coefficients generally take only primary and secondary fluorescence emissions into consideration, a practice retained in this chapter. In a critical survey of mathematical matrix correction procedures for X-ray fluorescence analysis, Tertian (1986) noted that, from 1968 on, Criss and Birks' fundamental parameters method had prompted some workers to re-evaluate the older influence coefficient approach by progressively giving these coefficients a sound theoretical status. By the mid-1960s, theoretical expressions had been developed to calculate the influence coefficient for the effect of element j on analyte i, but their applications were limited to monochromatic incident sources for binary systems involving absorption only. The task at hand involved deriving expressions that define influence coefficients explicitly as a function of fundamental parameters in the context of polychromatic incident sources in multielement systems when both absorption and enhancement effects are present.

4.4 INITIAL DEVELOPMENTS

4.4.1 von Hamos

The theoretical aspects of the intensity emitted from pure element specimens and from 'diluted substances', e.g. Cu in Cu–Al and Ni in Ni–Fe, for monochromatic incident X-radiation were examined by von Hamos (1945), who derived the following expression

$$P_{i\lambda} = P_{(i)\lambda} C_i k_i \tag{4.5}$$

where

$$k_i = \frac{a}{b + (a - b)C_i} \tag{4.6}$$

with

$$a = \frac{\mu_{i\lambda}}{\csc \Psi''} + \frac{\mu_{i\lambda_i}}{\csc \Psi'} \tag{4.7}$$

$$b = \frac{\mu_{j\lambda}}{\csc \Psi''} + \frac{\mu_{j\lambda_i}}{\csc \Psi'} \tag{4.8}$$

The conclusions reached were that nickel intensities in the presence of iron should be lower than those calculated by direct proportionality; on the other hand, for Cu in the presence of Al, copper intensities higher than those calculated by direct proportionality were predicted. Both predictions have since been confirmed in practice. It can be shown that Eqn (4.6) is equivalent to

$$k_i = \frac{\mu_i^*}{C_i \mu_i^* + C_j \mu_j^*} = \frac{\mu_i^*}{\mu_s^*} \tag{4.9}$$

with coefficients μ_i^* and μ_s^* as defined previously. Eqn (4.5) is equivalent to

$$C_i = K_i I_i M_i$$

$$= \frac{1}{P_{(i)\lambda}} \cdot P_{i\lambda} \cdot \frac{\mu_s^*}{\mu_i^*} \tag{4.10}$$

4.4.2 Beattie and Brissey

For a similar analytical context, Beattie and Brissey (1954) derived the following expression, one of the first models that defines influence coefficients as a function of two elements:

$$-\left(\frac{I_{(i)}}{I_i} - 1 \right) C_i + K_{ij} C_j + K_{ik} C_k + \cdots = 0 \tag{4.11}$$

It should be noted that the *relative intensity* expressed in the above is the reciprocal of R_i as defined in Eqn (4.1). Although it was shown that $K_{ij} = \mu_j^*/\mu_i^*$, which is valid for a monochromatic incident excitation source only, in practice the influence coefficients were computed using experimental intensity data for binary systems in the expression

$$K_{ij} = \frac{C_i}{C_j}\left(\frac{I_{(i)}}{I_i} - 1\right)$$ (4.12)

A drawback of the Beattie and Brissey model is that, for a system of N elements, there are N^2 possible solutions. An empirical procedure based on distinguishing the most abundant element was proposed as the best solution. That method was applied to the analysis of Cr–Fe–Ni–Mo alloys.

4.4.3 Lachance and Traill

Also working in the very restricted context of a monochromatic excitation source and absorption effects only, Lachance and Traill (1966) derived a model which overcomes the problem of overdefinition, namely

$$C_i = R_i\left[1 + \sum_j a_{ij\lambda}C_j\right]$$ (4.13)

referred to by Tertian (1986) as 'the canonical form . . . the simplest and most convenient one in practice'. The symbol $a_{ij\lambda}$ is used here *instead of the usual* α_{ij} to indicate that *the derivation relates to absorption effects only and to monochromatic excitation*. It is important to make this distinction at this time because the symbol α_{ij} is retained for that subset of influence coefficients which, although defined and calculated from theoretical principles, involves approximations and hence does not meet the criteria defining fundamental influence coefficients. The subset of α_{ij} influence coefficients is the topic of Chapter 5. It has been shown that

$$a_{ij\lambda} = \left(\frac{\mu_j^* - \mu_i^*}{\mu_i^*}\right)_\lambda = \left(\frac{\mu_j^*}{\mu_i^*}\right)_\lambda - 1$$ (4.14)

It is important to note that the influence coefficients relate to monochromatic excitation and to absorption effects only. Within those very strict limitations, a_{ij} coefficients are in the fundamental influence coefficient domain and remain constant in that context only.

4.5 INTRODUCING POLYCHROMATIC EXCITATION

4.5.1 Claisse and Quintin

Pursuing the study of influence coefficients and the fact that $a_{ij\lambda}$ can be considered as constant only in the limited case of a monochromatic incident source in systems that do not involve enhancement, Claisse and Quintin (1967) expanded Eqn (4.13) in order to take into consideration the context of a polychromatic incident source in the absence of enhancement. Based on the premise that the total emitted intensity is equal to the sum of the intensities at each monochromatic wavelength of the effective wavelength range, it was shown that R_i could be defined as the product of C_i and the quotient of two polynomials,

$$R_i = C_i \frac{p_0 + p_j C_j + p_{jj} C_j^2 + p_{jk} C_j C_k + \text{H.O.T.}}{q_0 + q_j C_j + q_{jj} C_j^2 + q_{jk} C_j C_k + \text{H.O.T.}} \qquad (4.15)$$

where H.O.T. represents the higher order terms of the series. Dividing the numerator by the denominator, and noting that p_0 and q_0 are equal, leads to

$$C_i = R_i \left[1 + \sum_j a_{ij} C_j + \sum_j a_{ijj} C_j^2 + \sum_j a_{ijjj} C_j^3 + \cdots \right.$$

$$\left. + \sum_j \sum_k a_{ijk} C_j C_k + \text{H.O.T.} \right] \qquad (4.16)$$

Considering that higher order coefficients such as a_{ijjj} and a_{ijkl} exist but are multiplied by the product of three concentrations or more, Claisse and Quintin concluded that these terms can probably be neglected; hence

$$C_i = R_i \left[1 + \sum_j a_{ij} C_j + \sum_j a_{ijj} C_j^2 + \sum_j \sum_k a_{ijk} C_j C_k \right] \qquad (4.17)$$

Claisse and Quintin showed that a_{ij} in Eqn (4.17) may be computed from

$$a_{ij} = \frac{\sum_\lambda P_{(i)\lambda} a_{ij\lambda}}{\sum_\lambda P_{(i)\lambda}} \qquad (4.18)$$

i.e., the a_{ij} coefficient is simply the average of the $a_{ij\lambda}$ coefficients for all incident monochromatic wavelengths when each is given a weight equal to $P_{(i)\lambda}$. The other coefficients can be defined by similar expressions. The fact that the values $P_{(i)\lambda}$ are constants for any given wavelength for fixed instrumental conditions results in the influence coefficients a_{ij}, a_{ijj} ... also being constants but, unfortunately, the exact matrix effect correction involves the sum of an infinite number of terms. The assumptions that Claisse and Quintin made to deal with this in practice if both absorption and enhancement are present are treated in Chapter 5.

4.5.2 Lachance

Given a reliable spectral distribution emanating from an X-ray tube (Gilfrich and Birks, 1968) and a fundamental parameters expression for primary and secondary fluorescence emission (Criss and Birks, 1968), it is possible to define and calculate fundamental influence coefficients for the total matrix effect, i.e. the combined effect of absorption and enhancement of element i by element j, for *binary* specimens. Lachance (1970) pointed out that Eqn (4.13) is beset by the spectre of not being theoretically valid in that the two basic assumptions made for its derivation are known to be sweeping simplification and physically unrealistic: (1) that a polychromatic incident beam can be represented by a unique wavelength and (2) that enhancement can be treated as negative absorption. Fundamental influence coefficients were calculated for polychromatic excitation in analytical contexts involving both absorption and enhancement but limited to *binary* systems. Using Criss and Birks' fundamental parameters equation to generate theoretical R_i values, it follows from Eqn (4.13) that

$$m_{ij} = \left(\frac{C_i - R_i}{C_j R_i}\right)_r = \frac{1}{C_j}\left(\frac{C_i}{R_i} - 1\right) \tag{4.19}$$

Thus the binary influence coefficient m_{ij} may quantify absorption only or absorption and enhancement, depending on the nature of elements i and j. The coefficient may be subscripted 'bin' to indicate that it is computed strictly from binary system data. If enhancement is not involved, then binary m_{ij} and a_{ij} influence coefficients are identical. Examples were given of the variations in the values of m_{ij} arrays for the analytes Fe and Mo alloyed with a number of other elements. Typical average values were also computed for the cross-influence coefficients in the Claisse–Quintin model.

4.6* FUNDAMENTAL INFLUENCE COEFFICIENT ALGORITHMS

The underlying concept in generating fundamental influence coefficients is that they be based on the theoretical principles of X-ray physics and hence defined explicitly in terms of fundamental and instrumental parameters. Every effort should be made to avoid unnecessary approximations or overlooking physical interactions that can be important. If fundamental influence coefficients are explicitly derived from physical theory, then the particular formalism used should not affect the final results. A major advantage of this approach is that it enables analysts to compute the values of influence coefficients *a priori*, i.e. independently of experimental data. Four fundamental influence coefficient models that have been proposed will now be examined. In order to show their common origin, all four will be derived from Eqn (3.45) expressed in analytical formalism

$$P_{(i)}C_i = P_i + S_i + \sum_j A_{ij}C_j - \sum_j E_{ij}C_j \tag{4.20}$$

which is in turn a combination of Eqns (3.29) and (3.41)

$$P_i = P_{(i)}C_i - \sum_j A_{ij}C_j \qquad (4.21)$$

$$P_i + S_i = P_i + \sum_j E_{ij}C_j \qquad (4.22)$$

4.6.1   Tertian and Broll–Tertian formalisms

The theoretical contributions of Tertian (1973, 1974) to the study of matrix effect corrections were carried out in the context of the Beattie and Brissey model expressed in the form

$$C_i = R_i\left[C_i + \sum_j K_{ij}C_j\right] \qquad (4.23)$$

where the K_{ij} coefficient was initially defined as $K_{ij} = \mu_j^*/\mu_i^*$. Subsequently, Tertian and Vié le Sage (1977) examined the underlying reasons for influence coefficient variation by distinguishing between the absorption case and the enhancement case *in the context of monochromatic incident radiation*. Considering binary systems, Eqn (4.23) was retained for the absorption case, namely

$$C_i = R_i[C_i + K_{ijp}C_j] \qquad (4.24)$$

where the subscript p is used to indicate that the influence coefficient relates to primary fluorescence emissions only. If enhancement is present, it was shown that K_{ijp} in the above expression must be replaced by

$$K_{ij} = \frac{K_{ijp} - e_{ij\lambda}C_i}{1 + e_{ij\lambda}(1 - C_i)} \qquad (4.25)$$

in which case the formalism of Eqn (4.23) is retained even if enhancement is present.

   Broll and Tertian (1983) extended the above relationship to encompass polychromatic incident radiation in multicomponent systems, based on the concept of Tertian's entities. Initially, consider analyte i in a multicomponent system in the context of polychromatic excitation. Fundamental absorption influence coefficients a_{ij}, a_{ik}, etc. for primary fluorescence can be derived from Eqn (4.21)

$$P_{(i)}C_i = P_i + \sum_j A_{ij}C_j \qquad (4.26)$$

Taking P_i in factor leads to

$$P_{(i)}C_i = P_i\left[1 + \sum_j \frac{A_{ij}}{P_i} C_j\right] \tag{4.27}$$

Solving for C_i yields

$$C_i = R_{ip}\left[1 + \sum_j a_{ijp}C_j\right] \tag{4.28}$$

where

$$R_{ip} = \frac{P_i}{P_{(i)}} \tag{4.29}$$

and

$$a_{ijp} = \frac{A_{ij}}{P_i} \tag{4.30}$$

Similarly, fundamental enhancement influence coefficients e_{ij}, e_{ik} can be derived from Eqn (4.22); considering the total intensity due to primary and secondary fluorescence emissions

$$P_i + S_i = P_i + \sum_j E_{ij}C_j \tag{4.31}$$

Taking P_i in factor at the right hand side of the equals sign leads to

$$P_i + S_i = P_i\left[1 + \sum_j \frac{E_{ij}}{P_i} C_j\right] \tag{4.32}$$

Dividing both sides by $P_{(i)}$ gives

$$R_i = R_{ip}\left[1 + \sum_j e_{ijp}C_j\right] \tag{4.33}$$

where

$$R_i = \frac{P_i + S_i}{P_{(i)}} \tag{4.34}$$

and

$$e_{ijp} = \frac{E_{ij}}{P_i} \tag{4.35}$$

Ratioing Eqns (4.28) and (4.33) results in

$$\frac{C_i}{R_i} = \frac{1 + \sum_j a_{ijp}C_j}{1 + \sum_j e_{ijp}C_j} \tag{4.36}$$

Cross multiplication and regrouping leads to

$$C_i = R_i \left[1 + \sum_j \left(a_{ijp} - e_{ijp} \frac{C_i}{R_i} \right) C_j \right] \tag{4.37}$$

resulting in the Broll–Tertian algorithm

$$C_i = R_i \left[1 + \sum_j m_{ijp} C_j \right] \tag{4.38}$$

where the effective influence coefficient m_{ijp} is defined:

$$m_{ijp} = \left(a_{ijp} - e_{ijp} \frac{C_i}{R_i} \right) \tag{4.39}$$

4.6.2 Rousseau formalism

Rousseau (1984a, b) observed that the large number of fundamental parameters involved in the classical equations proposed by Sherman (1955) together with the complexity of the primary beam emanating from X-ray tubes make these equations complex and difficult to handle. As a result, a new formalism was proposed to correct for matrix effects wherein explicit expressions were deduced for the calculation of absorption and enhancement for use in the Rousseau algorithm

$$C_i = R_i \left[\frac{1 + \sum_j a_{ijp} C_j}{1 + \sum_j e_{ijp} C_j} \right] \tag{4.40}$$

The process may be described as follows.
    Defining

$$a_{ij\lambda} = \left[\frac{\mu_j^*}{\mu_i^*} \right]_\lambda - 1 \tag{4.41}$$

$$W_{i\lambda} = \frac{\mu_{i\lambda} I_\lambda\, \Delta\lambda}{\mu_i^* \left[1 + \sum_j a_{ij\lambda} C_j \right]} \tag{4.42}$$

and making a few algebraic manipulations, the Sherman formulation takes the form

$$C_i = R_i \frac{\sum_\lambda W_{i\lambda} \left(1 + \sum_j a_{ij\lambda} C_j \right)}{\sum_\lambda W_{i\lambda} \left(1 + \sum_j e_{ij\lambda} C_j \right)} \tag{4.43}$$

Removing the brackets in both the numerator and the denominator of the above equation and dividing each term by $\Sigma_j W_{i\lambda}$ leads to Eqn (4.40) with

$$a_{ijp} = \frac{\sum_\lambda W_{i\lambda} a_{ij\lambda} \Delta\lambda}{\sum_\lambda W_{i\lambda} \Delta\lambda} \qquad (4.44)$$

and

$$e_{ijp} = \frac{\sum_\lambda W_{i\lambda} e_{ij\lambda} \Delta\lambda}{\sum_\lambda W_{i\lambda} \Delta\lambda} \qquad (4.45)$$

Commenting on that result, Rousseau stated that Eqn (4.40) is a fundamentally exact algorithm between concentration and intensity that has obvious physical meaning. It was pointed out that a_{ijp} through the $a_{ij\lambda}$ coefficients includes all the mass absorption coefficients (when there is no enhancement) in the Sherman equations. Similarly, the e_{ijp} coefficients correct for enhancement effects. It is stressed that the coefficients are variables, being a function of specimen composition, and hence are valid only for a given specimen. For practical applications, Rousseau noted that an iterative process could be applied directly to Eqn (4.1) in order to compute concentrations; for faster calculation, however, it is preferable in a first step to obtain an estimate of composition for each unknown, using influence coefficients of the type developed in the next chapter. These concentrations are then used to compute a_{ijp} and e_{ijp} values for each matrix element in each specimen.

Concerning the Broll–Tertian and the Rousseau formalisms, the equations defining the absorption and enhancement influence coefficients in these algorithms are equivalent; i.e. Eqn (4.30) is equivalent to Eqn (4.44) and Eqn (4.35) is equivalent to Eqn (4.45) because of the relationship

$$P_{i\lambda} = C_i W_{i\lambda} \qquad (4.46)$$

$$P_i = C_i \sum_\lambda W_{i\lambda} \qquad (4.47)$$

obtained by observing that the denominator in Eqn (4.42) is equal to $\Sigma\, C_i\mu_i^*$, which in turn is equal to μ_s^*, i. e.

$$\mu_i^* \left[1 + \sum_j \left\{ \frac{\mu_j^* - \mu_i^*}{\mu_i^*} \right\} C_j \right] = \mu_i^* + \sum_j (\mu_j^* - \mu_i^*) C_j$$

$$= \mu_i^* + \sum_j \mu_j^* C_j - \mu_i^* (1 - C_i)$$

$$= \mu_s^* \qquad (4.48)$$

Thus, in Eqns (4.44) and (4.45)

$$a_{ijp} = \frac{\sum_\lambda W_{i\lambda} a_{ij\lambda}}{\sum_\lambda W_{i\lambda}} = \frac{\sum_\lambda P_{i\lambda} a_{ij\lambda}}{\sum_\lambda P_{i\lambda}} = \frac{\sum_\lambda A_{ij\lambda}}{\sum_\lambda P_{i\lambda}} = \frac{A_{ij}}{P_i} \tag{4.49}$$

$$e_{ijp} = \frac{\sum_\lambda W_{i\lambda} e_{ij\lambda}}{\sum_\lambda W_{i\lambda}} = \frac{\sum_\lambda P_{i\lambda} e_{ij\lambda}}{\sum_\lambda P_{i\lambda}} = \frac{\sum_\lambda E_{ij\lambda}}{\sum_\lambda P_{i\lambda}} = \frac{E_{ij}}{P_i} \tag{4.50}$$

4.6.3 Lachance formalism

Recalling Eqn (3.45) as proposed by Lachance (1988) as an alternate formalism for Criss and Birks' fundamental parameters algorithm, namely

$$P_i + S_i = P_{(i)} C_i - \sum_j A_{ij} C_j + \sum_j E_{ij} C_j \tag{4.51}$$

transposition leads to

$$P_{(i)} C_i = P_i + S_i + \sum_j A_{ij} C_j - \sum_j E_{ij} C_j \tag{4.52}$$

which may be regarded as Criss and Birks' fundamental parameters algorithm expressed in a formalism more easily adapted to quantitative analysis. It can be interpreted as follows.

$P_{(i)} C_i$, viz. the fluorescence intensity of the analyte corrected for all matrix effects, *is equal to* $P_i + S_i$, viz. the total intensity due to primary fluorescence and secondary (enhancement) fluorescence, *plus* $\Sigma_j A_{ij} C_j$, viz. the sum of all absorption effects which compensates for 'lost' counts due to absorption, *minus* $\Sigma_j E_{ij} C_j$, viz. the sum of all enhancement effects which compensates for the 'extra' counts due to enhancement.

Taking $P_i + S_i$ in factor and dividing both sides of Eqn (4.52) by $P_{(i)}$ gives

$$C_i = \frac{P_i + S_i}{P_{(i)}} \left[1 + \sum_j \frac{A_{ij}}{P_i + S_i} C_j - \sum_j \frac{E_{ij}}{P_i + S_i} C_j \right] \tag{4.53}$$

Defining

$$a_{ij} = \frac{A_{ij}}{P_i + S_i} \tag{4.54}$$

$$e_{ij} = \frac{E_{ij}}{P_i + S_i} \tag{4.55}$$

$$m_{ij} = a_{ij} - e_{ij} \tag{4.56}$$

leads to

$$C_i = R_i \left[1 + \sum_j a_{ij} C_j - \sum_j e_{ij} C_j \right] \qquad (4.57)$$

$$= R_i \left[1 + \sum_j m_{ij} C_j \right] \qquad (4.58)$$

where a_{ij} and e_{ij} are fundamental influence coefficients that combine in a single term m_{ij} the quantification of the absorption and enhancement processes (examined in Chapters 2 and 3) relating to analyte i and elements $j, k, \ldots n$, in the context of polychromatic incident sources impinging on a multielement specimen.

Conversely, the simplest process to compute m_{ij} influence coefficients is to dispense with the concept of defining absorption and enhancement influence coefficients separately, i.e. dispense with the calculation of A_{ij}, E_{ij}, a_{ij} and e_{ij} and proceed directly from the expression

$$P_{(i)} C_i = P_i + S_i + \sum_j M_{ij} C_j \qquad (4.59)$$

where

$$M_{ij} = \sum_\lambda M_{ij\lambda} \, \Delta\lambda \qquad (4.60)$$

with

$$M_{ij\lambda} = C_i \frac{\mu_{i\lambda} I_\lambda}{\mu_s^*} \left[\frac{\mu_j^* - \mu_i^*}{\mu_i^*} - (e(e' + e''))_j \right] \qquad (4.61)$$

in which case m_{ij} in Eqn (4.58) is defined as

$$m_{ij} = \frac{M_{ij}}{P_i + S_i} \qquad (4.62)$$

4.6.4 De Jongh formalism

Noting that influence coefficients are often determined experimentally, thus requiring a large number of standard reference materials, and that, fortunately, they can also be obtained or at least estimated by fundamental parameters expressions, de Jongh (1973) extended the fundamental influence coefficient method to multielement systems. The conversion of intensities to concentrations is based on the following algorithm proposed by de Jongh (1979)

$$C_i = (D_i + K_i I_i) \left[1 + \sum_j m_{ijn} C_j \right] \qquad (4.63)$$

where D_i has the dimension weight fraction and accounts for residual background, and K_i is a calibration constant, dimension C/I.

The approach can be described as follows:

(a) the summation over j includes i and excludes a selected non-measured element or component n;

(b) the computation of m_{ijn} is summarized as consisting of three steps:

(1) d_{ij} coefficients are based on partial derivatives of Shiraiwa and Fujino's (1966) modification of Sherman's equations; they are calculated at an average specimen composition $(C_i, C_j, \ldots)_{avg}$ to satisfy the expression

$$\frac{C_i}{R_i} = \text{constant} \times [1 + d_{ij} \Delta C_j + d_{ik} \Delta C_k + \cdots + \text{H.O.T.}] \qquad (4.64)$$

where $\Delta C_j = C_j - C_{j,avg}$. It was observed that, in the absence of enhancement, the sum of the higher order terms is always negative and negligibly small. If the analyte intensity is strongly enhanced, the higher order terms can be neglected only for a limited (though still wide) range of concentrations;

(2) b_{ij} coefficients are calculated from d_{ij} coefficients by eliminating one of the influencing components, i.e. the major element for alloys, cellulose for plants etc. For example, if $d_{in} \Delta C_n$ is the term eliminated, then $b_{ij} = d_{ij} - d_{in}$.

(3) m_{ijn} coefficients are calculated so that they may be multiplied by C_j instead of ΔC_j.

It is very important to note at this point the distinguishing feature of m_{ijn} coefficients. Three subscripts are used for the influence coefficients in de Jongh's formalism in order to show their full physical definition, i.e. i is the analyte, j is the matrix element and n is the non-measured element or component. This is in contrast to the coefficients m_{ij} in Eqn (4.58) where the derivation is based on the concept that matrix elements substitute for the analyte. Equation (4.63) offers the advantage that the component chosen for elimination can be selected arbitrarily; It can be a constituent that cannot be determined by X-ray fluorescence.

The link between m_{ijn} and m_{ij} coefficients can be illustrated as follows. Expressing Eqn (4.58) in the form

$$C_i = R_i[1 + C_j m_{ij} + C_k m_{ik} + \cdots + C_n m_{in}] \qquad (4.65)$$

and substituting $(1 - C_i - C_j - C_k - \cdots)$ for C_n in the above leads to

$$C_i = R_i[1 + m_{in} + C_i(-m_{in}) + C_j(m_{ij} - m_{in}) + \cdots] \qquad (4.66)$$

Taking $(1 + m_{in})$ in factor results in

$$C_i = R_i(1 + m_{in})[1 + C_i m_{iin} + C_j m_{ijn} + C_k m_{ikn} + \cdots] \qquad (4.67)$$

where

$$m_{iin} = \frac{-m_{in}}{1 + m_{in}} \qquad (4.68)$$

$$m_{ijn} = \frac{m_{ij} - m_{in}}{1 + m_{in}} \qquad (4.69)$$

Expressed in de Jongh formalism

$$C_i = \left[D_i + \left(\frac{1 + m_{in}}{I_{(i)}} \right) I_i \right] \left[1 + \sum_{\substack{j=i,j,k,\dots \\ j \neq n}} m_{ijn} C_j \right] \tag{4.70}$$

Defining

$$K_i = \frac{1 + m_{in}}{I_{(i)}} \tag{4.71}$$

and substituting in the above gives

$$C_i = (D_i + K_i I_i) \left[1 + \sum_{\substack{j=i,j,k,\dots \\ j \neq n}} m_{ijn} C_j \right] \tag{4.72}$$

4.7 NUMERICAL EXAMPLES

The choice by analysts of different algebraic processes to transform the fundamental intensity–concentration expression of Criss and Birks (1968) to influence coefficient formalism has resulted in a number of models being proposed. In all cases the influence coefficient are defined explicitly in terms of fundamental and instrumental parameters, and hence are interrelated. This is illustrated numerically in the following examples relating to the system Y–Mo–Ag. Table 4.1 lists the analytical contexts and fundamental parameters used to generate the data in the other tables in this chapter.

Tables 4.2–4.5 provide a detailed illustration of the numerical data that enter into the computation of a fundamental absorption influence coefficient for polychromatic incident sources. Using Table 4.2 as reference, the effects

Table 4.1 Instrumental and fundamental parameters for the system Y–Mo–Ag used for computations in subsequent tables in this chapter

Table	Incident source	Analytical contexts ψ'	ψ''	C_{Ag}	C_{Mo}
4.2	W 45 kV (CP)	$90°$	$30°$	0.30	0.70
4.3	Cr 45 kV (CP)	$90°$	$30°$	0.30	0.70
4.4	W 45 kV (CP)	$90°$	$40°$	0.30	0.70
4.5	W 45 kV (CP)	$90°$	$30°$	0.10	0.90

	λ_i (Å)[a]	C_K	n_K	λ_K (Å)	C_{KL}	n_{KL}	p_K	f	ω_K
Y	0.831	273.2	2.85	0.728	38.04	2.73	0.854	1.0	0.715
Mo	0.710	330.1	2.85	0.620	47.30	2.73	0.856	1.0	0.765
Ag	0.561	440.0	2.85	0.486	65.63	2.71	0.848	1.0	0.827

[a] $\lambda_i = \lambda_{K-L_{2,3}}$

Table 4.2   Tabulation of I_λ, $P_{(i)\lambda}$, $P_{i\lambda}$, $a_{ij\lambda}$ and $A_{ij\lambda}$ values computed for 0.02 Å elements of the effective wavelength excitation range for analyte Ag (30%) in the presence of Mo

Analytical context
Excitation source: W target tube; 45 kV (CP)
Instrument geometry: $\psi' = 90°$, $\psi'' = 30°$
System: Mo–Ag; i = Ag, j = Mo; $C_{Ag} = 0.30$, $C_{Mo} = 0.70$

λ (Å)	I_λ	$P_{(Ag)}$	P_{Ag}	$a_{AgMo\lambda}$	A_{AgMo}
0.29	15.5	4.97	0.56	2.40	1.34
0.31	36.6	13.31	1.56	2.23	3.48
0.33	56.8	23.05	2.83	2.07	5.84
0.35	76.6	34.32	4.40	1.91	8.39
0.37	96.2	46.77	6.30	1.75	11.05
0.39	111.1	58.18	8.21	1.61	13.20
0.41	116.4	65.07	9.62	1.47	14.14
0.43	114.6	67.87	10.50	1.34	14.08
0.45	109.9	68.48	11.08	1.22	13.52
0.47	104.5	68.10	11.51	1.11	12.75
$\sum$		450.03	66.56		97.78

a_{AgMo} (polychromatic source) = 97.78/66.56 = 1.4691

of a different excitation source, different instrumental geometry and different specimen composition on the value of a_{ij} are shown. Columns in Table 4.2 list in turn:

- the ten effective incident wavelength elements, i.e. $\lambda < 0.486$ Å (the *K* absorption edge of Ag);
- the incident intensities for each wavelength element; see Table II.1.7;
- the intensities for the pure analyte Ag;
- the intensities for specimen $C_{Ag} = 0.3$, $C_{Mo} = 0.7$;
- the values of the fundamental influence coefficient for the absorption of Ag by Mo;
- the values of A_{AgMo}, i.e. the product $P_{Ag}a_{AgMo\lambda}$.

The relative intensity is given by ratioing the two sums

$$R_{Ag} = P_{Ag}/P_{(Ag)} = 66.56/450.03 = 0.1479$$

and the fundamental influence coefficient a_{AgMo} for the polychromatic source is given by

$$a_{AgMo} = A_{AgMo}/P_{Ag} = 97.78/66.56 = 1.4691$$

The intensity–concentration relationship expressed in terms of absorption by the matrix element is given by

$$P_{Ag} = P_{(Ag)}C_{Ag} - A_{AgMo}C_{Mo}$$
$$= (450.03 \times 0.3) - (97.78 \times 0.7)$$
$$- 66.56$$

and in influence coefficient formalism

$$C_{Ag} = R_{Ag}[1 + a_{AgMo}C_{Mo}]$$
$$= 0.1479[1 + (1.4691 \times 0.7)]$$
$$= 0.3000$$

4.7.1 Effect of target material

Conditions in Table 4.3 are identical to those in Table 4.2 except that the incident source is a tube with a chromium target (see Table II.1.8) instead of tungsten. It is noted that the ten $a_{AgMo\lambda}$ values are identical in the two tables, as expected. Conversely, the values in all the other columns are quite different, but the resulting relative intensity 0.1473 and the influence coefficient 1.4812 are similar to their W target counterparts. This is also predictable because in

Table 4.3 Tabulation of I_λ, $P_{(i)\lambda}$, $P_{i\lambda}$, $a_{ij\lambda}$ and $A_{ij\lambda}$ values as in Table 4.2 except that the excitation source is a Cr target tube

Analytical context
Excitation source: Cr target tube; 45 kV (CP)
Instrument geometry: $\psi' = 90°$, $\psi'' = 30°$
System: Mo–Ag; i = Ag, j = Mo; $C_{Ag} = 0.30$, $C_{Mo} = 0.70$

λ (Å)	I_λ	$P_{(Ag)}$	P_{Ag}	$a_{AgMo\lambda}$	A_{AgMo}
0.29	3.00	0.963	0.11	2.40	0.26
0.31	6.60	2.400	0.28	2.23	0.63
0.33	8.80	3.571	0.44	2.07	0.91
0.35	9.94	4.441	0.57	1.91	1.09
0.37	11.0	5.348	0.72	1.75	1.26
0.39	12.0	6.284	0.89	1.61	1.43
0.41	12.9	7.212	1.07	1.47	1.57
0.43	13.5	7.995	1.24	1.34	1.66
0.45	13.9	8.661	1.40	1.22	1.71
0.47	13.8	8.994	1.52	1.11	1.68
$\sum$		55.87	8.23		12.19

a_{AgMo} (polychromatic source) $= 12.19/8.23 = 1.4812$

Table 4.4   Tabulation of values as in Table 4.2 except that
the emergence angle is $40°$ instead of $30°$

Analytical context
Excitation source: W target tube; 45 kV (CP)
Instrument geometry: $\psi' = 90°$, $\psi'' = 40°$
System: Mo–Ag; i = Ag, j = Mo; $C_i = 0.30$, $C_j = 0.70$

λ (Å)	I_λ	$P_{(Ag)}$	P_{Ag}	$a_{AgMo\lambda}$	A_{AgMo}
0.29	15.5	5.86	0.70	2.18	1.52
0.31	36.6	15.50	1.94	2.00	3.87
0.33	56.8	26.56	3.50	1.83	6.39
0.35	76.6	39.02	5.41	1.66	9.00
0.37	96.2	52.80	7.70	1.51	11.63
0.39	111.1	65.07	9.98	1.37	13.63
0.41	116.4	72.14	11.62	1.23	14.32
0.43	114.6	74.63	12.60	1.11	13.99
0.45	109.9	74.73	13.20	1.00	13.17
0.47	104.5	73.81	13.61	0.90	12.19
	$\sum$	500.11	80.25		99.69

a_{AgMo} (polychromatic source) = 99.69/80.25 = 1.2422

Table 4.5   Tabulation of values as in Table 4.2 except that
the concentration of Ag is 10% instead of 30%

Analytical context
Excitation source: W target tube; 45 kV (CP)
Instrument geometry: $\psi' = 90°$, $\psi'' = 30°$
System: Mo–Ag; i = Ag, j = Mo; $C_i = 0.10$, $C_j = 0.90$

λ (Å)	I_λ	$P_{(Ag)}$	P_{Ag}	$a_{AgMo\lambda}$	A_{AgMo}
0.29	15.5	4.97	0.16	2.40	0.38
0.31	36.6	13.31	0.44	2.23	0.99
0.33	56.8	23.05	0.81	2.07	1.67
0.35	76.6	34.22	1.26	1.91	2.40
0.37	96.2	46.77	1.81	1.75	3.18
0.39	111.1	58.18	2.38	1.61	3.82
0.41	116.4	65.07	2.80	1.47	4.12
0.43	114.6	67.87	3.08	1.34	4.12
0.45	109.9	68.48	3.26	1.22	3.98
0.47	104.5	68.10	3.41	1.11	3.78
	$\sum$	450.03	19.41		28.44

a_{AgMo} (polychromatic source) = 28.44/19.41 = 1.4652

neither case are characteristic target lines present in the effective wavelength range.

4.7.2 Effect of instrumental geometry

The analytical context for Table 4.4 is identical to that of Table 4.2 except that the angle of emergence is $40°$ instead of $30°$. The relative intensity in this case is 0.1605 and the absorption influence coefficient equals 1.2422. The latter is lower than in Table 4.2 owing to the shorter path length of the emerging radiation; hence there is less absorption of the characteristic Ag radiations.

Table 4.6 Tabulation of $P_{(i)\lambda}$, $P_{i\lambda}$, $a_{ij\lambda}$ and $A_{ij\lambda}$ values in a context wherein the characteristic lines emitted by the target element are shorter than the absorption edge of the analyte

Analytical context
Excitation source: Rh target tube; 45 kV (CP); see Table II.1.9
Instrument geometry: $\psi' = 90°$, $\psi'' = 30°$
System: Y–Mo; i = Mo, j = Y; $C_{Mo} = 0.20$, $C_Y = 0.80$

λ (Å)	$P_{(Mo)}$	P_{Mo}	a_{MoY}	A_{MoY}	R_{Mo}
0.29	1.66	0.086	3.57	0.31	0.052
0.31	6.00	0.322	3.41	1.10	0.054
0.33	9.20	0.511	3.25	1.66	0.056
0.35	11.78	0.679	3.09	2.10	0.058
0.37	14.14	0.847	2.93	2.48	0.060
0.39	16.40	1.022	2.76	2.82	0.062
0.41	18.29	1.187	2.60	3.09	0.065
0.43	19.94	1.349	2.44	3.30	0.068
0.45	21.57	1.522	2.29	3.49	0.071
0.47	22.77	1.676	2.15	3.60	0.074
0.49	23.88	1.833	2.01	3.68	0.077
0.51	24.68	1.975	1.87	3.70	0.080
0.53	27.27	2.274	1.75	3.97	0.083
0.55	97.14	8.435	1.63	13.74	0.087
0.57	33.37	3.014	1.52	4.57	0.090
0.59	34.39	3.230	1.41	4.56	0.094
0.61	340.08	33.16	1.31	43.57	0.098
Σ	722.56	63.12		101.74	

R_{Mo} (polychromatic source) $= 63.12/722.56 = 0.0874$
a_{MoY} (polychromatic source) $= 101.74/63.12 = 1.6118$

4.7.3 Effect of specimen composition

Compared with Table 4.2, only the specimen composition has changed in Table 4.5; hence the columns headed I_λ and $a_{AgMo\lambda}$ remain identical. It is noted that the change in specimen composition has led to a slightly lower value for a_{AgMo}, namely 1.4652.

4.7.4 Effect of characteristic tube target lines

Two cases are considered, analyte Mo in the presence of Y, Table 4.6, in which case absorption only is involved, and analyte Mo in the presence of Ag, Table 4.7, in which case absorption and enhancement are involved. The target tube material is Rh (see Table II.1.9); the two K lines of rhodium, 0.615 Å and 0.546 Å, are shorter than the K absorption edge of Mo, 0.620 Å, and therefore both contribute strongly to the emission of Mo $K-L_3$ radiations. The K line intensities of rhodium have been added to the continuum; hence the values 97.14 and 340.08 at $\lambda = 0.55$ Å and 0.61 Å, respectively. Table 4.6 lists data for each of the 17 effective wavelength elements that can excite molybdenum atoms. From the summation values, one obtains the value 0.0874 for the relative intensity and 1.6118 for a_{MoY}. It is noted that the two characteristic lines of rhodium account for more than 50% of the characteristic fluorescence emission of molybdenum. Intensity data are plotted in Figure 4.1 to illustrate the relative importance of incident wavelength on emitted intensities.

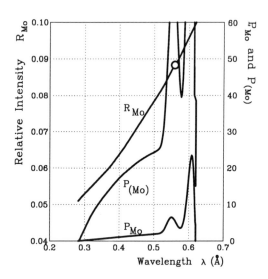

Figure 4.1 Plot of P_{Mo}, $P_{(Mo)}$ and R_{Mo} for each effective $\Delta\lambda$ interval. Context: Table 4.6. The circle denotes the relative intensity R_{Mo} generated by the summation

The analytical context for Table 4.7 is identical to that of Table 4.6, except that Ag is now the binary component with Mo. Although both K lines of silver can excite Mo atoms, the computations were carried out as though both effects occur at the Ag $K-L_{2,3}$ wavelength, as indicated by equating $f_{K-L_{2,3}}$ to unity in Table 4.1. The term A_{MoAg} is the product of $P_{\text{Mo}}a_{\text{MoAg}}$, and the term E_{MoAg} is the product of $P_{\text{Mo}}e_{\text{MoAg}}$. A glance at Table 4.7 and Figure 4.2 clearly shows the abrupt change that takes place at the K absorption edge of silver. The monochromatic absorption coefficient A_{MoAg} and influence coefficient a_{MoAg} change algebraic sign and, since all incident wavelength elements above 0.486 Å cannot excite silver atoms, enhancement takes place over the range 0.27 to 0.486 Å only. The relative intensity R_{Mo} is given by $214.78/722.56 = 0.2972$, in sharp contrast to the value of 0.0874 for the analytical context of Table 4.6.

Table 4.7 Tabulation of $(P_i + S_i)_\lambda$, $P_{i\lambda}$, $a_{ij\lambda}$, $A_{ij\lambda}$, $e_{ij\lambda}$ and $E_{ij\lambda}$ values in a context wherein the characteristic lines are shorter than the absorption edge of the analyte

Analytical context
Identical to Table 4.6 except that matrix element is Ag

λ (Å)	$P_{\text{Mo}} + S_{\text{Mo}}$	P_{Mo}	a_{MoAg}	A_{MoAg}	e_{MoAg}	E_{MoAg}	m_{MoAg}
0.29	0.60	0.25	0.382	0.10	1.693	0.43	-1.311
0.31	2.14	0.92	0.380	0.35	1.656	1.52	-1.276
0.33	3.24	1.41	0.378	0.53	1.618	2.29	-1.240
0.35	4.10	1.81	0.376	0.68	1.580	2.86	-1.204
0.37	4.86	2.18	0.374	0.81	1.542	3.36	-1.169
0.39	5.57	2.53	0.372	0.94	1.505	3.80	-1.133
0.41	6.14	2.82	0.370	1.04	1.468	4.14	-1.099
0.43	6.61	3.08	0.367	1.13	1.432	4.41	-1.065
0.45	7.07	3.34	0.365	1.22	1.397	4.67	-1.032
0.47	7.38	3.53	0.364	1.28	1.364	4.81	-1.000
0.49	5.90	5.90	-0.237	-1.40	0.0	0.0	-0.237
0.51	6.30	6.30	-0.271	-1.71	0.0	0.0	-0.271
0.53	7.20	7.20	-0.303	-2.18	0.0	0.0	-0.303
0.55	26.51	26.51	-0.334	-8.85	0.0	0.0	-0.334
0.57	9.40	9.40	-0.362	-3.41	0.0	0.0	-0.362
0.59	9.99	9.99	-0.389	-3.89	0.0	0.0	-0.389
0.61	101.78	101.78	-0.415	-42.20	0.0	0.0	-0.415
Σ	214.78	188.95		-55.54		32.30	

$P_{(\text{Mo})} = 722.56$ (Table 4.6)
R_{Mo} (polychromatic source) $= 214.78/722.56 = 0.2972$
m_{MoAg} (polychromatic source) $= (-55.54 - 32.30)/214.78 = -0.4090$

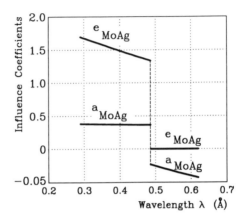

Figure 4.2 Plot of a_{MoAg} and e_{MoAg} for each effective $\Delta\lambda$ wavelength element. Context: Table 4.7

Summation data in Table 4.7 may be used to generate the fundamental absorption and enhancement influence coefficients in the Broll–Tertian, Rousseau and Lachance–Traill models, namely

$$a_{MoAgp} = A_{MoAg}/P_{Mo} = -55.54/188.95 = -0.2939$$

$$e_{MoAgp} = E_{MoAg}/P_{Mo} = 32.30/188.95 = 0.1709$$

$$m_{MoAg} = (A_{MoAg} - E_{MoAg})/(P_{Mo} + S_{Mo})$$
$$= (-55.54 - 32.30)/214.78$$
$$= -0.4090$$

Thus, expressed in Broll–Tertian formalism Eqn (4.37)

$$C_{Mo} = R_{Mo}\left[1 + \left\{a_{MoAgp} - \left(e_{MoAgp}\frac{C_{Mo}}{R_{Mo}}\right)\right\}C_{Ag}\right]$$

$$= 0.2972\left\{1 + \left[-0.2939 - \left(0.1709 \times \frac{0.2}{0.2972}\right)\right] \times 0.80\right\}$$

$$= 0.2000$$

Expressed in Rousseau formalism Eqn (4.40)

$$C_{Mo} = R_{Mo}\left[\frac{1 + a_{MoAgp}C_{Ag}}{1 + e_{MoAgp}C_{Ag}}\right]$$

$$= 0.2972\left[\frac{1 + (-0.2939 \times 0.8)}{1 + 0.1709 \times 0.8}\right]$$

$$= 0.2000$$

and in Lachance–Traill formalism Eqn (4.58)

$$C_{Mo} = R_{Mo}[1 + m_{MoAg}C_{Ag}]$$
$$= 0.2972[1 + (-0.4090 \times 0.8)]$$
$$= 0.2000$$

4.7.5 Binary m_{ij} arrays

A more comprehensive illustration of the variation of binary fundamental influence coefficients as a function of composition is presented in Table 4.8. The analytical contexts are identical to those in Tables 4.6 and 4.7, except that data are generated for 13 Mo–Y and 13 Mo–Ag binaries, the concentration of analyte C_{Mo} varying from 0.999 to 0.001. The data for the binaries $C_{Mo} = 0.2$ presented in greater detail in Tables 4.6 and 4.7 are labelled with superscripts in Table 4.8.

It is observed that m_{MoY} and m_{MoAg} vary smoothly between two limiting values, a property that will be made use of in the next chapter. This variation is representative of binary m_{ij} arrays, but in many cases the range of values is somewhat larger.

Table 4.8 Arrays of fundamental binary influence coefficients

Analytical context
Excitation source: Rh target tube; 45 kV (CP); see Table II.1.9
Instrument geometry: $\psi' = 90°$, $\psi'' = 30°$
Systems: Y–Mo; analyte = Mo; j = Y
         Mo–Ag; analyte = Mo; j = Ag

C_{Mo}	C_j	j = Y		j = Ag	
		R_{Mo}	m_{MoY}	R_{Mo}	m_{MoAg}
0.999	0.001	0.9973	1.688	0.9994	−0.381
0.99	0.01	0.9736	1.686	0.9938	−0.381
0.90	0.10	0.7715	1.666	0.9359	−0.384
0.80	0.20	0.6014	1.651	0.8671	−0.387
0.70	0.30	0.4692	1.640	0.7927	−0.390
0.60	0.40	0.3631	1.632	0.7119	−0.393
0.50	0.50	0.2759	1.625	0.6236	−0.397
0.40	0.60	0.2029	1.620	0.5264	−0.400
0.30	0.70	0.1408	1.615	0.4184	−0.404
0.20	0.80	0.0874[a]	1.612[a]	0.2972[b]	−0.409[b]
0.10	0.90	0.0409	1.609	0.1595	−0.415
0.01	0.99	0.0039	1.606	0.0172	−0.421
0.001	0.999	0.0004	1.606	0.0017	−0.422

[a] See Table 4.6 and text for detailed calculation.
[b] See Table 4.7 and text for detailed calculation.

4.7.6 Fundamental influence coefficients, multielement specimens

For space considerations, the computation of fundamental influence coefficients in multielement specimens will be limited to the same three elements as in previous tables in this chapter. The analytical context of Tables 4.6, 4.7 and 4.8 is retained, with Mo as the analyte, $C_{Mo} = 0.2$, with $C_Y = 0.3$ and $C_{Ag} = 0.5$, a situation involving strong absorption and strong enhancement effects. As in Tables 4.6 and 4.7, data are listed in Table 4.9 for all 17 effective incident wavelength increments.

The following points may be noted:

- S_{Mo} data (not tabulated) are given by $S_{Mo} = P_{Mo} e_{MoAg} C_{Ag}$.
- a_{MoY} and a_{MoAg} data are not tabulated; they are identical to those listed in Tables 4.6 and 4.7. Thus, data in the column headed A_{MoY} are the product of P_{Mo} and the value of a_{MoY} in Table 4.6, and the data in the column headed A_{MoAg} are the product of P_{Mo} and the value of a_{MoAg} in Table 4.7.

Thus, in Tables 4.6 and 4.9, the following fundamental entities are generated from theoretical principles and will be used to generate fundamental influence coefficients (applicable to the three element specimen under consideration) in the Broll–Tertian, Rousseau, Lachance–Traill and de Jongh formalisms.

$P_{(Mo)} = 722.56$ is the calculated emitted intensity for a specimen of pure molybdenum. It is computed following Criss and Birks (1968).

$P_{Mo} + S_{Mo} = 115.24$ is the calculated emitted intensity for specimen $C_{Mo} = 0.2$, $C_Y = 0.3$ and $C_{Ag} = 0.5$, calculated following Criss and Birks, Eqn (3.10).

$P_{Mo} = 107.89$ is a hypothetical intensity, i.e. the calculated intensity emitted if enhancement is not taken into consideration, Eqn (3.9). The entity P_{Mo} cannot be measured experimentally if enhancement is present. Its value is necessary to compute the fundamental influence coefficients in the Broll–Tertian and Rousseau formalisms.

$A_{MoY} = 172.64$ is the pro-rated sum of the net matrix effect of yttrium for each effective incident wavelength element.

$M_{MoY} = 172.64$, i.e. there being no enhancement, the matrix effect is equal to the absorption effect A_{MoY}.

$A_{MoAg} = -30.33$ is the pro-rated sum of the absorption effect of silver for each effective incident wavelength element.

$E_{MoAg} = 14.72$ is the pro-rated sum of the enhancement effect of silver for each effective incident wavelength element.

$M_{MoAg} = A_{MoAg} - E_{MoAg} = -45.04$ is the pro-rated sum of the total matrix effect of silver for each incident wavelength element.

The above entities are used to compute the fundamental influence coefficients in the following algorithms.

Table 4.9 Tabulation of $(P_i + S_i)_\lambda$, $P_{i\lambda}$, $e_{ik\lambda}$, $A_{ij\lambda}$, $A_{ik\lambda}$, $E_{ik\lambda}$ and $M_{ik\lambda}$ in the context of strong absorption and enhancement

Analytical context
Identical to Tables 4.6–4.8 except that:
System: Y–Mo–Ag: $i = Mo$ $j = Y$ $k = Ag$
$C_i = 0.20$ $C_j = 0.30$ $C_k = 0.50$

λ (A)	$P_{Mo} + S_{Mo}$	P_{Mo}	e_{MoAg}	A_{MoY}	A_{MoAg}	E_{MoAg}	M_{MoAg}
0.29	0.23	0.15	1.142	0.52	0.06	0.17	−0.11
0.31	0.85	0.54	1.124	1.85	0.21	0.61	−0.40
0.33	1.32	0.85	1.104	2.76	0.32	0.94	−0.62
0.35	1.72	1.11	1.085	3.44	0.42	1.21	−0.79
0.37	2.10	1.37	1.064	4.01	0.51	1.46	−0.95
0.39	2.48	1.63	1.043	4.50	0.61	1.70	−1.09
0.41	2.81	1.86	1.022	4.84	0.69	1.90	−1.21
0.43	3.12	2.08	1.001	5.09	0.76	2.08	−1.32
0.45	3.44	2.31	0.980	5.29	0.84	2.26	−1.42
0.47	3.69	2.49	0.959	5.35	0.91	2.39	−1.48
0.49	3.22	3.22	0.0	6.46	−0.76	0.0	−0.76
0.51	3.46	3.46	0.0	6.48	−0.94	0.0	−0.94
0.53	3.97	3.97	0.0	6.94	−1.21	0.0	−1.21
0.55	14.70	14.70	0.0	23.95	−4.90	0.0	−4.90
0.57	5.24	5.24	0.0	7.95	−1.90	0.0	−1.90
0.59	5.60	5.60	0.0	7.90	−2.18	0.0	−2.18
0.61	57.31	57.31	0.0	75.30	−23.76	0.0	−23.76
$\sum$	115.24	107.89		172.64	−30.33	14.72	−45.04

$P_{(Mo)} = 722.56$ (Table 4.6)
R_{Mo} (polychromatic source) $= 115.24/722.56 = 0.1595$
m_{MoY} (polychromatic source) $= 172.64/115.24 = 1.4981$
m_{MoAg} (polychromatic source) $= -45.04/115.24 = -0.3908$

For the Broll–Tertian and Rousseau formalisms:

$a_{ijp} = 172.64/107.89 = 1.6001$

$e_{ijp} = 0.0$

$m_{ijp} = 1.6001 - 0.0 = 1.6001$

$a_{ikp} = -30.33/107.89 = -0.2811$

$e_{ikp} = 14.72/107.89 = 0.1364$

$m_{ikp} = -0.2811 - \left(0.1364 \times \dfrac{0.2}{0.1595}\right) = -0.4521$

Thus the correction for matrix effect in the Broll–Tertian algorithm is expressed as

$$C_{Mo} = 0.1595\,[1 + (1.6001 \times 0.3) + (-0.4521 \times 0.5)]$$

$$= 0.1595\,[1.2540]$$

$$= 0.2000$$

The correction for matrix effects in the Rousseau algorithm takes the form

$$C_{Mo} = 0.1595 \left[\frac{1 + (1.6001 \times 0.3) + (-0.2811 \times 0.5)}{1 + \qquad\qquad (0.1364 \times 0.5)} \right]$$

$$= 0.1595\,[1.2540]$$

$$= 0.2000$$

The influence coefficients m_{ij} in the Lachance–Traill formalism are

$$m_{MoY} = 172.64/115.24 = 1.4981$$

$$m_{MoAg} = (-30.33 - 14.72)/115.24 = -0.3909$$

and the correction for matrix effects takes the form

$$C_{Mo} = 0.1595\,[1 + (1.4981 \times 0.3) + (-0.3909 \times 0.5)]$$

$$= 0.1595\,[1.2540]$$

$$= 0.2000$$

The influence coefficients Eqns (4.68) and (4.69) for the de Jongh formalism are:
(a) if yttrium is chosen as the eliminated element

$$m_{MoMoY} = \frac{-1.4981}{1 + 1.4981} = -0.5997$$

$$m_{MoAgY} = \frac{-0.3909 - 1.4981}{1 + 1.4981} = -0.7562$$

and the correction for matrix effects takes the form, Eqn (4.70)

$$C_{Mo} = 0.1595(1 + 1.4981)\,[1 + (-0.5997 \times 0.2) + (-0.7562 \times 0.5)]$$

$$= 0.3984\,[0.5020]$$

$$= 0.2000$$

(b) if silver is chosen as the eliminated element

$$m_{MoMoAg} = \frac{-(-0.3909)}{1 + (-0.3909)} = 0.6418$$

$$m_{MoYAg} = \frac{1.4981 - (-0.3909)}{1 + (-0.3909)} - 3.1013$$

Table 4.10 Comparison of fundamental influence coefficients defined as a function of the total emitted intensity and of the primary fluoresence intensity only

Analytical context
Excitation source: Rh target tube; 45 kV (CP)
Instrument geometry: $\psi' = 90°$, $\psi'' = 30°$
System: Mo–Ag; i = Mo, j = Ag

C_{Mo}	R_{Mo}	a_{MoAg}	e_{MoAg}	m_{MoAg}	a_{MoAgp}	e_{MoAgp}
0.999	0.9994	−0.2313	0.1492	−0.3805	−0.2314	0.1492
0.99	0.9938	−0.2318	0.1489	−0.3808	−0.2322	0.1492
0.90	0.9359	−0.2366	0.1470	−0.3836	−0.2401	0.1491
0.80	0.8671	−0.2414	0.1452	−0.3867	−0.2486	0.1496
0.70	0.7927	−0.2458	0.1441	−0.3898	−0.2569	0.1506
0.60	0.7119	−0.2496	0.1435	−0.3931	−0.2648	0.1523
0.50	0.6236	−0.2528	0.1437	−0.3965	−0.2724	0.1548
0.40	0.5264	−0.2555	0.1447	−0.4002	−0.2798	0.1584
0.30	0.4184	−0.2575	0.1468	−0.4043	−0.2870	0.1636
0.20	0.2972[a]	−0.2586	0.1504	−0.4090[a]	−0.2940	0.1709
0.10	0.1595	−0.2585	0.1562	−0.4147	−0.3008	0.1818
0.01	0.0172	−0.2568	0.1645	−0.4213	−0.3067	0.1965
0.001	0.0017	−0.2565	0.1655	−0.4220	−0.3073	0.1983

[a] See Table 4.7 for detailed calculation.

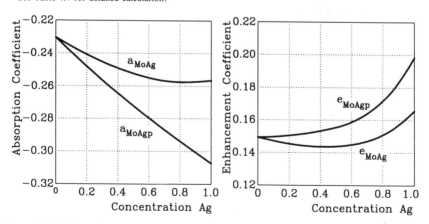

Figure 4.3 Plot of a_{MoAg}, e_{MoAg}, m_{MoAg}, a_{MoAgp} and e_{MoAgp} as a function of composition. Context: Table 4.10

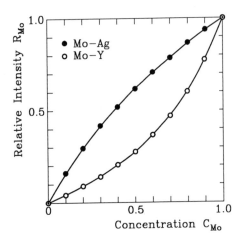

Figure 4.4 R_{Mo} plotted versus C_{Mo} for the systems Mo–Ag and Mo–Y

and the correction for matrix effects then takes the form

$$C_{Mo} = 0.1595(1 - 0.3909)\,[1 + (0.6418 \times 0.2) + (3.1013 \times 0.3)]$$
$$= 0.09715\,[2.05875]$$
$$= 0.2000$$

A comparison between a_{ij}, e_{ij}, m_{ij}, a_{ijp} and e_{ijp} as a function of composition in the system Mo–Ag is made in Table 4.10 and illustrated in Figure 4.3; plots of R_{Mo} versus C_{Mo} for matrix elements Ag and Y are shown in Figure 4.4.

4.8 DISCUSSION

According to Broll (1986, 1990), the reason why the concept of corrections based on influence coefficient models retains the attention of analysts is one of practicality. The attraction of the fundamental influence coefficient method is that it combines the theoretical exactness of the fundamental parameters approach while retaining the flexibility of the usual Lachance–Traill formulation. Thus, whether complex or comparatively simple, any given analytical context can be studied and the problem of converting intensities to concentrations solved in agreement with the accuracy required. It should be emphasized that theoretical exactness is retained only if there are no serious shortcomings in the quantification of the physical processes involved in defining the influence coefficients. In other words, the difference between the classical fundamental parameters algorithm and the fundamental influence coefficient algorithms is one of formalism only. The latter algorithms are derived explicitly from fundamental expressions and are therefore equivalent. For example, using a reference specimen instead of a pure element, the relative intensity R_i can be expressed as $R_{iu} = K_i I_{iu}$, where K is a constant determined from the reference specimen. In the classical fundamental parameters formalism

$$R_{iu} = \frac{I_{iu}}{I_{(i)}} = \left(\frac{P_i + S_i}{I_i P_{(i)}}\right)_r \cdot I_{iu} \tag{4.73}$$

whereas in the fundamental influence formalisms

$$R_{iu} = \frac{I_{iu}}{I_{(i)}} = \left(\frac{C_i}{I_i[\ldots]}\right)_r \cdot I_{iu} \tag{4.74}$$

leading to relationships which are most valuable for calibration

$$\frac{1}{I_{(i)}} = K_i = \left(\frac{P_i + S_i}{I_i P_{(i)}}\right)_r = \left(\frac{C_i}{I_i[\ldots]}\right)_r \tag{4.75}$$

where $[\ldots]$ represents the term in brackets in any one of Eqns (4.37), (4.40) or (4.58).

There are a number of options as to how fundamental influence coefficients are used in practice. The contention that the coefficients are 'exact' can be made only in the context of specimens of known composition; i.e. fundamental influence coefficients are a function of specimen composition, hence 'unique' to each specimen. One option consists in calculating the influence coefficients using the concentrations of a typical reference material on hand or an average composition. These are then treated as constants in the correction for matrix effects in unknowns, provided that the unknowns match the composition used in the original calculations fairly closely. A second option consists in proceeding as above, but in order to do so the departure of each unknown from the 'typical' reference material or average composition must be monitored. If it is found that a given unknown lies outside previously established concentration ranges, its concentrations are used to calculate a new set of fundamental influence coefficients to be used in the correction process in that particular case. The third option consists in calculating individual fundamental influence coefficients for each specimen, a method described in Chapter 9.

4.9 SUMMARY

The following equation quantifying the absorption effect was derived in Chapter 3, Eqn (3.29)

$$P_i = P_{(i)} C_i - \sum_j A_{ij} C_j$$

along with Eqn (3.41), quantifying enhancement due to secondary emission

$$P_i + S_i = P_i + \sum_j E_{ij} C_j$$

It was shown in this chapter that the two above expressions can be transformed to

$$C_i = \frac{P_i}{P_{(i)}} \left[1 + \sum_j \frac{A_{ij}}{P_i} C_j \right]$$

and

$$R_i = \frac{P_i}{P_{(i)}} \left[1 + \sum_j \frac{E_{ij}}{P_i} C_j \right]$$

respectively. Ratioing the above yielded

$$\frac{C_i}{R_i} = \frac{1 + \sum_j a_{ijp} C_j}{1 + \sum_j e_{ijp} C_j}$$

Solving for C_i led to the Broll–Tertian algorithm

$$C_i = R_i \left[1 + \sum_j \left\{ a_{ijp} - e_{ijp} \frac{C_i}{R_i} \right\} C_j \right]$$

or to the Rousseau algorithm

$$C_i = R_i \left[\frac{1 + \sum a_{ijp} C_j}{1 + \sum e_{ijp} C_j} \right]$$

Conversely, the absorption and enhancement effect can be quantified by the combined expressions, i.e.

$$P_{(i)} C_i = P_i + S_i + \sum A_{ij} C_j - \sum E_{ij} C_j$$

Regrouping and solving for C_i led to the Lachance–Traill algorithm

$$C_i = \frac{P_i + S_i}{P_{(i)}} \left[1 + \sum \left\{ \frac{A_{ij} - E_{ij}}{P_i + S_i} \right\} C_j \right]$$

$$= R_i \left[1 + \sum m_{ij} C_j \right]$$

with

$$m_{ij} = \frac{M_{ij}}{P_i + S_i} = \frac{A_{ij} - E_{ij}}{P_i + S_i}$$

It was also shown that the de Jongh algorithm could be derived from the same entities P_i, S_i, A_{ij} and E_{ij}, which are fundamental parameters functions, in which case

$$C_i = (D_i + K_i I_i)\left[+ \sum_{\substack{j=i,j,k\ldots \\ J \neq n}} m_{ijn} C_j \right]$$

This model allows any one element to be selected for elimination, usually the major constituent. Provided that the computation of the fundamental influence coefficients in all the above algorithms matches the appropriate physical model for the particular specimen composition, i.e. provided unnecessary approximations are avoided, then all are equivalent to the approach proposed by Criss and Birks (1968) from which the algorithms are derived.

5 ALPHA COEFFICIENT CONCEPT

It often happens when seeking the solution to a problem that developments proceed simultaneously along parallel lines, each in turn influencing the course of the other. The quest for a mathematical solution that compensates for the effect of absorption and enhancement is such a case. The concept of quantifying the total matrix effect as a sum of the products of the influence coefficient and concentration of each individual element pre-dates the fundamental parameters approach proposed by Criss and Birks (1968). In a survey of mathematical matrix correction procedures, Tertian (1986) states: 'From 1968 on, the emergence of the fundamental parameter method, a matrix correction procedure on its own, has revived interest in the theoretical equations of X-ray fluorescence emission, and prompted some workers to re-evaluate the older influence coefficient method, by progressively giving these coefficients a sound theoretical status.' This aspect was addressed in the previous chapter, in which it was shown that a_{ij}, e_{ij}, m_{ij}, m_{ijn}, a_{ijp}, e_{ijp} and m_{ijp} influence coefficients can be defined explicitly in terms of fundamental parameters. In a subsequent paper, Tertian (1988) referred to the alpha model as the simplest and most convenient one in practice, 'provided that the significance and properties of the alphas or "influence coefficients", are fully understood.' This chapter examines the concepts that underlie alpha coefficients, which may be considered as a subset of influence coefficients.

5.1 GENERAL CONSIDERATIONS

Initially, it must be pointed out that there is no inherent significance attached to the term 'alpha coefficient'. The symbol α_{ij} is often associated with the Lachance–Traill (1966) model, namely

$$C_i = R_i \left[1 + \sum_j \alpha_{ij} C_j \right] \tag{5.1}$$

which, when expressed in the $C = KIM$ formalism of Eqn (3.52) becomes

$$C_{iu} = \left(\frac{C_i}{I_i\left[1 + \sum_j \alpha_{ij}C_j\right]} \right)_r I_{iu}\left[1 + \sum_j a_{ij}C_j\right]_u \qquad (5.2)$$

where u stands for specimens of unknown composition and r stands for reference materials of known composition. *The symbol α_{ij} is retained in the present text to designate the subset of influence coefficients quantifying the matrix effect of element j on analyte i that are calculated from theory, although some approximation is involved in the process.* Alpha coefficients are therefore not deemed fundamental. On the other hand, alpha coefficients are not deemed *empirical*, as they are linked to m_{ij} influence coefficients, which are in turn linked explicitly to fundamental and instrumental parameters. This is pointed out in a comprehensive review of matrix correction methods by Tertian (1988). In the section titled Comprehensive study of alpha influence coefficients, Tertian states that the fundamental parameters approach 'emphasized the shortcomings of multiple regression correction coefficients, but gave little consideration to what we may call the "physical" alphas, i.e. those consistent with Beattie and Brissey's or Lachance and Traill's ideas'. The next section therefore outlines what type of influence coefficients the symbol α designates in this and subsequent chapters.

5.2 CONCEPT AND DOMAIN OF ALPHA COEFFICIENTS

At this stage, the question might very well be asked: if it is possible to calculate m_{ij} influence coefficients which are exact in the sense that they are derived explicitly from fundamental parameters expressions, why is there a need for α_{ij} in Eqn (5.1)? The answer lies in the fact that, strictly speaking, m_{ij} coefficients as defined in the previous chapter can only be computed for specimens of known or assumed composition; i.e. m_{ij} coefficients are variables and, being a function of composition, are unique to each specimen. By the same token, influence coefficients obtained from multiple regression of experimental data are also unique in the sense that they relate specifically to the number and nature of the materials used to generate the data. Quite often, these coefficients may be very different from those predicted by theory. For the sake of expediency, analysts may sometimes choose to forsake the theoretical exactness of the fundamental approach for the flexibility of the Lachance–Traill formulation without completely sacrificing the theoretical aspects of X-ray fluorescence emission. Thus, one may enumerate a number of characteristics associated with the concept of alpha coefficients:

• alpha coefficients are computed without reference to experimental data;

- alpha coefficients relate directly to data generated by fundamental parameters computations;
- depending on the model, there may be severe limits on its application imposed by the approximations inherent to the model.

Simply stated, the concept inherent to alpha coefficients is that they are calculated *a priori*, i.e. without prior knowledge of the elemental concentrations in the specimens being analysed. The following numerical examples provide an illustration of how the same analytical context may be envisaged from the fundamental influence coefficient and α coefficient concepts. Consider the numerical data listed for Fe48 and Fe52 in Table II.3.8. The former relate to the analytical context wherein the concentration of the analyte Fe is 10% and those of Cr and Ni are 60% and 30%, respectively, whereas Fe52 relates to Fe also present at 10%, but with the concentrations of Cr and Ni reversed; i.e. Cr and Ni are present at 30% and 60%, respectively. The pertinent data are $P_{(Fe)} = 1742.48$; $P_{Fe} + S_{Fe} = 84.27$, $M_{FeCr} = 160.30$ and $M_{FeNi} = -20.66$ for Fe48; $P_{Fe} + S_{Fe} = 133.47$, $M_{FeCr} = 220.96$ and $M_{FeNi} = -42.52$ for Fe52.

Thus, envisaged from the fundamental influence coefficient concept (m_{ij} formalism, Eqn (4.62)), $m_{FeCr} = (160.30/84.27) = 1.9022$ and $m_{FeNi} = (-20.66/84.27) = -0.2452$ in the case of Fe48, and $m_{FeCr} = (220.96/133.47) = 1.6555$ and $m_{FeNi} = (-42.52/133.47) = -0.3186$ in the case of Fe52. It therefore follows that, from a fundamental influence coefficient point of view, one is dealing with two distinct analytical contexts, and that the m_{ij} coefficients relating to Fe48 would be applicable only to specimens of composition similar to Fe48, and not to specimens of type Fe52. For example, if the fundamental influence coefficients computed for composition Fe48 are used with Fe52 data

$$C_{Fe} = \frac{133.47}{1742.48} [1 + (1.9022 \times 0.3) + (-0.2452 \times 0.6)]$$

$$= 0.0766[1.4236]$$

$$= 0.1090$$

and, conversely, if Fe52 coefficients are used with Fe48 data

$$C_{Fe} = \frac{84.27}{1742.48} [1 + (1.6555 \times 0.6) + (-0.3186 \times 0.3)]$$

$$= 0.0484[1.8977]$$

$$= 0.0918$$

These values are quite different from the known C_{Fe}, which is equal to 0.10 in both cases.

Now let us consider the same fundamental data from a different perspective, namely Eqn (5.1), which in the case of a three element system can be transformed and expressed as

$$\frac{C_i}{R_i} - 1 = C_j\alpha_{ij} + C_k\alpha_{ik} \tag{5.3}$$

Thus, in this context, the determination of influence coefficients can be envisaged as the algebraic solution of two equations in two unknowns, i.e. in the present case using Fe48 and Fe52 data

$$\frac{C_{Fe}}{R_{Fe}} - 1 = C_{Cr}\alpha_{FeCr} + C_{Ni}\alpha_{FeNi} \tag{5.4}$$

$$(0.1/0.04836) - 1 = 1.0678 = 0.6\alpha_{FeCr} + 0.3\alpha_{FeNi}$$

$$(0.1/0.0766) - 1 = 0.3055 = 0.3\alpha_{FeCr} + 0.6\alpha_{FeNi}$$

Solving for α_{FeCr} and α_{FeNi} results in $\alpha_{FeCr} = 2.0334$ and $\alpha_{FeNi} = -0.5077$. Substitution in Eqn (5.1) yields

$$C_{Fe} = 0.04836\,[1 + (2.0334 \times 0.6) + (-0.5077 \times 0.3)]$$
$$= 0.04836\,[2.0677]$$
$$= 0.1000$$

for Fe48 and

$$C_{Fe} = 0.0766\,[1 + (2.0334 \times 0.3) + (-0.5077 \times 0.6)]$$
$$= 0.0766\,[1.3054]$$
$$= 0.1000$$

in the case of Fe52. Thus, an identical pair of alpha coefficient values, namely 2.0334 and -0.5077, can be used to quantify the matrix effect of chromium (absorption) and nickel (absorption and enhancement), which is not the case if fundamental influence coefficients are used in this particular analytical context. As a matter of fact, the values 2.0334 and -0.5077 would probably be considered acceptable for all compositions within the range (10% Fe, 70% Cr, 20% Ni) and (10% Fe, 20% Cr, 70% Ni).

The above demonstration is presented not in defence of treating alpha coefficients as constants, but as a simple illustration of how the computation of a matrix effect correction term may be envisaged quite differently. The process of calculating alpha coefficients as in the above example breaks down if carried out for multielement systems with large ranges of concentrations. The only criterion that may be adduced for labelling the coefficients alpha in such cases is that they are calculated without reference to experimental data; i.e. they are calculated a priori using R_i data resulting from a fundamental parameters expression.

5.3 CONCEPT OF THEORETICAL INFLUENCE COEFFICIENTS

Gillam and Heal (1952), in their investigation of the experimental and theoretical aspects of X-ray fluorescence analysis, derived an expression for emitted intensity for the limited context of absorption effects in a binary system and with monochromatic excitation. Although the expression *influence coefficient* was not used, the expression is equivalent to

$$\frac{P_{i\lambda}}{P_{(i)\lambda}} = \frac{C_i \mu_i^*}{C_i \mu_i^* + C_j \mu_j^*} \tag{5.5}$$

from which they derived

$$\frac{P_{i\lambda}}{P_{(i)\lambda}} = \frac{C_i}{C_i + C_j(\mu_j^*/\mu_i^*)} \tag{5.6}$$

and showed that this expression gave a qualitative account of the experimental intensity measurements reported by Friedman and Birks (1948) for iron in the presence of silver.

Beattie and Brissey (1954) expanded Eqn (5.6) so as to incorporate absorption effects in multielement systems for monochromatic excitation. From the following expression

$$\frac{I_{(i)}}{I_i} C_i = C_i + \sum_j \frac{\mu_j^*}{\mu_i^*} C_j \tag{5.7}$$

a set of simultaneous equations was derived

$$
\begin{aligned}
-(R_{iR} - 1)C_i + & \quad K_{ij}C_j + & \quad K_{ik}C_k + \cdots = 0 \\
K_{ji}C_i & \quad - (R_{jR} - 1)C_j + & \quad K_{jk}C_k + \cdots = 0 \\
K_{ki}C_i & \quad + \quad K_{kj}C_j - (R_{kR} - 1)C_k + \cdots = 0
\end{aligned} \tag{5.8}
$$

where $R_{iR} = I_{(i)}/I_i$, i.e. the reciprocal of R_i as previously defined, and $K_{ij} = \mu_j^*/\mu_i^*$. Lacking the means to incorporate the effect of polychromatic excitation and enhancement effects, advantage was taken of the fact that K_{ij} can be calculated from the relationship for a binary system, namely

$$K_{ij} = \frac{C_i}{C_j} (R_{iR} - 1) \tag{5.9}$$

Relative intensities were measured for binary systems involving enhancement; thus, in those cases K_{ij} represents the influence coefficient for the total matrix effect (absorption and enhancement) of element j on analyte i. The fact that the set of Eqns (5.8) is further governed by the constraint $C_i + C_j + \cdots = 1$ results in a problem of over determination; i.e., for a system of N elements, N^2 independent solutions are theoretically possible. To circumvent this,

Beattie and Brissey proposed a calculation procedure that involved distinguishing one element from the others, and found that the best solution was obtained by distinguishing the most abundant element in the sample analysed. A detailed numerical example was given for a specimen in the Cr–Fe–Ni–Mo system.

Although the Lachance–Traill model, Eqn (5.1), solved the overdetermination problem associated with the Beattie and Brissey model, its derivation was also related to the context of monochromatic incident sources involving absorption only. Lacking a practical process for incorporating the effect of a polychromatic incident source, alpha coefficients were computed based on the concept of an equivalent wavelength λ_{eqv}; i.e. assuming that a single wavelength could represent a polychromatic source. $I_{(i)}$ values were then calculated using the relationship

$$I_{(i)} = \frac{I_i\left[1 + \sum_j \alpha_{ij\lambda}C_j\right]}{C_i} \tag{5.10}$$

where $\alpha_{ij\lambda}$ was calculated from the theoretical expression

$$\alpha_{ij\lambda} = \left(\frac{\mu_j^* - \mu_i^*}{\mu_i^*}\right)_{\lambda_{eqv}} \tag{5.11}$$

Equation (5.10) yielded fairly constant values for the pure analytes which were averaged to yield $I_{(i)}$. In a few cases it became evident that the calculated $\alpha_{ij\lambda}$ coefficients undercorrected the matrix effect (giving lower $I_{(i)}$ values than the general trend), in which case the coefficients were adjusted accordingly. The $I_{(i)}$ values were then used to compute R_i for a separate set of specimens treated as unknowns, and concentrations were generated by an iterative process. It was also shown that the same $\alpha_{ij\lambda}$ values could be applied to previously published experimental data. Both the Beattie and Brissey and the Lachance and Traill approaches were based on gross approximations (e.g. theoretical calculations made assuming monochromatic excitation, binary systems, and dealing with the absorption effect only), assumptions which certainly limited their application when applied to experimental data involving polychromatic incident sources, multielement specimens and enhancement. In contrast, subsequent developments have shown that the Lachance–Traill formalism may be retained wherein the α coefficients are much more closely linked to fundamental theory. These developments are examined next.

5.4* EXTENSION TO POLYCHROMATIC EXCITATION

Given the Claisse and Quintin (1967) theoretical demonstration and experimental confirmation that influence coefficients vary as a function of specimen composition for polychromatic incident sources, and the provision by Criss and Birks (1968) of a workable fundamental expression relating emitted intensity and concentration, the procedures for computing alpha coefficients developed mainly along the following lines.

(a) Fundamental parameters expressions were used to generate arrays of binary m_{ij} coefficients, the purpose being to establish the limits within which they vary and the pattern of the variation.

(b) α models were proposed based on the premise that the variations of binary m_{ij} could be approximated by a linear expression.

(c) Subsequently, α models were proposed based on the premise that the variations were better approximated by a hyperbolic expression.

(d) In both (b) and (c), it was shown that the total effect is not strictly equal to the sum computed using binary coefficients. This led to the concept of introducing an additional correction term for *third element effects*.

Table 5.1 Typical $m_{Mnj,bin}$ arrays

Analytical context
Excitation source: W target tube, 45 kV (CP); see Table II.1.7
Instrument geometry: $\psi' = 63°$, $\psi'' = 33°$
Analyte: Mn in various binary systems

C_{Mn}	C_j	j =	Mg	Si	Ti	V	Cr
					$m_{Mnj,bin}$		
0.999	0.001		−0.385	−0.045	2.119	2.506	−0.109
0.99	0.01		−0.385	−0.046	2.111	2.496	−0.109
0.90	0.10		−0.388	−0.053	2.053	2.425	−0.109
0.80	0.20		−0.392	−0.060	2.011	2.377	-0.109
0.50	0.50		−0.404	−0.082	1.943	2.303	−0.109
0.20	0.80		−0.418	−0.101	1.909	2.268	−0.109
0.10	0.90		−0.423	−0.108	1.901	2.260	−0.109
0.01	0.99		−0.429	−0.113	1.895	2.253	−0.109
0.001	0.999		−0.429	−0.114	1.894	2.253	−0.109

C_{Mn}	C_j	j =	Fe	Co	Ni	Zn	Y
0.999	0.001		0.046	−0.159	−0.111	−0.102	0.887
0.99	0.01		0.045	−0.161	−0.113	−0.104	0.884
0.90	0.10		0.041	−0.181	−0.134	−0.120	0.860
0.80	0.20		0.035	−0.202	−0.159	−0.137	0.837
0.50	0.50		0.010	−0.277	−0.237	−0.185	0.789
0.20	0.80		−0.033	−0.375	−0.329	−0.233	0.753
0.10	0.90		−0.056	−0.416	−0.367	−0.252	0.742
0.01	0.99		−0.086	−0.462	−0.406	−0.273	0.733
0.001	0.999		−0.090	−0.467	−0.411	−0.275	0.732

(e) Comprehensive alpha coefficient algorithms were derived.

These will now be examined in turn.

5.4.1 Binary m_{ij} influence coefficient arrays

The ability to generate binary m_{ij} influence coefficient $m_{ij,bin}$ arrays for any two elements in any given analytical context in order to establish the limiting values that the coefficients may take is a very important step in the process of using α coefficients. Another is the ability to study the variation patterns as the atomic number of the matrix elements increases or decreases. Tables 5.1 and 5.2 give the analytical context and list the fundamental parameters used to generate the $m_{ij,bin}$ arrays for analytes Mn and Co, respectively.

Let us first examine the data in Table 5.1, as plotted in Figure 5.1. The calculation of each $m_{Mnj,bin}$ coefficient involves the summation of 81 discrete wavelength elements. The characteristic lines of the elements Mg, Si, Ti, V and Cr, i.e. atomic number j < 25, cannot eject Mn K shell electrons and hence cannot cause enhancement of Mn spectral lines. The elements Fe, Co, Ni, Zn and Y, on the other hand, can cause enhancement of Mn. In the case of Mg and Si, the overall absorption effect is negative; i.e, μ_{Mg}^* and μ_{Si}^* are both less than μ_{Mn}^*, so that all m_{MnMg} and m_{MnSi} values are negative. The fact that the elements Ti and V are strong absorbers of Mn $K-L_{2,3}$ radiations is reflected in the magnitude of the positive values for m_{MnTi} and m_{MnV}. The element Cr represents a special case, in which the wavelength of the absorption edge of

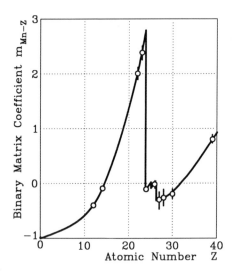

Figure 5.1 Plot of $m_{Mnj,bin}$ versus the atomic number of matrix element j. Value at 50% Mn is shown as circle

the matrix element is longer than the K absorption edge of the analyte and shorter than the emitted line of the analyte. The values of m_{ij} remain essentially constant and are slightly negative under these circumstances. The element Fe also represents a special case, in this instance because only the $K-M_3$ line of iron can eject Mn K level electrons. This results in the $m_{MnFe,bin}$ array ranging from very low positive to very low negative values. Cobalt and nickel, on the other hand, are both strong enhancers of Mn; zinc is somewhat less so, and although yttrium does cause some enhancement, absorption has gradually become the dominant matrix effect as the atomic number of the matrix element increases. This trend continues until the L absorption edges of the matrix element are shorter than the K absorption edge of Mn, as shown in Figure 5.1.

The $m_{Coj,bin}$ data in Table 5.2, as plotted in Figure 5.2, reflect the same trends as those observed for $m_{Mnj,bin}$ in Figure 5.1, and in this case the

Table 5.2   Typical $m_{Coj,bin}$ arrays

Analytical context
Excitation source: W target tube, 45 kV (CP); see Table II.1.7
Instrument geometry: $\psi' = 63°$, $\psi'' = 33°$
Analyte: Co in binaries

					$m_{Coj,bin}$		
C_{Co}	C_j	j =	Mg	Si	Ti	Mn	Fe
0.999	0.001		−0.552	−0.302	1.295	2.186	−0.100
0.99	0.01		−0.552	−0.302	1.289	2.176	−0.100
0.90	0.10		−0.554	−0.307	1.245	2.100	−0.100
0.80	0.20		−0.557	−0.313	1.211	2.048	−0.100
0.50	0.50		−0.566	−0.330	1.148	1.968	−0.100
0.20	0.80		−0.579	−0.349	1.114	1.930	−0.100
0.10	0.90		−0.584	−0.355	1.106	1.922	−0.100
0.01	0.99		−0.590	−0.361	1.099	1.915	−0.100
0.001	0.999		−0.590	−0.362	1.099	1.915	−0.100
C_{Co}	C_j	j =	Ni	Cu	Zn	Rb	Y
0.999	0.001		0.026	−0.307	−0.315	0.161	0.323
0.99	0.01		0.026	−0.308	−0.317	0.159	0.321
0.90	0.10		0.020	−0.324	−0.330	0.144	0.304
0.80	0.20		0.014	−0.341	−0.345	0.130	0.288
0.50	0.50		−0.013	−0.398	−0.389	0.095	0.250
0.20	0.80		−0.058	−0.464	−0.436	0.064	0.219
0.10	0.90		−0.082	−0.490	−0.455	0.054	0.209
0.01	0.99		−0.112	−0.519	−0.475	0.044	0.200
0.001	0.999		−0.116	−0.522	−0.478	0.043	0.199

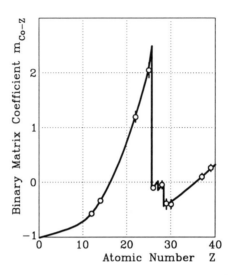

Figure 5.2   Plot of $m_{Coj,bin}$ versus the atomic number of matrix element j. Value at 50% Co is shown as circle

calculation of each m_{Coj} value is the resultant of the summation of 66 terms. Again the influence coefficients for matrix elements Mg and Si are all negative and, as expected from the larger difference in μ_j^* and μ_i^* values, are slightly more so than for Mn as analyte. The $m_{CoFe,bin}$ and $m_{MnCr,bin}$ arrays show that iron behaves towards cobalt as chromium does towards manganese; similarly the $m_{CoNi,bin}$ and $m_{MnFe,bin}$ arrays show that nickel behaves towards cobalt in a similar fashion to that of iron towards manganese. Enhancement is the dominant matrix effect when the matrix elements are copper and zinc and, although enhancement is still present as the atomic number of j increases, the absorption effect gradually becomes dominant, as indicated in the positive values of the $m_{CoRb,bin}$ and $m_{CoY,bin}$ arrays.

There is no doubt that the characteristic lines of the target element have an influence on the values of the m_{Mnj} and m_{Coj} coefficients in the above two tables. However, the pattern of the variations is representative of $m_{ij,bin}$ arrays in general except for shifts to lower or higher numerical values, depending on the atomic number of the elements.

5.5*   LINEAR APPROXIMATIONS OF BINARY m_{ij} ARRAYS

Although their theoretical study was limited to absorption effects, Claisse and Quintin (1967) were the first to propose an algorithm in which the coefficient α_{ij} is made to vary as a linear function of the concentration of C_j. Based on experimental data involving enhancement, the authors also proposed that the linear approximation be extended to include both absorption and enhancement. The theoretical aspect of Claisse and Quintin's contribution that pertains to fundamental influence coefficients is covered in detail in Section 4.5 and is only briefly recalled here. For the binary system i–j, the model proposed is expressed as

$$C_i = R_i[1 + \alpha_j C_j + \alpha_{jj} C_j^2] \qquad (5.12)$$

$$C_j = R_j[1 + \alpha_i C_i + \alpha_{ii} C_i^2] \qquad (5.13)$$

which may also be expressed in the conventional formalism

$$C_i = R_i[1 + \alpha_{ij,lin} C_j] \qquad (5.14)$$

$$C_j = R_j[1 + \alpha_{ji,lin} C_i] \qquad (5.15)$$

where

$$\alpha_{ij,lin} = \alpha_j + \alpha_{jj} C_j \qquad (5.16)$$

$$\alpha_{ji,lin} = \alpha_i + \alpha_{ii} C_i \qquad (5.17)$$

This formalism brings out more clearly that, at the binary level, the Claisse–Quintin model is equivalent to alpha coefficients in the Lachance–Traill model that vary linearly as a function of C_j. This extends the concentration range over which the model is applicable. If m_{ij} is plotted as a function of C_j, the intercept of a linear least squares fit on the Y axis gives α_j, and α_{jj} is given by the slope of the line. Values for α_j and α_{jj} may also be obtained algebraically, given m_{ij} values for two binaries. The two equations to be solved are

$$m_{ij1} = \alpha_j + C_{j1} \alpha_{jj} \qquad (5.18)$$

$$m_{ij2} = \alpha_j + C_{j2} \alpha_{jj} \qquad (5.19)$$

where the subscripts 1 and 2 refer to binaries 1 and 2, respectively. By Cramer's Rule

$$\alpha_j = \frac{m_{ij1} C_{j2} - m_{ij2} C_{j1}}{C_{j2} - C_{j1}} \qquad (5.20)$$

and

$$\alpha_{jj} = \frac{m_{ij2} - m_{ij1}}{C_{j2} - C_{j1}} \qquad (5.21)$$

To approximate the complete binary array, i.e. $C_i = 1.0$ to $C_i = 0.0$, Rousseau and Claisse (1974) recommended using the m_{ij} values for $C_j = 0.8$ and $C_j = 0.2$ in solving Eqns (5.20) and (5.21). This is shown graphically in Figure 5.3. If the concentration range is limited to $C_i = 0.0$–0.5, then one may choose the m_{ij} values for $C_j = 0.4$ and $C_j = 0.1$ instead; i.e. the linear fit is limited to the

Table 5.3 Comparison of $m_{ij,bin}$ array and α_{ij} array generated by the Claisse–Quintin algorithm. Recalculated C_i values using the Claisse–Quintin and Lachance–Traill algorithms

Analytical context
Excitation source: W target tube, 45 kV (CP); see Table II.1.7
Instrument geometry: $\psi' = 63°$, $\psi'' = 33°$
System: Cr–Co; analyte: Co

C_{Co}	C_{Cr}	R_{Co}	$m_{CoCr,bin}$	α_{CoCr}	$C_{Co1}{}^a$	$C_{Co2}{}^a$
0.999	0.001	0.9971	1.8821	1.7998	0.9989	0.9988
0.99	0.01	0.9718	1.8735	1.7981	0.9893	0.9882
0.90	0.10	0.7621	1.8092	1.7814	0.8979	0.8909
0.80	0.20	0.5915	1.7629	(1.7629)	(0.8000)	0.7912
0.70	0.30	0.4608	1.7307	1.7443	0.7019	0.6942
0.60	0.40	0.3566	1.7069	1.7258	0.6027	0.5974
0.50	0.50	0.2711	(1.6885)	1.7073	0.5026	(0.5000)
0.40	0.60	0.1996	1.6738	1.6887	0.4018	0.4018
0.30	0.70	0.1387	1.6617	1.6702	0.3008	0.3026
0.20	0.80	0.0862	1.6517	(1.6517)	(0.2000)	0.2026
0.10	0.90	0.0403	1.6432	1.6332	0.0996	0.1015
0.01	0.99	0.0038	1.6366	1.6165	0.0099	0.0102
0.001	0.999	0.0004	1.6359	1.6148	0.0010	0.0011

^{a} $C_{Co1} = R_{Co}[1 + \alpha_{CoCr}C_{Cr}]$ with $\alpha_{CoCr} = 1.7999 + (-0.1853C_{Cr})$.
$C_{Co2} = R_{Co}[1 + 1.6885C_{Cr}]$

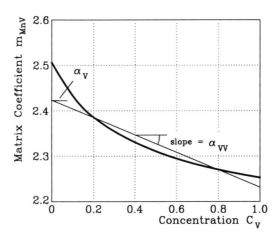

Figure 5.3 Illustration of a graphical method for obtaining α_j and α_{jj} for the Claisse–Quintin algorithm

range under consideration. Table 5.3 illustrates the 'goodness of fit' between the m_{CoCr} array and $\alpha_{CoCr,lin}$ array computed using a linear approximation, and gives the recalculated C_{Co} values using $\alpha_{CoCr,lin}$ (column headed C_{Co1}), while the column headed C_{Co2} illustrates the much larger error introduced if α_{CoCr} is simply equated to m_{CoCr} for the binary $C_i = 0.5$ and treated as a constant.

Table 5.4 is similar to Table 5.3 except that the analyte is chromium and the matrix element is cobalt, a strong enhancer of chromium. In this case, a linear fit is less representative of the m_{CrCo} array and, as column C_{Cr2} indicates, equating α_{CrCo} to m_{CrCo} for $C_{Cr} = 0.5$ cannot be considered an option for quantitative analysis. Similar evidence that equating α_{ij} to m_{ij} for the binaries $C_i = 0.5$ and applying the resultant values over wide ranges of elemental concentrations is not a valid procedure, especially when enhancement is the dominant matrix effect, was shown by Lachance (1970) and confirmed experimentally by Rasberry and Heinrich (1974).

For reasons that will be made evident under the treatment of the 'third element effect', it was shown by Tertian (1975) that it is preferable to define the variation of the coefficient α_{ij} as a function of C_i rather than C_j. Thus, a

Table 5.4 Comparison of $m_{ij,bin}$ and α_{ij} arrays generated by the Claisse–Quintin algorithm. Recalculated C_i values using the Claisse–Quintin and Lachance–Traill algorithms

Analytical context
Excitation source: W target tube, 45 kV (CP); see Table II.1.7
Instrument geometry: $\psi' = 63°$, $\psi'' = 33°$
System Cr–Co; analyte: Cr

C_{Cr}	R_{Cr}	a_{CrCo}	e_{CrCo}	$m_{CrCo,bin}$	α_{CrCo}	C_{Cr1}[a]	C_{Cr2}[a]
0.999	0.9991	0.2545	0.3157	−0.0609	−0.0501	0.9990	0.9908
0.99	0.9906	0.2529	0.3162	−0.0633	−0.0529	0.9901	0.9887
0.90	0.9078	0.2360	0.3216	−0.0806	−0.0845	0.9005	0.8901
0.80	0.8182	0.2168	0.3282	−0.1114	(−0.1114)	(0.8000)	0.7863
0.70	0.7302	0.1973	0.3354	−0.1381	−0.1421	0.6991	0.6874
0.60	0.6427	0.1775	0.3434	−0.1659	−0.1729	0.5982	0.5925
0.50	0.5541	0.1573	0.3525	(−0.1952)	−0.2037	0.4977	(0.5000)
0.40	0.4628	0.1367	0.3630	−0.2263	−0.2344	0.3977	0.4086
0.30	0.3666	0.1158	0.3754	−0.2596	−0.2652	0.2986	0.3165
0.20	0.2620	0.0947	0.3906	−0.2960	(−0.2960)	(0.2000)	0.2211
0.10	0.1435	0.0733	0.4103	−0.3370	−0.3267	0.1013	0.1183
0.01	0.0160	0.0540	0.4345	−0.3805	−0.3544	0.0104	0.0129
0.001	0.0016	0.0521	0.4375	−0.3854	−0.3572	0.0010	0.0013

[a] $C_{Cr1} = R_{Cr}[1 + \alpha_{CrCo}C_{Co}]$ with $\alpha_{CrCo} = -0.0498 + (-0.3077C_{Co})$;
$C_{Cr2} = R_{Cr}[1 + (-0.1952C_{Co})]$

linear fit to a plot of m_{ij} versus C_i, would yield the coefficients m_{jT} and m_{jjT} by solving the two equations

$$m_{ij1} = \alpha_{jT} + C_{i1}\alpha_{jjT} \tag{5.22}$$

$$m_{ij2} = \alpha_{jT} + C_{i2}\alpha_{jjT} \tag{5.23}$$

for α_{jT} and α_{jjT} which are then used to define

$$\alpha_{ij,\text{lin}} = \alpha_{jT} + \alpha_{jjT}C_i \tag{5.24}$$

The coefficients are subscripted T in order to differentiate them from their equivalent in the Claisse–Quintin model; the interrelation is given by

$$\alpha_{jT} = \alpha_j + \alpha_{jj} \tag{5.25}$$

$$\alpha_{jjT} = -\alpha_{jj} \tag{5.26}$$

The advantage of the Tertian formalism is significant when multielement systems are being considered, by far the more common situation in practice. This advantage may be retained in a modified Claisse–Quintin model, namely

$$C_i = R_i\left[1 + \sum_j (\alpha_j + \alpha_{jj}C_M)C_j\right] \tag{5.27}$$

where

$$C_M = C_j + C_k + \cdots C_n = 1 - C_i \tag{5.28}$$

5.5.1 Numerical example

The following data are taken from Table 5.3, in which $i = \text{Co}$, $j = \text{Cr}$, $m_{\text{CoCr}} = 1.6517$ at $C_{\text{Co}} = 0.2$ and $m_{\text{CoCr}} = 1.7629$ at $C_{\text{Co}} = 0.8$. The application of the Claisse–Quintin algorithm expressed in the Lachance–Traill formalism of Eqn (5.14) for the Co–Cr binary system requires the successive calculation of α_{Cr}, α_{CrCr} and $\alpha_{\text{CoCr,lin}}$ using Eqns (5.20), (5.21) and (5.16):

$$\alpha_{\text{Cr}} = \frac{(1.7629 \times 0.8) - (1.6517 \times 0.2)}{0.8 - 0.2}$$

$$= 1.8000$$

$$\alpha_{\text{CrCr}} = \frac{1.6517 - 1.7629}{0.8 - 0.2}$$

$$= -0.1853$$

and, at $C_{\text{Co}} = 0.10$

$$\alpha_{\text{CoCr,lin}} = 1.8000 + (-0.1853 \times 0.9)$$

$$= 1.6332$$

Calculation of C_{Co} from Eqn (5.14) gives

$$C_{Co} = 0.0403\,[1 + (1.6332 \times 0.9)]$$
$$= 0.0995$$

whereas if the m_{ij} value for $C_i = 0.5$ is used in Eqn (5.14)

$$C_{Co} = 0.0403\,[1 + (1.6885 \times 0.9)]$$
$$= 0.1015$$

Alternatively, $\alpha_{CoCr,lin}$ can be determined using the Tertian formalism, Eqns (5.25), (5.26) and (5.24):

$$\alpha_{Cr} = \frac{(1.6517 \times 0.8) - (1.7629 \times 0.2)}{0.8 - 0.2}$$

$$= 1.6147$$

$$\alpha_{CrCr} = \frac{1.7629 - 1.6517}{0.8 - 0.2}$$

$$= 0.1853$$

$$\alpha_{CoCr,lin} = 1.6147 + (0.1853 \times 0.1)$$
$$= 1.6332$$

5.6* HYPERBOLIC APPROXIMATIONS OF BINARY m_{ij} ARRAYS

Studies were pursued by a number of workers in order to find an algorithm that would provide a better fit of α_{ij} arrays to the fundamental $m_{ij,bin}$ arrays. Attempts to transform the mathematical expressions in power series proved unsuccessful and doomed to failure, as shown by Rousseau and Claisse (1974). The concept of subjectively selecting higher order terms in the Claisse–Quintin model, as proposed by Lachance and Claisse (1980), did not fare much better, as shown by Rousseau (1987) and Tertian (1987). A number of three coefficient hyperbolic functions as better approximations for the variation observed in $m_{ij,bin}$ arrays have been postulated:

● Lachance (1981) proposed the model

$$\alpha_{ij,hyp} = \left(\alpha_1 + \frac{\alpha_2 C_M}{1 + (1 - C_M)\alpha_3}\right)_{ij} \tag{5.29}$$

where the subscript hyp is used to indicate that the $m_{ij,bin}$ arrays are approximated by a hyperbolic function, i.e.

$$\alpha_{ij,hyp} \cong m_{ij,bin} \tag{5.30}$$

In practice

$$\alpha_1 = m_{ij,bin} \text{ calculated for } C_i = 0.999, \ C_j = 0.001 \tag{5.31}$$

$$\alpha_2 = (m_{ij,bin} \text{ calculated for } C_i = 0.001, \ C_j = 0.999) - \alpha_1 \tag{5.32}$$

$$\alpha_3 = \frac{\alpha_2}{(m_{ij,bin} \text{ calculated for } C_i = 0.5) - \alpha_1} - 2 \tag{5.33}$$

- Tertian and Claisse (1982, p. 136) and Broll and Tertian (1983) proposed the expression

$$\alpha_{ij,hyp} = k_1 + \frac{k_2 C_i}{C_i + k_3(1 - C_i)} \tag{5.34}$$

where k_1 is the limiting value at $C_i = 0$, k_2 is (the limiting value at $C_i = 1.0$) $- k_1$ and

$$k_3 = 1 + \frac{k_1 + k_2}{2} \tag{5.35}$$

It was observed that k_3 was a correct value (especially when considering polychromatic excitation) in nearly all binary systems. In other cases, a suitable value for k_3 was obtained by trial and error.

- Tertian (1987, 1988) defined effective $\alpha_{ij,hyp}$ influence coefficients

$$\alpha_{ij,hyp} = \left(\gamma_1 + \frac{\gamma_2 C_i}{1 + (1 - C_i)\gamma_3} \right)_{ij} \tag{5.36}$$

where, in practice

$$\gamma_1 = m_{ij,bin} \text{ calculated for } C_i = 0.001, \ C_j = 0.999 \tag{5.37}$$

$$\gamma_2 = (m_{ij,bin} \text{ calculated for } C_i = 0.999, \ C_j = 0.001) - \gamma_1 \tag{5.38}$$

$$\gamma_3 = \frac{(\gamma_1 + \gamma_2) - (2 \times m_{ij,bin} \text{ calculated for } C_i = 0.5)}{(m_{ij,bin} \text{ calculated for } C_i = 0.5) - \gamma_1} \tag{5.39}$$

Equations (5.29) and (5.36) are equivalent given that

$$\gamma_1 = \alpha_1 + \alpha_2 \tag{5.40}$$

$$\gamma_2 = -\alpha_2 \tag{5.41}$$

$$\gamma_3 = \frac{-\alpha_3}{1 + \alpha_3} \tag{5.42}$$

5.6.1 Numerical example

Table 5.5 illustrates the close agreement between the $m_{CoV,bin}$ and $\alpha_{CoV,hyp}$ arrays (V is a strong absorber) when the latter is computed using Eqn (5.29) or (5.36); Table 5.6 illustrates the close agreement between the $m_{VCo,bin}$ and

Table 5.5 Comparison of $m_{ij,bin}$ array and $\alpha_{ij,hyp}$ array generated by Eqn (5.29) or (5.36). Recalculated C_i values using $\alpha_{ij,hyp}$ and $\alpha_{ij,lin}$ influence coefficients

Analytical context
Excitation source: W target tube, 45 kV (CP); see Table II.1.7
Instrument geometry: $\psi' = 63°$, $\psi'' = 33°$
System: Co–V; analyte: Co

C_{Co}	C_V	R_{Co}	$m_{CoV,bin}$	$\alpha_{CoV,hyp}$	$C_{Co}{}^a$	$C_{Co}{}^b$
0.999	0.001	0.9974	1.5791	1.5783	(0.9990)	0.9989
0.99	0.01	0.9747	1.5723	1.5720	0.9900	0.9894
0.90	0.10	0.7813	1.5188	1.5199	0.9001	0.8982
0.80	0.20	0.6174	1.4784	1.4794	0.8001	(0.8000)
0.70	0.30	0.4879	1.4494	1.4500	0.7001	0.7017
0.60	0.40	0.3819	1.4274	1.4276	0.6000	0.6025
0.50	0.50	0.2933	1.4100	1.4100	(0.5000)	0.5024
0.40	0.60	0.2177	1.3959	1.3958	0.4000	0.4017
0.30	0.70	0.1524	1.3843	1.3841	0.3000	0.3008
0.20	0.80	0.0953	1.3745	1.3743	0.2000	(0.2000)
0.10	0.90	0.0449	1.3661	1.3660	0.1000	0.0996
0.01	0.99	0.0043	1.3595	1.3595	0.0100	0.0099
0.001	0.999	0.0004	1.3588	1.3589	(0.0010)	0.0010

$^a C_{Co} = R_{Co}[1 + \alpha_{CoV,hyp}C_V]$ with

$$\alpha_{CoV,hyp} = 1.5791 + \frac{(-0.2203C_V)}{1 + (-0.6972C_{Co})}$$

$^b C_{Co} = R_{Co}[1 + \alpha_{CoV,lin}C_V]$ with

$$\alpha_{CoV,lin} = 1.5131 + (-0.1733 \times C_V)$$

$\alpha_{VCo,hyp}$ arrays (Co is a strong enhancer) when $\alpha_{VCo,hyp}$ is computed using Eqn (5.29) or (5.36). In both tables, concentrations calculated using $\alpha_{ij,hyp}$ coefficients are subscripted a and those using the Claisse–Quintin model are subscripted b.

The following data taken from Table 5.6, in which $i = V$ and $j = Co$, are used to demonstrate the calculation of $\alpha_{ij,hyp}$. Substitution in expressions defining α coefficients relating to Eqn (5.29) from $C_V = 0.999$

$$\alpha_1 = 0.0452$$

from $C_V = 0.001$ and α_1

$$\alpha_2 = -0.2951 - 0.0452$$
$$= -0.3403$$

from $C_V = 0.50$, α_1 and α_2

$$\alpha_3 = \frac{-0.3403}{-0.1079 - 0.0452} - 2$$

$$= 0.2227$$

Table 5.6 Comparison of $m_{ij,bin}$ array and $\alpha_{ij,hyp}$ array generated by Eqn (5.29) or (5.36). Recalculated C_i values using $\alpha_{ij,hyp}$ and $\alpha_{ij,lin}$ influence coefficients

Analytical context
Excitation source: W target tube, 45 kV (CP); see Table II.1.7
Instrument geometry: $\psi' = 63°$, $\psi'' = 33°$
System: Co–V; analyte: V

C_V	R_V	a_{VCo}	e_{VCo}	$m_{VCo,bin}$	$\alpha_{VCo,hyp}$	$C_V{}^a$	$C_V{}^b$
0.999	0.9990	0.3635	0.3187	0.0452	0.0449	(0.9990)	0.9990
0.99	0.9896	0.3609	0.3188	0.0421	0.0424	0.9900	0.9900
0.90	0.8986	0.3350	0.3199	0.0151	0.0169	0.9002	0.9002
0.80	0.8024	0.3065	0.3216	−0.0151	−0.0126	0.8004	(0.8000)
0.70	0.7097	0.2783	0.3238	−0.0455	−0.0431	0.7005	0.6996
0.60	0.6189	0.2504	0.3268	−0.0764	−0.0748	0.6004	0.5991
0.50	0.5285	0.2228	0.3306	−0.1079	−0.1079	(0.5000)	0.4987
0.40	0.4368	0.1953	0.3356	−0.1403	−0.1422	0.3995	0.3987
0.30	0.3417	0.1681	0.3422	−0.1742	−0.1780	0.2991	0.2992
0.20	0.2404	0.1410	0.3512	−0.2102	−0.2154	0.1990	(0.2000)
0.10	0.1290	0.1139	0.3637	−0.2498	−0.2544	0.0995	0.1008
0.01	0.0140	0.0896	0.3802	−0.2906	−0.2909	0.0100	0.0103
0.001	0.0014	0.0871	0.3822	−0.2951	−0.2947	(0.0010)	0.0010

^{a} $C_V = R_V[1 + \alpha_{VCo,hyp}C_{Co}]$ with

$$\alpha_{VCo,hyp} = 0.0452 + \frac{(-0.3403C_{Co})}{1 + (0.2227C_V)}$$

^{b} $C_V = R_V[1 + \alpha_{VCo,lin}C_{Co}]$ with

$$\alpha_{VCo,lin} = 0.0499 + (-0.3252 \times C_{Co})$$

Substitution in Eqn (5.29) for the case $C_V = 0.2$ yields

$$\alpha_{ij,hyp} = 0.0452 + \frac{-0.3403 \times 0.8}{1 + (1 - 0.8) \times 0.2227}$$

$$= -0.2154$$

Similarly, substitution in expressions relating to Eqn (5.36) from $C_V = 0.001$

$$\gamma_1 = -0.2951$$

from $C_V = 0.999$ and γ_1

$$\gamma_2 = 0.0452 - (-0.2951)$$

$$= 0.3403$$

from $C_V = 0.50$, γ_1 and γ_2

$$\gamma_3 = \frac{(-0.2951 + 0.0452) - (2 \times (-0.1079))}{-0.1079 - (-0.2951)}$$

$$= -0.1822$$

Substitution in Eqn (5.36) for the case $C_V = 0.2$ yields

$$\alpha_{ij,hyp} = -0.2951 + \frac{0.3403 \times 0.2}{1 + (1 - 0.2) \times (-0.1822)}$$

$$= -0.2154$$

It therefore follows that knowledge of $(\alpha_1, \alpha_2, \alpha_3)_{ij}$ or $(\gamma_1, \gamma_2 \text{ and } \gamma_3)_{ij}$ coefficients can be used as constants to generate very close approximations of binary m_{ij} values over the whole range $C_i = 1-0$.

5.7* THIRD ELEMENT EFFECT

The topic is best introduced by examining a simple numerical example. The analytical context and fundamental parameters relating to Tables 5.1 and 5.2 are retained; the relative intensities computed from fundamental parameters are listed in Table 5.7.

From data relating to specimens 1 and 2, it is possible to compute the 'exact' $m_{ij,bin}$ influence coefficients from Eqn (4.19)

$$m_{ij,bin} = \frac{C_i - R_i}{C_j R_i} \tag{5.43}$$

Table 5.7 Numerical example of the third element effect

Analytical context
Excitation source: W target tube, 45 kV (CP); see Table II.1.7
Instrument geometry: $\psi' = 63°$, $\psi'' = 33°$
System: Mn–Ti–Co; analyte: Mn

Specimen	C_{Mn}	C_{Ti}	C_{Co}	R_{Mn}	$m_{ij,bin}$	[FP]^a	[$m_{ij,bin}$]^b
1	0.10	0.90	0.00	0.03690	1.9008	2.7100	(2.7100)
2	0.10	0.00	0.90	0.15992	−0.4164	0.6563	(0.6563)
3	0.10	0.60	0.30	0.05103		1.9596	2.0156
4	0.10	0.30	0.60	0.07925		1.2619	1.3204
5	0.20	0.80	0.00	0.07910	1.9087	2.5284	(2.5284)
6	0.20	0.00	0.80	0.28559	−0.3784	0.7003	(0.7003)
7	0.20	0.60	0.20	0.09834		2.0337	2.0704
8	0.20	0.20	0.60	0.17879		1.1186	1.1568
9	0.50	0.50	0.00	0.25360	1.9427	1.9716	(1.9716)
10	0.50	0.00	0.50	0.58038	−0.2771	0.8615	(0.8615)
11	0.50	0.40	0.10	0.28757		1.7387	1.7494
12	0.50	0.10	0.40	0.46581		1.0734	1.0835

^{a} [FP] correction term calculated from fundamental parameters.
^{b} [$m_{ij,bin}$] correction term calculated using $m_{ij,bin}$ coefficients.

where R_i is computed from the fundamental parameters expression. Thus

$$m_{\text{MnTi,bin}} = \frac{0.1 - 0.03689}{0.9 \times 0.03689}$$

$$= 1.9008$$

$$m_{\text{MnCo,bin}} = \frac{0.1 - 0.15993}{0.9 \times 0.15993}$$

$$= -0.41636$$

The term 'exact' is used to indicate that the values 1.9008 and -0.4164 have been obtained from fundamental parameters expressions and are the only influence coefficient values that generate the value $C_{\text{Mn}} = 0.1$ in binaries 1 and 2. However, substitution of these values in the alpha model Eqn (5.1) for the two ternary specimens yields

$$C_{\text{Mn}} = 0.05103 [1 + (1.9008 \times 0.6) + (-0.41636 \times 0.3)]$$

$$= 0.0510 [2.01557]$$

$$= 0.1029$$

$$C_{\text{Mn}} = 0.07925 [1 + (1.9008 \times 0.3) + (-0.41636 \times 0.6)]$$

$$= 0.0793 [1.3204]$$

$$= 0.1047$$

To the extent that the $\alpha_{ij,\text{hyp}}$ are very good approximations of $m_{ij,\text{bin}}$ influence coefficients and interchangeable in practice, one must therefore conclude that the following three expressions are incompatible

$$C_i = R_i [1 + \alpha_{ij,\text{hyp}} C_j] \tag{5.44}$$

$$C_i = R_i [1 + \alpha_{ik,\text{hyp}} C_k] \tag{5.45}$$

$$C_i = R_i [1 + \alpha_{ij,\text{hyp}} C_j + \alpha_{ik,\text{hyp}} C_k] \tag{5.46}$$

even when dealing with a situation where the concentration of analyte i is equal in all three specimens. If the values of $\alpha_{ij,\text{hyp}}$ and $\alpha_{ik,\text{hyp}}$ that satisfy Eqns (5.44) and (5.45) are substituted in Eqn (5.46), an erroneous value for C_i is found, which is ascribed to *third element effects*. Tertian and Claisse (1982) observed that enhancement is the chief source of the third element effect and that two models have been proposed to compensate for this effect, namely cross-coefficient α_{ijk} models and a model based on the correction factor $1 + \varepsilon_i$. These are examined next.

5.7.1 Cross-coefficient formalisms

As mentioned in the previous chapter, Claisse and Quintin's derivation of an influence coefficient algorithm for multielement systems yielded an expression involving the products of influence coefficients times the product of the concentration of two matrix elements, i.e.

$$C_i = R_i \left[1 + \sum_j \alpha_j C_j + \sum_j \alpha_{jj} C_j^2 + \sum_j \sum_k \alpha_{ijk} C_i C_k \right] \qquad (5.47)$$

The coefficients α_{ijk}, α_{ijl}, ... were termed *crossed coefficients* and, subsequently, Tertian and Claisse (1982) referred to them as a *correction for third element effects*. It should be emphasized that third element effects are in no way related to the process of tertiary fluorescence described in Section 2.5.1. It is noted that Claisse and Quintin (1967) dealt only with the absorption process and calculated α_j and α_{jj} (both theoretical and experimental) in binary systems only. Crossed coefficients were not determined because it was felt that the approximations on α_j and α_{jj} coefficients were too large to obtain α_{ijk} coefficients with sufficient accuracy.

Given that hyperbolic functions ($\alpha_{ij,hyp}$ coefficients) yield a better approximation of binary m_{ij} arrays than linear functions, a hyperbolic function, combined with cross-coefficients, is retained for the COLA algorithm (Tao et al., 1985), resulting in the expression

$$C_i = R_i \left[1 + \sum_j \alpha_{ij} C_j \right] \qquad (5.48)$$

where

$$\alpha_{ij} = \left(\alpha_1 + \frac{\alpha_2 C_M}{1 + (1 - C_M)\alpha_3} \right)_{ij} + \sum_k \alpha_{ijk} C_k \qquad (5.49)$$

$$= \alpha_{ij,hyp} + \sum_k \alpha_{ijk} C_k \qquad (5.50)$$

Thus, the COLA model combines the concept of approximating the variation in binary m_{ij} arrays by a hyperbolic function and the concept of cross-coefficients to compensate for third element effects. The coefficients α_1, α_2 and α_3 are constants for a given analytical context, and α_{ijk} coefficients vary within well defined limits. Cross-coefficient α_{ijk} values can be computed from data generated by fundamental parameters expressions *for three element* systems, i.e.

$$\frac{C_i}{R_i} - 1 - \alpha_{ij,hyp} C_j - \alpha_{ik,hyp} C_k = \alpha_{ijk} C_j C_k \qquad (5.51)$$

or, given that

$$\frac{C_i}{R_i} = \left[1 + \sum_j m_{ij} C_j \right] \qquad (5.52)$$

and defining

$$[FP] = \left(\frac{C_i}{R_i}\right)_r = \left(\frac{P_{(i)}C_i}{P_i + S_i}\right)_r = \left[1 + \sum_j m_{ij}C_j\right]_r \tag{5.53}$$

$$[m_{ij,\text{bin}}] = \left[1 + \sum_j m_{ij,\text{bin}}C_j\right] \cong \left[1 + \sum_j \alpha_{ij,\text{hyp}}C_j\right] \tag{5.54}$$

$$\Delta = [FP] - [m_{ij,\text{bin}}] = \alpha_{ijk}C_jC_k \tag{5.55}$$

$$\alpha_{ijk} = \frac{\Delta}{C_jC_k} \tag{5.56}$$

If Δ is plotted versus the product C_jC_k, the slope of a least squares fit through the origin provides an effective value for α_{ijk}. Conversely, one may simply average a number of α_{ijk} values obtained from Eqn (5.56). Rousseau and Claisse (1974) proposed the computation of a single value α_{ijk} at a composition where the correction term $R_iC_jC_k$ is close to its maximum value (namely $C_i = 0.30$, $C_j = C_k = 0.35$) that is valid in practical applications.

5.7.2 Tertian formalism

Whereas the cross-coefficient formalism retained in Eqn (5.50) compensates for third element effects following Claisse and Quintin (1967) by introducing an *additive term* on the right hand side of the equal sign in Eqn (5.47), i.e. *plus* $\Sigma_{j,k}\,\alpha_{ijk}C_jC_k$, Tertian (1973) proposed a model that compensates for third element effects by the introduction of *a divisor*, namely

$$C_i = \frac{R_i}{1 + \varepsilon_i}\left[1 + \sum_j \alpha_{ij,\text{hyp}}C_j\right] \tag{5.57}$$

where the term $1 + \varepsilon_i$ represents the third element effect correction factor. Subsequently, Tertian (1976) derived an expression for $1 + \varepsilon_i$ which may be defined as

$$1 + \varepsilon_i = \frac{R_i}{C_M} \sum_j \left(\frac{C_j}{R_{ij}}\right)_\text{bin} \tag{5.58}$$

where $R_{ij,\text{bin}}$ is the relative intensity for the binary i–j AT THE SAME CONCENTRATION LEVEL as C_i in the multielement specimen. The factor may also be computed from the relationship

$$1 + \varepsilon_i = \frac{\left[1 + \sum_j m_{ij,\text{bin}}C_j\right]}{\left[1 + \sum_j m_{ij}C_j\right]} \tag{5.59}$$

It was noted that $(\gamma_1, \gamma_2$ and $\gamma_3)_{ij}$ used to calculate $\alpha_{ij,hyp}$ are constants for fixed instrumental operating conditions and can be stored for any application in any other analytical system. The following practical procedure was proposed by Tertian.

(a) a comparison-standard technique is recommended. As shown previously this obviates the necessity for recourse to the measurement of pure elements, which for many reasons is inexpedient. It is only necessary for the unknown and the standard to be more or less similar.

(b) If the unknown u is categorized and can be related to some definite reference r, the working equation is

$$C_{iu} = C_{ir} \frac{I_{iu}}{I_{ir}} \frac{1 + \varepsilon_{ir}}{1 + \varepsilon_{iu}} \frac{1 + \sum_j \alpha_{iju,hyp}C_{ju}}{1 + \sum_j \alpha_{ijr,hyp}C_{jr}} \tag{5.60}$$

There is a good chance that the $1 + \varepsilon_i$ correction terms are little different and can be cancelled and that α_{iju} can be equated to α_{ijr}; hence

$$C_{iu} = C_{ir} \frac{I_{iu}}{I_{ir}} \frac{1 + \sum_j \alpha_{ijr,hyp}C_{ju}}{1 + \sum_j \alpha_{ijr,hyp}C_{jr}} \tag{5.61}$$

which is equivalent to

$$C_{iu} = \left(\frac{C_i}{I_i \left[1 + \sum_j \alpha_{ij,hyp}C_j \right]} \right)_r I_{iu} \left[1 + \sum_j \alpha_{ijr,hyp}C_{ju} \right] \tag{5.62}$$

Applying the usual iterative procedure generally yields an excellent estimate of the unknown composition. If deemed necessary, new α_{iju} and $1 + \varepsilon_{iu}$ values corresponding to the $C_{iu}, C_{ju}, \ldots$ values obtained can be computed and used to generate a more accurate analysis of the unknown.

(c) If the unknown is poorly characterized, except qualitatively, a reference specimen is selected with the only condition that it contain all elements of interest in sufficient proportions. The calculations are unchanged. The estimated composition from Eqn (5.61) is not as valid as in (b), but it is sufficient to calculate new α_{ij} values and $1 + \varepsilon_i$ corresponding to the estimated value and to obtain a correct analysis in a second step, Eqn (5.60).

5.7.3 Numerical examples

Examples of the magnitude of cross-coefficient α_{ijk} and of the correction factor $1 + \varepsilon_i$ are shown in Tables 5.8–5.10, wherein the theoretical data were

Table 5.8   Example of third element effect when matrix element j and k are both absorbers. Both α_{ijk} and $1 + \varepsilon_i$ terms are negligible

Analytical context
Excitation source: W target tube, 45 kV (CP); see Table II.1.7
Instrument geometry: $\psi' = 63°$, $\psi'' = 33°$
System: Mn–Mg–Ti; i = Mn, j = Mg, k = Ti

	C_{Mn}	C_{Mg}	C_{Ti}	R_{Mn}	$[FP]^a$	$[m_{ij,bin}]^b$	Δ^c	α_{ijk}^d	$1 + \varepsilon_i^e$
1	0.10	0.90	0.00	0.1616	0.6188				
2	0.10	0.00	0.90	0.0369	2.7100				
3	0.10	0.60	0.30	0.0760	1.3164	1.3162	0.0002	0.001	1.000
4	0.10	0.30	0.60	0.0497	2.0136	2.0135	0.0001	0.001	1.000
5	0.20	0.80	0.00	0.3005	0.6656				
6	0.20	0.00	0.80	0.0791	2.5284				
7	0.20	0.60	0.20	0.1767	1.1318	1.1310	0.0008	0.007	0.999
8	0.20	0.20	0.60	0.0970	2.0621	2.0618	0.0003	0.003	1.000
9	0.50	0.50	0.00	0.6265	0.7981				
10	0.50	0.00	0.50	0.2536	1.9716				
11	0.50	0.40	0.10	0.4831	1.0350	1.0328	0.0022	0.055	0.998
12	0.50	0.10	0.40	0.2877	1.7380	1.7367	0.0013	0.033	0.999

^{a} [FP] = correction term calculated from fundamental parameters.
^{b} $[m_{ij,bin}]$ = correction term calculated using $m_{ij,bin}$ coefficients.
^{c} Δ = [FP] − $[m_{ij,bin}]$.
^{d} $\alpha_{ijk} = \Delta/(C_j C_k)$.
^{e} $1 + \varepsilon_i = [m_{ij,bin}]/[FP]$.

generated for the same analytical context as in the previous tables in this chapter. Data for the six binaries may be used directly to compute values of the $\alpha_{ij,hyp}$ coefficients, which are then used to compute the correction terms for the six three element specimens. Equation (5.56) is then used to compute α_{ijk} values.

Details relating to the computations for specimens 5–8 in Table 5.9 are as follows.

Substitution in Eqn (5.53)

Specimen 5 [FP] = 0.20/0.07910 = 2.5284

Specimen 6 [FP] = 0.20/0.28559 = 0.7003

Specimen 7 [FP] = 0.20/0.09834 = 2.0337

Specimen 8 [FP] = 0.20/0.17879 = 1.1186

Substitution in Eqn (5.43)

Specimen 5 $m_{ij,bin} = \dfrac{0.2 - 0.07910}{0.8 \times 0.07910} = 1.9106$

Table 5.9 Example of third element effect when one matrix element is a strong absorber and the other is a strong enhancer. $1 + \varepsilon_i$ and α_{ijk} terms are significant

Analytical context
Excitation source: W target tube, 45 kV (CP); see Table II.1.7
Instrument geometry: $\psi' = 63°$, $\psi'' = 33°$
System: Mn–Ti–Co; $i = Mn$, $j = Ti$, $k = Co$

	C_{Mn}	C_{Ti}	C_{Co}	R_{Mn}	$[FP]^a$	$[m_{ij,bin}]^b$	Δ^c	$\alpha_{ijk}^{\ d}$	$1 + \varepsilon_i^{\ e}$
1	0.10	0.90	0.00	0.0369	2.7100				
2	0.10	0.00	0.90	0.1599	0.6253				
3	0.10	0.60	0.30	0.0510	1.9596	2.0156	−0.0560	−0.311	1.029
4	0.10	0.30	0.60	0.0793	1.2619	1.3204	−0.0585	−0.325	1.046
5	0.20	0.80	0.00	0.0791	2.5284				
6	0.20	0.00	0.80	0.2856	0.7003				
7	0.20	0.60	0.20	0.0983	2.0337	2.0714	−0.0377	−0.314	1.019
8	0.20	0.20	0.60	0.1788	1.1186	1.1574	−0.0388	−0.323	1.035
9	0.50	0.50	0.00	0.2536	1.9716				
10	0.50	0.00	0.50	0.5804	0.8615				
11	0.50	0.40	0.10	0.2876	1.7387	1.7494	−0.0107	−0.268	1.006
12	0.50	0.10	0.40	0.4658	1.0734	1.0835	−0.0101	−0.253	1.009
Avg								−0.299	1.024

Footnotes $a–e$ have the same meaning as in Table 5.8.

Specimen 6 $m_{ik,bin} = \dfrac{0.2 - 0.28559}{0.8 \times 0.28559} = -0.3746$

Substitution in Eqn (5.54)

Specimen 7 $[m_{ij,bin}] = [1 + (1.9106 \times 0.6) + (-0.3746 \times 0.2)]$
$= 2.0714$

Specimen 8 $[m_{ij,bin}] = [1 + (1.9106 \times 0.2) + (-0.3746 \times 0.6)]$
$= 1.1574$

Substitution in Eqn (5.55)

Specimen 7 $\Delta = 2.0337 - 2.0714 = -0.0377$

Specimen 8 $\Delta = 1.1186 - 1.1574 = -0.0388$

Substitution in Eqn (5.56)

Specimen 7 $\alpha_{ijk} = -0.0377/0.12 = -0.314$

Specimen 8 $\alpha_{ijk} = -0.0382/0.12 = -0.323$

Substitution in Eqn (5.59)

Table 5.10   Example of third element effect when matrix element j and k enhance
analyte i. $1 + \varepsilon_i$ and α_{ijk} terms are significant

Analytical context
Excitation source: W target tube, 45 kV (CP); see Table II.1.7
Instrument geometry: $\psi' = 63°$, $\psi'' = 33°$
System: Mn–Co–Zn; i = Mn, j = Co, k = Zn

	C_{Mn}	C_{Co}	C_{Zn}	R_{Mn}	$[FP]^a$	$[m_{ij,bin}]^b$	Δ^c	α_{ijk}^d	$1 + \varepsilon_i^e$
1	0.10	0.90	0.00	0.1599	0.6253				
2	0.10	0.00	0.90	0.1294	0.7728				
3	0.10	0.60	0.30	0.1428	0.7003	0.6746	0.0257	0.143	0.963
4	0.10	0.30	0.60	0.1320	0.7576	0.7237	0.0339	0.188	0.955
5	0.20	0.80	0.00	0.2856	0.7003				
6	0.20	0.00	0.80	0.2459	0.8133				
7	0.20	0.60	0.20	0.2678	0.7468	0.7286	0.0182	0.152	0.976
8	0.20	0.20	0.60	0.2469	0.8100	0.7851	0.0249	0.207	0.969
9	0.50	0.50	0.00	0.5804	0.8615				
10	0.50	0.00	0.50	0.5508	0.9078				
11	0.50	0.40	0.10	0.5693	0.8783	0.8707	0.0076	0.190	0.991
12	0.50	0.10	0.40	0.5511	0.9073	0.8985	0.0088	0.220	0.990
Avg.								0.183	0.974

Footnotes $a-e$ having the meanings given in Table 5.8

Specimen 6 $1 + \varepsilon_i = 2.0714/2.0337 = 1.018$

Specimen 7 $1 + \varepsilon_i = 1.1574/1.1186 = 1.035$

Substitution in Eqn (5.58)

Specimen 6 $1 + \varepsilon_i = \dfrac{(0.6/0.0791) + (0.2/0.2856)}{(0.8/0.0983)} = 1.019$

Specimen 7 $1 + \varepsilon_i = \dfrac{(0.2/0.0791) + (0.6/0.2856)}{0.08/0.1788} = 1.035$

A limited comparison of the error introduced if: (a) the term for third
element effects is omitted from the calculation of the matrix effect correction;
(b) an average value for α_{ijk} is used to compensate for third element effects
using the COLA algorithm, Eqns (5.48) and (5.50); or (c) an average value for
$1 + \varepsilon_i$ is used to compensate for third element effects using the Tertian
algorithm, Eqn (5.57); is presented in Table 5.11. The data relate to the
calculation of C_{Mn} for the analytical contexts of Tables 5.9 and 5.10.

Table 5.11 Comparison of recalculated Mn concentrations

System C_{Mn}	C_{Mn} (recalculated)					
	Mn–Ti–Co (Table 5.9)			Mn–Co–Zn (Table 5.10)		
3	0.10	0.1028^a	0.1001^b	0.1004^c	0.0963^a	0.1010^b	0.0989^c
4	0.10	0.1047	0.1005	0.1022	0.0955	0.0999	0.0980
7	0.20	0.2036	0.2001	0.1988	0.1950	0.2010	0.2002
8	0.20	0.2068	0.2005	0.2020	0.1939	0.1993	0.1991
11	0.50	0.5032	0.4997	0.4914	0.4957	0.4998	0.5089
12	0.50	0.5047	0.4992	0.4929	0.4952	0.4992	0.5084

^{a} Recalculated using $\alpha_{ij,hyp}$ coefficients only.
^{b} Recalculated using $\alpha_{ij,hyp}$ and averaged α_{ijk} coefficients.
^{c} Recalculated using $\alpha_{ij,hyp}$ and averaged $(1 + \varepsilon_i)$ terms.

Details of the calculations of C_{Mn} relating to Specimen 7 (system Mn–Ti–Co) in Table 5.9 are as follows:
Substitution in Eqn (5.46)

$$\text{Specimen 7 } C_i^a = 0.0983\,[1 + (1.9106 \times 0.6) + (-0.3746 \times 0.2)]$$
$$= 0.0983\,[2.0714]$$
$$= 0.2036$$

Substitution in Eqn (5.48)

$$C_i^b = 0.0983\,[1 + \{1.9106 + (-0.299 \times 0.2)\} \times 0.6$$
$$+ (-0.3746 \times 0.2)]$$
$$= 0.0983\,[2.0356]$$
$$= 0.2001$$

or

$$C_i^b = 0.0983\,[1 + (1.9106 \times 0.6)$$
$$+ \{-0.3746 + (-0.298 \times 0.6)\} \times 0.2]$$
$$= 0.0983\,[2.0356]$$
$$= 0.2001$$

Substitution in Eqn (5.57)

$$C_i^c = \frac{0.0983}{1.024}\,[2.0714]$$
$$= 0.1988$$

5.8 DISCUSSION

There can be little doubt that the evolution of the concept of influence coefficients and particularly alpha coefficients has been such that any comparison between the entity labelled α_{ij} in this chapter and the influence coefficients defined by Beattie and Brissey (1954) or by Lachance and Traill (1966) would be barely recognizable. The two factors mainly responsible for this are:

- the formulation of a fundamental parameters approach by Criss and Birks (1968) which made it possible to evaluate models in the light of the fundamental principles of X-ray physics, and
- the spectacular progress in computer technology which has been such that computations that were considered insurmountable in the early 1960s became current practice in the late 1970s.

In hindsight, it is not so much that the models or concepts first proposed were wrong but that they were incomplete, hence more limited than current models. Thus, as a result of a revived interest in the theoretical aspects of X-ray fluorescence emission, alpha coefficient models, although based on approximations, have been gradually given a sounder link to X-ray fluorescence emission theory.

The state of the art regarding influence coefficients in the mid 1960s was such that three major barriers had to be overcome.

(a) Derivations of expressions from fundamental principles were all limited to dealing with absorption effects only. Enhancement was not taken into consideration and yet, in practice, enhancement is involved in the majority of analytical contexts.

(b) The derivations were based on the assumption that the incident excitation radiation was monochromatic. In practice, the excitation source is generally an X-ray tube and hence polychromatic.

(c) The definition of the influence coefficients was envisaged in terms of two element systems only.

5.8.1 The equivalent wavelength concept

To avoid having to deal with the integral over the effective wavelength range λ_{min} to λ_{abs} for polychromatic incident radiation, analysts were prompted to seek and define a single wavelength that is representative of the whole effective polychromatic excitation source. Given that emitted primary fluorescence intensities vary as a function of specimen composition, a quantitative definition of the term 'equivalent wavelength' cannot be expressed in terms of I_i. The solution lies in defining 'equivalent' in terms of relative intensity R_i. Simply stated, the equivalent wavelength is that specific incident wavelength which would yield the same relative intensity as that for the same specimens irradiated with a polychromatic beam. It has been shown (Lachance, 1970; Tertian, 1973; and Vié le Sage and Tertian, 1976) that, even if the analytical context is limited to absorption only, in practice one should consider an infinity of such equivalent wavelengths. Although they all fall within a fairly narrow band, the band shifts as a function of instrument geometry and specimen composition. This phenomenon, combined with the fact that the concept breaks down if enhancement is involved, led Tertian and Claisse (1982) to conclude that equivalent wavelengths can seldom be calculated in practice.

5.8.2 Variation of α_{ij} with specimen composition

As shown by Claisse and Quintin (1967), even in the very limited context of binary systems in which absorption is the only matrix effect, the coefficients in Eqn (5.1) vary as a function of specimen composition if the excitation is polychromatic. Not only is this also the case if enhancement is present but, in

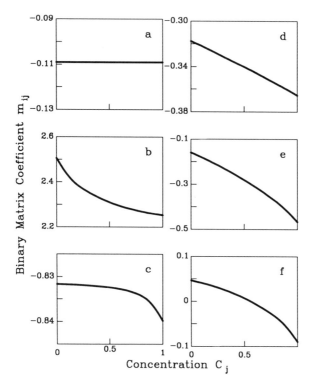

Figure 5.4 Illustration of the various shapes of plots of m_{ij} versus C_j in binary systems (see Section 5.8.2 for identifications)

the main, the variations are relatively even larger. Typical curves of the variation of the coefficient $m_{ij,bin}$ as a function of C_j are illustrated in Figure 5.4 where:

(a) is representative of $m_{MnCr,bin}$ in Table 5.1, i.e. element Z_j equal to $Z_i - 1$ in that region of the Periodic Table;

(b) is representative of strong matrix absorption, $m_{MnV,bin}$ in Table 5.1;

(c) is representative of a low Z matrix element, i.e. the effective absorption by the matrix element is less than that of the analyte, e.g. $Z_j \cong 5-8$, $Z_i \cong 24-50$.

(d) is representative of $m_{CoSi,bin}$ in Table 5.2, i.e. an intermediate absorption matrix element. The strong absorption curve (c) is gradually transformed to curve (f), which is representative of low absorption. For some intermediate absorption elements the curve is almost linear;

(e) is representative of strong enhancement, $m_{MnCo,bin}$ in Table 5.1;

(f) is representative of $m_{MnFe,bin}$ in Table 5.1, i.e. enhancement by the $K-M_3$ line only.

Figure 5.4 (a–f) also provides a clear illustration of how the three coefficients α_1, α_2 and α_3 combine to approximate very closely the curvature in all cases, depending on the algebraic sign and magnitude of the coefficients as shown below.

	α_1	α_2	α_3
(a)	negative	zero	zero
(b)	positive	negative	negative
(c)	negative	negative	positive
(d)	negative	negative	positive
(e)	negative	negative	positive
(f)	positive	negative	positive

Thus the three α coefficients may be visualized as describing the following:

- α_1 is the value of $m_{ij,bin}$ in the context where a minute amount of element j is substituted for an equal amount of element i in a specimen of pure analyte i, i.e. the $m_{ij,bin}$ value at the origin in Figure 5.4(a–f), the *initial* value of α_{ij};
- α_2 (the difference between the two limiting values of $m_{ij,bin}$) represents the *maximum* variation of α_{ij};
- α_3 controls the rate and mode of curvature, i.e. nil, concave (−) or convex (+).

5.8.3 Limited theoretical justification of alpha coefficient models

The concern of analysts regarding the theoretical exactness or the theoretical justification of influence coefficient models is ever present. For example, given the expression

$$C_i = R_i \left[1 + \sum_j (\text{influence coefficient})_{ij} C_j \right] \tag{5.63}$$

the question is: 'can that model be derived and the influence coefficients defined explicitly from fundamental X-ray fluorescence emission theory'? It was shown in the previous chapter that the answer is yes, *provided that* it is recognized that the fundamental influence coefficients m_{ij}, m_{ik}, ... are variables that depend on specimen composition; i.e.

$$m_{ij} = \frac{M_{ij}}{P_i + S_i} \tag{5.64}$$

where both M_{ij} and $P_i + S_i$ are unquestionably functions of specimen composition. On the other hand, Tertian (1987) and Lachance (1988) have shown that a theoretical justification of the Claisse–Quintin and COLA models is not feasible. This does not deter one from using those models in practice. The concept underlying α influence coefficients is not to emulate m_{ij} coefficients but to provide a flexible alternative to calculate a valid matrix effect correction term. It therefore follows that the definition of a given influence coefficient in the fundamental influence domain is different from and incompatible with that in the α coefficient domain.

Consider for example the definition of the coefficient α_{jj} in the Claisse–Quintin model. Rousseau (1984a) derived the following expression explicitly from fundamental theory:

$$\alpha_{jj} = - \frac{a_{ijp} e_{ijp}}{1 + \sum_j e_{ijp} C_j} \tag{5.65}$$

which leads to the conclusion that α_{jj} is equal to zero if there is no enhancement. Recalling Eqn (5.21), the α_{jj} coefficient in the α coefficient domain is defined quite differently, namely

$$\alpha_{jj} = \frac{a_{ij2} - a_{ij1}}{C_{j2} - C_{j1}} \tag{5.66}$$

where a_{ij} has been substituted for m_{ij} to indicate clearly that enhancement is not involved in the case under consideration. Thus, in the absence of enhancement, α_{jj} can only be equal to zero if a_{ij1} and a_{ij2} are equal, which implies *monochromatic* excitation. It is well documented that binary a_{ij} coefficients are very seldom constants (see, for example, m_{MnCr} and m_{CoFe} arrays in Tables 5.1 and 5.2); this is a clear indication that Eqns (5.65) and (5.66) refer to two entirely different entities.

A similar reasoning is applicable in the case of the COLA model, where both the $\alpha_{ij,hyp}$ and α_{ijk} coefficients are based on approximating patterns observed in data generated from fundamental parameters expressions, but nonetheless the coefficients remain approximations. The α domain provides a useful intermediate between the explicitly defined fundamental m_{ij} or a_{ijp} and e_{ijp} influence coefficients and those influence coefficients that are strictly empirical. Thus the 'advanced' α models rely on simultaneously adjusting the value of α_{ij} based on the latest estimate of composition, and getting a better estimate of composition based on the latest α_{ij} values as part of the iteration process. Alternatively, the models can serve to generate an estimate of composition for subsequent treatment by any fundamental algorithm.

5.9 SUMMARY

Because m_{ij} and α_{ij} are used in identical algorithms, it is important to demarcate clearly the domain of fundamental influence coefficients m_{ij} and that of semi-theoretical influence coefficients α_{ij}. Both m_{ij} and α_{ij} coefficients are variable as a function of specimen composition. Only in the very limited analytical contexts in which unknowns and reference materials can be well categorized in restricted concentration ranges can α_{ij} be considered as constant in the expression based on Eqn (5.48).

$$C_{iu} = K_{(i)}I_{iu}\left[1 + \sum_j \alpha_{ij}C_{ju}\right]$$

$$= R_{iu}\left[1 + \sum_j \alpha_{ij}C_{ju}\right]$$

Then, the recommended way to obtain α_{ij} is to equate it to m_{ijr}, (i.e. to treat m_{ijr} as an arbitrary constant), where the reference may be either an average composition or the composition of any real typical reference on hand. Alternatively, one can define the variable α_{ij} for use in the modified Claisse–Quintin model

$$\alpha_{ij} = \alpha_j + \alpha_{jj}C_M + \sum_k \alpha_{ijk}C_k$$

$$= \alpha_{ij,lin} + \sum_k \alpha_{ijk}C_k$$

or preferably, for application to even wider ranges of concentrations, define α_{ij} for use in the COLA algorithm

$$\alpha_{ij} = \alpha_{ij,hyp} + \sum_k \alpha_{ijk}C_k$$

Conversely, the third element effect may be compensated for by the introduction of the $1 + \varepsilon_i$ correction term in the expression

$$C_{iu} = C_{ir} \frac{I_{iu}}{I_{ir}} \frac{1 + \varepsilon_{ir}}{1 + \varepsilon_{iu}} \frac{1 + \sum_{j} \alpha_{iju,hyp}C_{ju}}{1 + \sum_{j} \alpha_{ijr,hyp}C_{jr}}$$

If the unknown is categorized and can be related to a similar standard, the simpler model

$$C_{iu} = C_{ir} \frac{I_{iu}}{I_{ir}} \frac{1 + \sum_{j} \alpha_{ijr}C_{ju}}{1 + \sum_{j} \alpha_{ijr}C_{jr}}$$

with application of the usual iterative process should generally yield an excellent estimate of composition. If the unknown is only known qualitatively, then the above equation yields an estimate of composition which enables the calculation of the appropriate $1 + \varepsilon_i$ value corresponding to this estimate for use in Eqn (5.60).

6 EMPIRICAL INFLUENCE COEFFICIENTS

In contrast to the two previous chapters, in which influence coefficients were based on theoretical principles, this chapter examines the concepts underlying influence coefficient models wherein the values of the coefficients are the resultant of regression techniques. Generally, data from a large number of reference materials are treated by a method of least squares in order to generate a 'best fit' set of values for the coefficients. As stated previously, in order to avoid confusion created by the fact that identical models are sometimes used in the fundamental, alpha and empirical influence coefficient domains, the symbol r_{ij} will be used exclusively to represent influence coefficients obtained by regression methods. It is also noted that binary specimens present a special case because by definition, for any given binary specimen, the three influence coefficients m_{ij}, α_{ij} and r_{ij} are equal.

6.1 GENERAL CONSIDERATIONS

From the very first, X-ray fluorescence analysts realized that concentration in unknowns could not be obtained by simply measuring the intensity of a characteristic line in a specimen of unknown composition and comparing it directly with the measured intensity of the same line in a reference material of known composition. It was also realized that one could not always circumvent the non-linear relationship between concentration and intensity by simply using higher order expressions of the type

$$C_i = a + bI_i + cI_i^2 + \cdots \tag{6.1}$$

and determining the coefficients a, b, c, ... by regression of a large amount of experimental data. The computation of a correction term based on empirical influence coefficients offered a possible solution.

6.2 CONCEPT AND DOMAIN OF EMPIRICAL INFLUENCE COEFFICIENTS

As mentioned previously, it was in the limited context of a monochromatic incident source that the theoretical aspects of primary fluorescence emission were developed by von Hamos (1945). This was also the case for Gillam and Heal (1952), who developed an expression for secondary fluorescence emission. The problems that led analysts to contemplate the use of empirical influence coefficients to correct for the combined effects of absorption and enhancement for polychromatic excitation were outlined by Sherman (1954):

- the relationship becomes much more complex if the incident beam is polychromatic;
- the computational labour involved makes the process too complicated for general use;
- additional terms must be added if enhancement is present and the treatment becomes much more complicated in a mathematical sense.

These point to the conclusion that:

- for limited ranges in composition, a close linear approximation may be made using a system of equations involving influence coefficients computed from observations on samples of known composition;
- applications depend wholly on the particular elements involved and the ranges of their concentrations.

Thus one may visualize a correction term for matrix effects given intensity data for two reference materials. Firstly, if an apparent concentration is obtained for material 1

$$C_{i(app)} = \left(\frac{C_i}{I_i}\right)_{r1} I_i \tag{6.2}$$

it is likely that the ratio $C_i/C_{i(app)}$ will be either smaller or larger than unity owing to matrix effects. Thus one must visualize a correction term involving the summation of the products of matrix element concentrations times their respective influence coefficients, i.e.

$$\left(\frac{C_i}{C_{i(app)}}\right)_{r1} = [\cdots + C_j r_{ij} + C_k r_{ik} + \cdots] \tag{6.3}$$

where the influence coefficients r_{ij}, r_{ik}, ... compensate for the matrix effects of elements j, k, ... on analyte i, whether the dominant effect is absorption or enhancement. The fact that the magnitude of r_{ij} can be computed strictly from experimentally measured intensities by regression methods implies that the spectral distribution of the incident beam, instrumental geometry, mass attenuation coefficients etc. do not enter into the calculations and need not be known. Thus, the two criteria that demarcate empirical influence coefficients are:

- they are based strictly on experimental data;
- fundamental and instrumental parameters do not figure in the computations.

This is in sharp contrast to m_{ij} and α_{ij} coefficients, which are computed directly or indirectly from fundamental parameters expressions.

6.3* CRISS–BIRKS AND LACHANCE–TRAILL FORMALISMS

Based on the knowledge that, for the restricted analytical context of a monochromatic incident excitation source and the absence of enhancement, intensities can be converted to concentrations using the expression

$$C_i = K_i I_i [C_i \mu_i^* + C_j \mu_j^* + \cdots] \tag{6.4}$$

Birks (1959) noted that:

- similar equations will result for each effective wavelength in the incident source but with different values for μ^*;
- it can be imagined intuitively that μ_i^* may include enhancement;
- μ^* may be expressed in terms of new coefficients r_{ij} representing the overall effect of element j on element i;
- unfortunately, the coefficients r_{ij} are not really constants;
- there are situations, however, where the r_{ij} may be considered constant to a first or second order approximation.

Hence, to the extent that a polychromatic incident source can be approximated by a singly equivalent wavelength, and that enhancement can be visualized as

Table 6.1 Comparison of empirical influence coefficients r_{ij} used for the determination of Cr, Fe and Ni in the systems Cr–Fe–Ni and Cr–Fe–Ni–Co–Mo[a]

i		Cr	Fe	Ni	Co	Mo
				j		
Cr	Test 1	1.559	0.516	0.670	0.971	1.731
	Test 2	1.801	1.608	2.616		
Fe	Test 1	2.419	1.023	0.850	1.143	1.286
	Test 2	3.391	1.997	2.215		
Ni	Test 1	15.428	−1.110	−0.990	−7.721	5.336
	Test 2	−1.050	3.651	5.611		

[a] Data taken from Criss and Birks (1968).

negative absorption (both sweeping approximations), then Eqn (6.4) may be written as

$$C_i = K_i I_i [C_i r_{ii} + C_j r_{ij} + C_k r_{ik} + \cdots] \qquad (6.5)$$

leading to

$$\frac{C_i}{R_i} = [C_i r_{ii} + C_j r_{ij} + C_k r_{ik} + \cdots] \qquad (6.6)$$

This algorithm was used by Criss and Birks (1968) to illustrate the incongruity of empirical influence coefficient methods. Values were obtained by regression of experimental data for the systems Cr–Fe–Co–Ni–Mo and Cr–Fe–Ni (see Table 6.1 for a comparison of coefficient values). As noted by Criss and Birks, the coefficients obtained in this way do not represent the effect of one element on another, as was hypothesized in the derivation of Eqn (6.6); rather, they are simply the best set of numbers to describe the intensities measured. This is quite evident when the coefficients are of opposite algebraic sign; e.g. when comparing r_{ij} values for identical systems but using different reference

Table 6.2 Comparison of chemical and X-ray analyses for four stainless steel specimens using the r_{ij} coefficients listed in Table 6.1[a]. Application of Eqn (6.5)

Alloy		Cr	Fe	Ni	Mn
			Content, wt. %		
303	Chem	17.2	71.2	8.7	1.3
	Test 1	17.8	73.8	8.4	
	Test 2	used as reference			
304	Chem	18.6	69.5	9.4	1.4
	Test 1	18.8	71.0	9.7	
	Test 2	19.0	71.7	9.3	
316	Chem	17.7	64.8	12.8	1.8
	Test 1	17.8	66.0	13.2	
	Test 2	used as reference			
321	Chem	17.8	68.2	10.8	1.6
	Test 1	18.5	71.7	10.9	
	Test 2	18.4	70.6	11.0	
347	Chem	17.7	67.9	10.7	1.6
	Test 1	18.5	71.7	11.7	
	Test 2	18.4	70.2	11.4	
430	Chem	17.5	81.3		
	Test 1	17.6	79.3		
	Test 2	17.5	81.4		

[a] Data taken from Criss and Birks (1968).

materials. However, it was demonstrated that quite different sets of coefficients led to approximately the same calculated concentrations as shown in Table 6.2.

Criss and Birks noted that if the reference standard used in defining relative intensity for each measurement is a pure element, then the derivation of Eqn (6.6) demands that $r_{ii} = 1$ for each element. It was suggested that this may not be true in practice, hence r_{ii} values were computed. Criss and Birks also noted that it is possible to compensate somewhat for uncertainties in measuring intensities by using more than the minimum number of standards.

If the Lachance–Traill algorithm is chosen as the model for the correction of matrix effects, then empirical influence coefficients can be computed from the expression

$$\left(\frac{C_i}{R_i} - 1\right)_r = C_{jr}r_{ij} + C_{kr}r_{ik} + \cdots \tag{6.7}$$

using a large suite of accurately analysed reference materials that includes specimens of the pure analytes. The coefficients may be obtained by standard techniques of regression analysis that require solving for r in expressions of the type

$$\sum_n C_j^2 r_{ij} + \sum_n C_j C_k r_{ik} + \cdots = \sum_n \left(C_j \frac{C_i}{R_i} - 1\right)$$

$$\sum_n C_j C_k r_{ij} + \sum_n C_k^2 r_{ik} + \cdots = \sum_n \left(C_k \frac{C_i}{R_i} - 1\right) \tag{6.8}$$

where n includes all reference specimens. If specimens of the pure analytes are not available, values for $I_{(i)}$, r_{ij}, $r_{ik}, \ldots$ may be computed by solving the set of linear equations

$$I_{i1} = C_{i1}I_{(i)} - I_{i1}C_{j1}r_{ij} - I_{i1}C_{k1}r_{ik} - \cdots$$
$$I_{i2} = C_{i2}I_{(i)} - I_{i2}C_{j2}r_{ij} - I_{i2}C_{k2}r_{ik} - \cdots \tag{6.9}$$
$$I_{i3} = C_{i3}I_{(i)} - I_{i3}C_{j3}r_{ij} - I_{i3}C_{k3}r_{ik} - \cdots$$

for $I_{(i)}$, r_{ij}, $r_{ik}, \ldots$ given concentrations and net measured intensities.

6.4* METHOD OF RASBERRY AND HEINRICH

Based on a study of an alloy system in which the matrix effects are severe (Cr–Fe–Ni, over a range of composition from 0 to 100%), Rasberry and Heinrich (1974) proposed an empirical method for the calibration of X-ray fluorescence analysis. A suite of 36 alloys of known composition plus three pure elements provided the relative intensities measured at 45 and 20 kV; 31 alloys were used for calibration, and the remaining five were treated as unknowns in order to evaluate the method.

It was observed that theory and experiment indicate that a plot of relative intensity versus concentration for binaries can be approximated adequately by a hyperbolic function when the matrix effect consists of absorption only:

$$R_i = \frac{C_i}{1 + r_{ij}(1 - C_i)} \qquad (6.10)$$

where r_{ij} is a constant. It was noted that, in the case of binary alloys that did not involve enhancement, the hyperbolic curves could be studied more incisively if C_i/R_i is plotted as a function of C_j or $1 - C_i$; hyperbolas are then represented as straight lines with slope r_{ij}. However, in the three cases where enhancement was predominant, namely analyte Cr in Fe–Cr, analyte Cr in Cr–Ni, and analyte Fe in Fe–Ni, it was observed that R versus C curves were nonhyperbolic and that a plot of C/R versus C falls below the unity level and *shows marked curvature*. This led the authors to propose the equation

$$C_i = R_i \left[1 + \sum_j r_{ij}C_j + \sum_k \frac{r_{ik}}{1 + C_i} C_k \right] \qquad (6.11)$$

where the coefficients r_{ij} are used for elements j, for which the more significant effect on analyte i is absorption. The coefficients r_{ik} are used for elements k for which the predominant effect on analyte i is secondary fluorescence. It was noted that, in general, analysts can easily predict which elements will cause fluorescence and thus require use of r_{ik} coefficients; alternatively, this prediction can be made by directly examining a graph of (C/R) as a function of C. It was also noted that the divisor $(1 + C_i)$ in the fluorescence term was selected on the basis that the fluorescence curve had a slope which was about twice as large when C_i approached zero as when it was near unity. It was noted that in some cases it may be necessary to apply within one equation the equivalent of both r_{ij} and r_{ik} coefficients to treat adequately the effects caused by a given element; i.e.

$$r_{ij} = r_{ij,a} + \frac{r_{ij,e}}{1 + C_i} \qquad (6.12)$$

where $r_{ij,a}$ and $r_{ij,e}$ relate to absorption and enhancement, respectively.

Given the fact that $(n - 1)$ coefficients are required for the determination of each element in a specimen containing n elements, the minimum number of coefficients required for calibration of the complete composition is $n(n - 1)$. An alternative method to plotting binary data consists in solving a set of simultaneous equations for each element using multielement reference materials. Given a suite of C_i and R_i data, the only unknown variables are the empirical influence coefficients. The values of the six influence coefficients for the system Cr–Fe–Ni are listed in Table 6.3. Compared with the equivalent coefficients listed in Table 6.1, test 2, the coefficients determined using Eqn (6.11) are all positive when absorption is the only matrix effect, and all negative when the

Table 6.3 Empirical influence coefficients r_{ij} used for the determination of Cr, Fe and Ni in the Cr–Fe–Ni system[a]: (a) from graphs, and (b) by equations

	r_{ij}		r_{ik}	
Analyte	r_{CrFe}		r_{CrNi}	
Cr (a)	− 0.46		− 0.27	
Cr (b)	− 0.419		− 0.234	
	r_{FeCr}		r_{FeNi}	
Fe (a)	2.10		− 0.47	
Fe (b)	2.099		− 0.465	
	r_{NiCr}	r_{NiFe}		
Ni (a)	1.20	1.71		
Ni (b)	1.226	1.711		

[a] Data taken from Rasberry and Heinrich (1974).

predominant effect is enhancement. Data relating to the five reference materials which were measured and treated as unknowns in order to evaluate the accuracy of the method are listed in Table 6.4.

In order to evaluate the effect of operating voltage on the magnitude of the influence coefficients, intensities were measured using a tungsten tube operated at 45 and 20 kV. It was noted that, at the lower operating voltage, the r_{ik}

Table 6.4 Comparison of chemical and X-ray analyses for five alloy specimens using the r_{ij} and r_{ik} coefficients listed in Table 6.3[a]. Application of Eqn (6.11)

	Content, wt. %					
	Cr		Fe		Ni	
Sample	Chem	XRF	Chem	XRF	Chem	XRF
5054	25.77	25.22	72.50	72.29	0.15	0.15
5202	21.30	20.90	63.03	62.77	14.80	15.07
5364	27.84	26.75	47.21	47.50	23.57	24.09
3987	0.00	0.00	34.31	33.78	65.52	65.12
1188	15.40	14.21	6.60	6.55	72.65	72.36

[a] Data taken from Rasberry and Heinrich (1974).

coefficients (enhancement effect) were essentially unchanged, whereas there was an appreciable decrease in the r_{ij} coefficients (absorption effect), as expected.

6.5* METHOD OF LUCAS-TOOTH AND PRICE

The mathematical model proposed by Lucas-Tooth and Price (1961) is inherently different from all other influence coefficients algorithms examined so far in that the correction term involves the summation of the products of empirical influence coefficients and *intensities* rather than *concentrations*. Fully aware of the theoretical aspects of X-ray fluorescence emission, the authors proposed to make the following sweeping assumptions and demonstrated that the practical results indicate their merit.

(1) In addition to absorption effects, enhancement also may occur. This is regarded as negative absorption and assumed to obey the same laws.

(2) The absorption of element i by element j is linearly proportional to C_j. Thus, in terms of intensity, the actual departure from the linear curve is assumed to be proportional to $C_j I_i$.

It was pointed out that the second assumption is not unreasonable given that, over small ranges, twice as much C_j might well be expected to have twice as great an effect on the emitted intensity I_i. Based on the above assumptions, the concentration in a multicomponent system is deemed to be given by

$$C_i = r_0 + r_i I_i \left[1 + \sum_j r_{ij} C_j \right] \tag{6.13}$$

This expression was considered very inconvenient to use, as the concentrations of the other elements appear in the right hand side of the equation. As they are of course related by similar equations, the solution of such equations would be involved. Rather than transform the above equations into an expansion involving higher powers, the authors considered that it was not unreasonable to write the expression

$$C_i = r_0 + I_i [r_i + r_{ii} I_i + r_{ij} I_j + r_{ik} I_k + \cdots] \tag{6.14}$$

It was pointed out that, to get the best possible fit, it is important to choose $r_0, r_i, r_{ii}, r_{ij}, \ldots$ so that $\Sigma \Delta_i^2$ is minimum, where

$$\Delta_i = \frac{C_i - C_{ix}}{I_i} \tag{6.15}$$

where C_{ix} is the concentration calculated by Eqn (6.14). In order to test the various assumptions made in deriving Eqn (6.14), a suite of alloys (brasses, bronzes and gun-metals, and pure copper) was prepared and Cu intensities were measured. These alloys were considered to represent a severe test for the

Table 6.5 Numerical example of empirical influence coefficients used for determination of Cu for the system Cu–Zn–Sn[a]

Analyte: Cu
Specimen No. 15
$C_{Cu} = 84.00\%$; $C_{Zn} = 4.99\%$; $C_{Sn} = 11.01\%$

Coefficients	× Intensities	= Product
r_0	0.78	= 0.78
r_{Cu}	$2.798 \times 10^{-3} \times 19261$	= 53.892
r_{CuCu}	$4.84 \times 10^{-8} \times 19261 \times 19261$ =	17.956
r_{CuZn}	$-0.64 \times 10^{-9} \times 19261 \times 7151$ =	-0.088
r_{CuSn}	$1.32 \times 10^{-8} \times 19261 \times 44965$ =	11.432
	Σ =	83.96

[a] Data taken from Lucas-Tooth and Price (1961).

Table 6.6 Comparison of chemical and X-ray analyses for 15 copper alloys using the influence coefficients listed in Table 6.5[a]

	Cu, %	Zn, %	Sn, %	I_{Cu}	I_{Zn}	I_{Sn}	Cu, %[b]	Cu %[c]
1	100.0	0	0	24724	349	800	100.0	99.82
2	96.62	0	3.34	23323	334	14325	94.33	96.78
3	94.54	0	5.42	22426	334	22470	90.71	94.53
4	92.62	0	7.35	21667	313	30490	87.64	92.85
5	90.61	0	9.36	20870	295	38330	84.41	90.82
6	65.27	34.73	0	17740	50795	790	71.75	65.27
7	62.95	37.05	0	17247	53618	775	69.76	63.03
8	67.61	32.39	0	18244	47799	785	73.79	67.58
9	69.98	30.02	0	18750	44680	840	75.84	69.94
10	87.88	5.01	7.11	20771	7778	29030	84.01	87.64
11	87.93	2.96	9.11	20376	4552	37175	82.43	87.83
12	87.93	0.83	11.24	20031	1659	45015	81.02	87.94
13	89.85	3.06	7.09	21220	4845	28515	85.83	89.88
14	86.73	1.99	11.28	19738	3028	45215	79.83	86.60
15	84.00	4.99	11.01	19261	7151	44965	77.90	83.97

[a] Data taken from Lucas-Tooth and Pyne (1964).
[b] Calculated by direct proportionality to pure Cu.
[c] Calculated from Eqn (6.14).

evaluation of the method as, in the worst case, a 7% shift in copper is required to compensate for absorption and enhancement effects. Table 6.5 lists data relating to the application of Eqn (6.14) to one of the alloys, and Table 6.6 provides an evaluation of the method for the whole suite of alloys.

6.6* METHOD OF LUCAS-TOOTH AND PYNE

Lucas-Tooth and Pyne (1964) found that dealing with corrections based on $I_j, I_k \ldots$ as proposed in Eqn (6.14) was inconvenient for two reasons:

(a) measured intensity values are usually large numbers which cause the various constants to have unwieldy powers of ten associated with them;

(b) not only does the same specimen not give exactly the same intensity on different spectrometers, but subsequent readings on the same specimen on the same spectrometer may be different owing to drift; hence the necessity and inconvenience of recalibration for all the coefficients.

Lucas-Tooth and Pyne also noted that it is much more convenient to express the correction for matrix effects in terms of the product of empirical influence coefficients and *apparent concentrations*. Since the intensity is roughly proportional to concentration, the authors proposed to define an apparent concentration $C_{i,app}$

$$C_{i,app} = K_i I_i \approx C_i \tag{6.16}$$

Thus, by adjustment of K_i (e.g., slope of linear least squares fit of plot C_i versus I_i), any specimen could be made to yield an equivalent identical apparent concentration at any time. The equivalent expression to Eqn (6.14) then becomes

$$C_i = r_0 + C_{i,app}[r_i + r_{ii}C_{i,app} + r_{ij}C_{j,app} + \cdots] \tag{6.17}$$

It was noted that there is a certain redundancy in this last equation given that the sum of all the constituents equals 1. Thus, any one element that is particularly intractable or undesirable can be omitted in the correction term; i.e. if one assumes that $r_{in} = 0$, then all the other coefficients can be readjusted to suit the data. The application of Eqn (6.17) for the determination of a major constituent was illustrated in the analytical context of chromium as analyte (7–32%) in high-alloy steels. Other constituents were present in the following ranges: Ni 0.2–24%; Mn 0.35–10%; Al 0.07–5.0%; Mo 0.1–4.4%; Cu 1.3–3.6%; W 0.25–3.0%; Si, 0.2–2.4%; lesser concentrations of Ti, V and Nb were also present, the balance being essentially iron. The procedure used was as follows:

- iron was chosen as the eliminated element;
- I_{Cr} was plotted versus C_{Cr}, showing wide scatter;
- similar plots for the other constituents also showed some scatter (Ni and Mn), the remaining elements less so;
- apparent concentrations were obtained for all elements by linear least squares fit;
- influence coefficients were determined by least squares fitting.

Table 6.7 illustrates the process involved in a typical calculation, and Table 6.8

Table 6.7 Numerical example of the use of empirical influence coefficients for the determination of Cr in high-temperature alloy steels.[a] Specimen 7542; $C_{Cr(chem)} = 16.93\%$. Application of Eqn (6.17)[b]

Coefficient	$\times C_{j,app} \times C_{i,app} =$		
r_0	-0.217		$= -0.22$
r_{Cr}	0.738	$\times 14.46$	$= 10.68$
r_{CrCr}	0.014	$\times 14.46 \times 14.46 =$	2.93
r_{CrMo}	0.029	$\times \ 4.38 \times 14.46 =$	1.82
r_{CrNi}	0.003	$\times 19.87 \times 14.46 =$	0.86
r_{CrTi}	0.046	$\times \ 0.56 \times 14.46 =$	0.38
r_{CrCu}	$0.0033 \times$	$3.57 \times 14.46 =$	0.17
r_{CrMn}	$0.0046 \times$	$2.19 \times 14.46 =$	0.15
r_{CrSi}	$0.0071 \times$	$1.12 \times 14.46 =$	0.11
		$\Sigma \ \ =$	$16.9 \ \%$

[a] Data taken from Lucas-Tooth and Pyne (1964).
[b] Coefficients values rounded off, chemical values used for $C_{j,app}$.

Table 6.8 Comparison of chemical and X-ray analyses for 15 alloys using the influence coefficients listed in Table 6.7[a]

	Content, wt. %		
Sample	$C_{Cr(chem)}$	$C_{Cr,app}$	$C_{Cr,XRF}$
7401	7.43	8.45	7.39
7402	8.33	9.48	8.36
7407	11.83	12.87	11.87
7408	12.02	12.91	12.00
7417	16.70	16.63	16.68
7418	19.21	17.87	19.29
7426	15.82	15.76	15.87
7429	31.67	28.08	31.93
7506	14.87	14.18	15.00
7536	19.95	18.74	19.95
7541	18.28	15.96	18.27
7542	16.93	14.46	16.99
7543	25.74	23.24	25.77
7545	19.73	17.81	19.71
8230	15.35	14.40	15.28

[a] Data taken from Lucas-Tooth and Pyne (1964).

is an abridged listing (15 from a total of 60) of results obtained on some of the samples containing both chromium and matrix elements in very wide ranges of content.

The authors noted that the use of $C_{j,app}$, $C_{k,app}$, ... is confusing if those elements are absent in a particular sample; i.e. since backgrounds are not subtracted, $C_{j,app}$, $C_{k,app}$, etc. are not zero. This implies that the correction includes one or more elements that are not present in the specimen. It was concluded that this does not lead to errors, since the r_i for the particular determination will be, so to speak, expecting the addition of these constant terms.

6.7 THE JAPANESE INDUSTRIAL STANDARD METHOD

Concerned about the problems with obtaining constants for matrix correction models and the desirability of specifying X-ray intensities in a common unit, Ito *et al.* (1981) proposed the Japanese Industrial Standard (JIS) method as an alternative to the α-correction method of Eqn (5.1). The principle of the JIS method is as follows:

- the calibration curve of the binary system consisting of the analyte and the principal constituent element is assumed to be parabolic and is expressed as

$$C_{i,app} = a + bI_i + cI_i^2 \tag{6.18}$$

- this apparent or tentative analytical value is corrected using the expression

$$C_i = C_{i,app}\left[1 + \sum_j r_{ij}C_j\right] \tag{6.19}$$

where the summation does not include the analyte i or the major or principal constituent, of the binary system used in the calibration.

Combining Eqns (6.18) and (6.19) leads to

$$C_i = (a + bI_i + cI_i^2)\left[1 + \sum_j r_{ij}C_j\right] \tag{6.20}$$

in which correction for the absorption–enhancement effects of the analyte and the major constituent is included in the X-ray intensity factor $(a + bI_i + cI_i^2)$, i.e. the calibration, and the correction for the other matrix elements is included in the correction factor $[1 + \sum_j r_{ij}C_j]$.

For a comparison of the accuracy of the JIS expression, Eqn (6.20), and α methods, Ito *et al.* selected ferrous alloys; hence iron was the element selected for elimination, i.e., j $\neq$ Fe in

$$C_i = (a + bI_i + cI_i^2)\left[1 + \sum_j r_{ij}C_j\right]$$ (6.21)

More than 30 reference alloys were used for calibration and 27 different alloys were used for comparing the accuracy of both methods. X-ray intensities were measured using 14 spectrometers in 12 laboratories. The calibration and influence coefficients were *determined by multiple regression*. Precise analysis using one calibration curve over a wide range of concentrations proved difficult; hence it was necessary to divide the concentration range into smaller ranges. The authors concluded that the correction coefficients were very similar in both the JIS and the α models and found no difference between the models as to their accuracy.

6.8* DISCUSSION

Confronting empirical influence coefficient approaches with the strict tenets of fundamental X-ray fluorescence emission theory may be regarded as a non-issue, inasmuch as the proponents of empirical methods knowingly acknowledge and accept the fact that they are based strictly on establishing statistical correlation between experimental data. Briefly stated, whereas the aim of methods based on fundamental theory is to be *universal* in their application, the raison d'être of empirical methods is more modest, namely to provide valid compositional data for *limited and well defined* analytical contexts. A criterion for the adequacy of any method is whether it fulfils the task at hand. This does not free the analyst from being knowledgeable (possibly concerned) about the premises, concepts and implications underlying all methods. This aspect is all the more important when identical algorithms are used with fundamental influence coefficients in one instance and with empirical coefficients in another. Thus one should examine first those aspects of empirical methods that are of a general nature and inherent to this approach, and secondly those that can be considered as more dependent on the experimental procedure followed.

A severe limitation inherent in any influence coefficient model relates to whether, directly or indirectly, the influence coefficients are considered constants. That influence coefficients are not constants can be shown experimentally by carefully measuring intensities for a suite of accurately known binary systems, especially if strong enhancement is involved. Thus, any model that implies constant influence coefficients is limited in its application over wide concentration ranges, regardless of whether the coefficients are computed from theory or by regression. Conversely, the concentration ranges depend in part on whether the coefficients have a semblance of physical meaning on the one hand, or are totally different from what is predictable from basic principles on the other. The following numerical example, Table 6.9—using theoretical data, is presented to illustrate the former case. The analytical context chosen is identical to that of Table 5.7. Apparent Mn concentrations $C_{Mn(app)}$ were computed from fundamental parameters R_{Mn} expressions for 15 hypothetical specimens. The data for the six ternaries in (a) were treated as relating to reference materials, and Eqn (6.7) was used to compute r_{MnTi} and r_{MnCo} influence coefficients, 1.853 and -0.425, respectively. Columns headed m_{MnTi} and m_{MnCo} (included for comparison) illustrate the wide variation in the value of the fundamental influence coefficients in this context. The column headed Mn(calc) lists the Mn concentrations obtained from the expression

$$C_{Mn} = C_{Mn,app}[1 + (1.853 \times C_{Ti}) + (-0.425 \times C_{Co})] \tag{6.22}$$

It is observed that, even though the value for r_{MnTi} is barely within the range of the six m_{MnTi} values and r_{MnCo} is completely outside the range of m_{MnCo} values, the re-calculated Mn concentrations, while not strictly quantitative, nevertheless provide reasonable estimates of C_{Mn}.

Although the above numerical example should only be considered as a fair indication of what one would expect were all 15 hypothetical specimens to be prepared and measured experimentally, it must be pointed out that the

Table 6.9 Numerical example illustrating the limited range of the Lachance–Traill model using influence coefficients computed by regression of theoretical intensity data

Concentration (%)					Coefficients	
Mn	Ti	Co	Mn(app)	Mn(calc)	m_{MnTi}	m_{MnCo}
(a) Used to compute influence coefficients by regression						
10	30	60	7.93	10.3	1.4949	-0.3110
10	60	30	5.10	10.1	1.7212	-0.2438
20	20	60	17.88	20.0	1.5146	-0.3072
20	60	20	9.83	19.9	1.7962	-0.2202
50	10	40	46.58	47.3	1.7351	-0.2502
50	40	10	28.76	48.9	1.8943	-0.1905
(b) Treated as unknowns						
5	15	80	5.60	5.25	(1.853)	(-0.425)
5	45	50	3.20	5.19		
5	80	15	2.01	4.85		
15	15	70	15.38	15.1		
15	70	15	6.69	14.9		
25	10	65	26.61	24.2		
25	65	10	11.42	24.7		
35	15	50	31.66	33.7		
35	50	15	18.50	34.5		

experimental route would probably introduce a number of difficulties. The 15 hypothetical specimens all totalled exactly 100% and were quite dissimilar in concentration range. In actual practice, this is very seldom if ever achieved. Thus, it is not only the number of reference materials that is important, but also their relevancy to the task at hand. Each analyte should be present in combination with fairly wide concentration ranges of each matrix element. Reference materials in which two or more constituents occur in constant ratios are of little use, as it is likely that a single one will be targeted as the main contributor to the matrix effect and thus pave the way for generating influence coefficients that have little if any physical meaning.

Last but not least, the objective evaluation of a method implies that a fairly large number of specimens of equally reliable composition *that were not used to generate the coefficients* should be used to check the method's validity. This may prove onerous in the case of empirical methods because it is highly recommended that as many reference materials as possible be used to generate the influence coefficients in the first place. This implies that ever larger numbers of reference materials are needed as the number of analytes increases.

6.9 SUMMARY

With the exception of the Lucas-Tooth and Price model, in which the correction term is a sum of products involving intensities, in all empirical influence coefficient models examined the correction term for matrix effects is a function of concentrations. Six algorithms have been examined, namely:

(1) $\quad C_i = R_i [C_i r_{ii} + C_j r_{ij} + C_k r_{ik} + \cdots]$

wherein Criss and Birks opted to include the r_{ii} coefficient in the regression, whereas Beattie and Brissey opted to equate r_{ii} to unity from theoretical principles.

(2) the Lachance–Traill model:

$$C_i = R_i [1 + r_{ij}C_j + r_{ik}C_k + \cdots]$$

(3) the Rasberry–Heinrich model:

$$C_i = R_i \left[1 + \sum_j r_{ij}C_j + \sum_k \left(\frac{r_{ik}}{1 + C_i} \right) C_k \right]$$

or

$$C_i = R_i \left[1 + \sum_j \left(r_{ij,a} + \frac{r_{ij,e}}{1 + C_i} \right) C_j \right]$$

(4) the Lucas-Tooth and Price model:

$$C_i = r_0 + I_i [r_i + r_{ii} I_i + r_{ij} I_j + r_{ik} I_k + \cdots]$$

(5) the Lucas-Tooth and Pyne model:

$$C_i = r_0 + C_{i,app} [r_i + r_{ii} C_{i,app} + r_{ij} C_{j,app} + \cdots]$$

(6) the JIS model:

$$C_i = (a + b I_i + c I_i^2) \left[1 + \sum_j r_{ij} C_j \right]$$

7 GLOBAL MATRIX EFFECT CORRECTION APPROACHES

The aim of this chapter is to examine a number of methods that have been proposed to compensate for matrix effects, based on evaluating the extent or magnitude of the total matrix effect indirectly. This is in contrast to methods treated in previous chapters, wherein the concept was to apply a correction for matrix effects based on the summation of each individual effect. It therefore follows that the concepts underlying *global* methods (global in the sense that the magnitude of the matrix effect is quantified as a whole) are quite different, and their application is therefore generally restricted to specific analytical contexts. The analytical context is all important, because compensation for matrix effects is based not on knowledge of fundamental and instrumental parameters but on a combination of specimen preparation and experimental intensity data. Another distinction is that, whereas the so-called mathematical correction methods imply comprehensive analysis (i.e. determination of all elements present), global methods are amenable to the determination of only one or a few elements.

7.1 GENERAL CONSIDERATIONS

For continuity, the general expressions for the conversion of intensities to concentrations adopted in the previous three chapters, namely

$$C = KIM \tag{7.1}$$

which led to Eqn (3.52)

$$C_{iu} = \left(\frac{C_i}{I_i M_i}\right)_r I_{iu} M_{iu} \tag{7.2}$$

are retained but the terms M_{iu} and M_{ir} are envisaged quite differently.

Consider, for example, the case where M_i was equated to μ_s^*, i.e.

$$M_i = \mu_s^* = \sum_i C_i \mu_i^* \qquad (7.3)$$

and the effective mass attenuation coefficients were used to correct for absorption effects. Obviously this procedure for computing μ_s^* implies knowledge of the total composition of the specimen. However, it can be shown that it is possible to quantify μ_s^* globally without any specific knowledge of sample composition. Theoretically, the mass attenuation coefficient μ_s can be measured experimentally by preparing very thin specimens and proceeding as in Section 1.8.1. A method that overcomes this onerous process is based on the fact that intensity measurements at the Compton peak are inversely proportional to the total absorption of the characteristic X-radiations of analyte i by the specimen.

Another example of how differences between M_{iu} and M_{ir} may be dealt with involves taking appropriate specimen preparation steps that minimize their difference. In these cases the values M_{iu} and M_{ir} are not quantified per se but simply made nearly equal, and so may be considered as cancelling out. Thus, the ultimate goal of global methods is simply to compensate for the difference between M_{ir} and M_{iu} by any valid process that does not involve specific knowledge of fundamental or instrumental parameters.

7.1.1 Analytical context

The fact that analysts have the capability to generate theoretical intensities from fundamental parameters makes it possible to evaluate the inherent limitations of global methods. This is therefore the option chosen for the numerical examples that will be used to illustrate the computational processes of the global methods examined. Consider, for example, a method in which the mathematical expression for converting intensities to concentrations was derived on the assumption that the incident excitation is monochromatic. Analysts may often use the method in a polychromatic context based on the concept of the equivalent wavelength. The fact that the 'equivalent wavelength' is actually a narrow band introduces an inherent error in the analytical results. Thus, by generating theoretical intensity data for polychromatic excitation and treating these data as if emanating from specimens of unknown composition, it is possible to assess the extent of the error introduced free of any experimental uncertainties. The same reasoning may be applied when dealing with methods based on analytical contexts that do not involve enhancement if they are applied in situations where some enhancement is present. Considerations of space preclude an in-depth numerical illustration of global methods that would take into consideration a number of combinations of target material, operating voltage, spectrometer geometry and specimen composition.

The analytical context retained to generate the theoretical intensity data is that often used in this text, namely: excitation source: X-ray tube operated at 45 kV; instrument geometry: $\psi' = 63°$, $\psi'' = 33°$

The choice of the systems is arbitrary but the elements are chosen so as to 'maximize the matrix effect to be minimized'. For example, the selection of Ni as the analyte in the presence of Fe is used to illustrate strong absorption. The converse, analyte Fe in the presence of Ni, represents cases of strong enhancement. A number of other binary or more complex systems could have been selected that would lead to the same general conclusion. In other words, the elements chosen are of secondary importance; the goal is to select systems that involve significant matrix effects and to show numerically to what degree the mathematical expressions compensate for the effects. This may then serve as a rough guideline for practical applications. The only exception to the use of theoretical data is the numerical example for the matrix matching method, in which case the data are hypothetical.

7.2* THE DOUBLE DILUTION METHOD

As expressed by Tertian and Claisse (1982), the goal of the double dilution method is to compensate for matrix effects, not by attenuation, but by measuring intensities on specimens at two levels of dilution. Based on principles first suggested by Sherman (1958) and on subsequent studies and development by Tertian (1969a, 1969b, 1972a, 1972b), the double dilution method provides a powerful tool for the correction of matrix effects, owing to its great flexibility in dealing with non-routine analytical requests.

7.2.1 Principle of the method

The intensity emitted by analyte i from a specimen of unknown composition in the context of *monochromatic* incident excitation and absence of enhancement is given by combining Eqns (2.1), (2.23) and (2.24), leading to

$$I_{iu} = \frac{Q_i C_{iu}}{\mu_u^*} \tag{7.4}$$

where

$$Q_i = g_i G_i I_\lambda \mu_{i\lambda} \tag{7.5}$$

Let us now consider a specimen as made up of two components, a portion C_s of the original sample, and the balance C_d of diluent. The emitted intensity I_{id} for that diluted specimen is given by

$$I_{id} = \frac{Q_i C_{iu} C_s}{C_s \mu_s^* + C_d \mu_d^*} \tag{7.6}$$

Given the fact that by definition $C_d = 1 - C_s$, substitution in the above equation yields

$$I_{id} = \frac{Q_i C_{iu} C_s}{C_s \mu_s^* + (1 - C_s)\mu_d^*} \tag{7.7}$$

$$= \frac{Q_i C_{iu} C_s}{\mu_d^* + C_s(\mu_s^* - \mu_d^*)} \tag{7.8}$$

Taking μ_d^* in factor in the denominator and defining

$$Q_d = Q_i / \mu_d^* \tag{7.9}$$

$$a_{sd} = \frac{\mu_s^* - \mu_d^*}{\mu_d^*} \tag{7.10}$$

leads to

$$I_{id} = \frac{Q_d C_{iu} C_s}{1 + a_{sd} C_s} \tag{7.11}$$

where a_{sd} is none other than the absorption influence coefficient quantifying the effect of substituting a portion of the sample by an equal portion C_d of the diluent (i.e., analogous to a_{ij}). Let us consider writing the above expression in the usual formalism, namely an unknown having analyte concentration C_{iu} and a reference material, analyte concentration C_{ir}. Both are diluted at two different dilution ratios, i.e. $C_{s1} + C_{d1}$ and $C_{s2} + C_{d2}$, where

$$C_{s2} = nC_{s1}; \; n < 1 \tag{7.12}$$

in the case of the unknown, and

$$C_{s2} = mC_{s1}; \; m < 1 \tag{7.13}$$

in the case of the reference material. Substitution in Eqn (7.11) yields the four relationships

$$I_{iu1} = \frac{Q_d C_{iu} C_{s1}}{1 + a_{sd} C_{s1}} \tag{7.14}$$

$$I_{iu2} = \frac{Q_d C_{iu} n C_{s1}}{1 + a_{sd} n C_{s1}} \tag{7.15}$$

$$I_{ir1} = \frac{Q_d C_{ir} C_{s1}}{1 + a_{sd} C_{s1}} \tag{7.16}$$

$$I_{ir2} = \frac{Q_d C_{ir} m C_{s1}}{1 + a_{sd} m C_{s1}} \tag{7.17}$$

Eliminating a_{sd} in Eqns (7.14) and (7.15) and Eqns (7.16) and (7.17) yields

$$Q_d C_{iu} n C_{s1} = (1 - n)\left(\frac{I_{i1} I_{i2}}{I_{i1} - I_{i2}}\right)_u \qquad (7.18)$$

$$Q_d C_{ir} m C_{s1} = (1 - m)\left(\frac{I_{i1} I_{i2}}{I_{i1} - I_{i2}}\right)_r \qquad (7.19)$$

In practice, computations are facilitated if $n = m$, in which case ratioing Eqns (7.18) and (7.19) results in

$$C_{iu} = \left(\frac{C_i}{D_i}\right)_r D_{iu} \qquad (7.20)$$

where

$$D_{ir} = \left(\frac{I_1 I_2}{I_1 - I_2}\right)_{ir} \qquad (7.21)$$

$$D_{iu} = \left(\frac{I_1 I_2}{I_1 - I_2}\right)_{iu} \qquad (7.22)$$

Thus, for either the unknown or the reference, the product of intensities measured on two diluted specimens divided by the difference between the two intensities is linearly proportional to the concentration of the analyte in the respective samples.

7.2.1.1 Numerical example

Equation (7.20) is exact under the premise that the incident excitation is *monochromatic* and enhancement is *not* present, because the influence coefficient a_{sd} is constant *only* in that context. When polychromatic excitation is involved, a slight error is introduced because a_{sd} is no longer constant. The following numerical example is based on theoretically calculated intensities for a polychromatic incident source in order to illustrate the extent of the error that is introduced in such a situation. Analyte i is nickel and the binary 40% Ni–60% Fe is considered as the unknown; the binary 60% Ni–40% Fe serves as the reference material. Theoretical emitted intensities for unknown and reference are listed in Table 7.1(a), along with intensities for both specimens diluted with the elements Si and Ti where $C_d = 0.50$ and 0.75. In Table 7.1(b), columns 1, 2 and 3 list C_{Ni} calculated by direct proportionality for the original sample and for the two diluted specimens, respectively. For example, C_{Ni} for the undiluted unknown is given by

$$C_{Ni} = \frac{0.6}{618.84} \times 344.22 = 0.3337$$

Table 7.1 Theoretical data used to illustrate the double dilution method context: analyte Ni in Fe–Ni system

(a) Intensities calculated from fundamental theory (polychromatic excitation)

Specimen	C_{Ni}	C_{Fe}	C_{dil}	ratio	none	Si	Ti
						I_{Ni} (nominal)	
						Diluent	
unk	0.4	0.6		nil	344.22		
ref	0.6	0.4		nil	618.84		
unk d1	0.2	0.3	0.5	1:1		272.69	184.66
unk d2	0.1	0.15	0.75	1:3		192.68	95.86
ref d1	0.3	0.2	0.5	1:1		470.32	303.58
ref d2	0.15	0.1	0.75	1:3		318.11	150.60
D_{unk}						656.7	199.3
D_{ref}						982.9	298.9

(b) Calculated concentrations[a]

	(1)	(2)	(3)	(4)
			C_{Ni}	
Dilution ratio	nil	1:1	1:3	
Diluent: Si	0.3337	0.3479	0.3634	0.4009
Diluent: Ti	0.3337	0.3650	0.3819	0.4002

[a] (1), (2) and (3) C_{Ni} calculated by direct proportionality; (4) C_{Ni} calculated by double dilution method, Eqn (7.20).

In all three columns the results are not quantitative. Concentrations in column 4 are calculated by the double dilution method (Eqn 7.20).

Firstly, substitution of data relating to dilutions with Si in Eqns (7.21) and (7.22) yields

$$D_{ir} = \frac{470.32 \times 318.11}{470.32 - 318.11} = 982.9$$

$$D_{iu} = \frac{272.69 \times 192.68}{272.69 - 192.68} = 656.7$$

Then, substitution of these two values in Eqn (7.20) results in

$$C_{iu} = \frac{0.6}{982.9} \times 656.7 = 0.4009$$

The difference in the results obtained by direct proportionality is a measure of the success of the concept of the double dilution method. If Ti, a stronger absorber of Ni $K-L_{2,3}$ radiation than Si (hence more similar to Fe, the matrix element), is used as the diluent under identical operating conditions, C_{iu} is then computed as equal to 0.4002. This is treated in work by Tertian and Claisse (1982), who showed that the error introduced when Eqn (7.20) is applied to polychromatic excitation depends on the choice of diluent.

7.2.2 Effect of enhancement

Enhancement effects are compensated for by the double dilution method only insofar as they can be visualized as negative absorption, which is a gross approximation. This is illustrated in the following numerical example, in which the only difference from the above numerical example is that Fe is now considered as the analyte and Ni as the matrix element. Table 7. 2(a) lists the calculated iron intensities for the undiluted specimens along with those for the unknown and the reference diluted with Si and Ti at $C_d = 0.50$ and 0.75. Also listed is a second set of intensities for $C_d = 0.6667$ and 0.8889 for dilution with Si. Table 7.2(b) lists C_{Fe} values calculated by direct proportionality and by the double dilution method based on data in Table 7.2(a).

The following calculations pertaining to intensity data for $C_d = 0.6667$ and 0.8889 exemplify the power of the double dilution method; i.e. using I_i data individually and combining them to calculate D_i. In the case of direct proportionality, C_{Fe} for the unknown is obtained by substituting I_i data directly into the expression

$$C_{iu} = \left(\frac{C_i}{I_i}\right)_r I_{iu} \tag{7.23}$$

$$= \frac{0.6}{445.97} \times 313.70 = 0.4220$$

$$= \frac{0.6}{154.10} \times 105.00 = 0.4088$$

for dilutions 1:2 and 1:8, respectively. If the same intensities I_1 and I_2 are combined and D_{ir} and D_{iu} are calculated

$$D_{ir} = \frac{445.97 \times 154.10}{445.97 - 154.10} = 235.5$$

$$D_{iu} = \frac{313.70 \times 105.00}{313.70 - 105.00} = 157.8$$

Table 7.2   Theoretical data used to illustrate the double dilution method. Context: analyte Fe in Fe–Ni system

(a) Intensities calculated from fundamental theory (polychromatic excitation)

| | | | | | I_{Fe} (nominal) | | |
| | | | | | | Diluent | |
Specimen	C_{Fe}	C_{Ni}	C_{dil}	ratio	none	Si	Ti
unk	0.4	0.6		nil	873.63		
ref	0.6	0.4		nil	1179.19		
unk d1	0.2	0.3	0.5	1:1		463.42	221.20
unk d2	0.1	0.15	0.75	1:3		236.41	87.12
ref d1	0.3	0.2	0.5	1:1		647.52	318.50
ref d2	0.15	0.1	0.75	1:3		339.62	128.29
D_{unk}						482.6	143.7
D_{ref}						714.2	313.7
unk d3	0.1333	0.2000	0.6667	1:2	313.70		
unk d4	0.0444	0.0667	0.8889	1:8	105.00		
ref d3	0.2000	0.1333	0.6667	1:2	445.97		
ref d4	0.0667	0.0444	0.8889	1:8	154.10		
D_{unk}					157.8		
D_{ref}					235.5		

(b) Calculated concentrations

| | | C_{Fe} | | |
| | | | Diluent | |
Ratio	Correction applied	none	Si	Ti
nil	direct proportionality	0.4445		
1:1	direct proportionality		0.4297	0.4167
1:3	direct proportionality		0.4176	0.4075
	double dilution (1:1 and 1:3)		0.4054	0.4014
1:2	direct proportionality		0.4220	
1:8	direct proportionality		0.4088	
	double dilution (1:2 and 1:8)		0.4022	

Substitution in Eqn (7.20) leads to

$$C_{iu} = \frac{0.6}{235.5} \times 157.8 = 0.4022$$

Although this represents a relative error of 0.55%, it must be pointed out that the above example involves (a) strong enhancement, (b) a polychromatic excitation source and (c) dilution with a constituent that has quite different absorbing properties to the analyte all of which were verified experimentally by Tertian and shown in Tertian and Claisse (1982, p. 249), who concluded that anomalous results occur only with exceptionally severe matrix effects.

7.2.3 Practical considerations

The power of the double dilution method lies in its quasi-universality; i.e. it is applicable to any element present in sufficient concentration in a soluble or mixable material. If the method is used in conjunction with mixtures, the problems associated with particle size effect, mineralogical effects and particle segregation may remain, and adequate mixing may also pose problems. These are best overcome by fusing with a flux as the diluent, in which case the diluted specimens are in the form of cast solid solution discs.

If the double dilution method is applied in the general analytical context of *polychromatic* excitation, and enhancement is present, then the influence coefficient a_{sd} of Eqn (7.10) is no longer a constant because of the polychromaticity of the excitation source and, especially, because of enhancement; i.e. m_{sd} would be a more appropriate symbol to define the matrix effect. As shown by Tertian (1969a,b, 1972a,b), the error introduced due to enhancement can be compensated in most cases if the dilution factors are fairly large. Unfortunately, higher dilution or the use of a stronger absorber as a diluent leads to decreases in emitted intensity. Whatever the analytical context, the dilutions must be such that I_1/I_2 ratios smaller than 2 should be avoided for fear of large relative errors in the denominator when computing D_i; i.e. $I_1 - I_2$ should not be the difference between two nearly equal values. It therefore follows that weighings and intensity measurements must be carried out somewhat more diligently than in most other methods.

7.3* COMPTON (INCOHERENT) SCATTER METHODS

The use of scattered radiation to provide a measure of the absorption effect of characteristic X-radiations in specimens, and hence as a method for compensating for absorption, is attractive because of its simplicity. The observation by Andermann and Kemp (1958) that the emitted intensities of fluorescence radiations and primary radiation scattered by a specimen at wavelengths near the fluorescence lines are affected in the same way, i.e. that their ratio is almost independent of specimen composition, is the basis of this method. Subsequent studies have generally been oriented towards the use of Compton (incoherent) scatter, preferably the scattered lines of the tube target or scattered background at low 2θ angles, where intensities are more intense, as shown in Figure 7.1. The following theoretical treatment is an outline of the more rigorous treatment of absorption corrections based on scattered

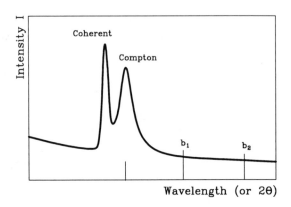

Figure 7.1   Typical line and background scattered radiations

radiation in Tertian and Claisse (1982). The limitations and pitfalls of using Compton scattered tube radiation directly to compensate for absorption, or indirectly for the determination of an effective mass absorption coefficient μ_s^* to be used in turn to compensate for absorption effects in practice, were examined in depth by Feather and Willis (1976) and by J. P. Willis (1991).

7.3.1   Theoretical considerations

Compton or incoherent scattering is the result of the impact of an incident photon on an electron loosely bound to the atom, by which the photon loses some energy and assumes new energy in the process. It is this process that results in the emission of broad peaks on the long wavelength side of characteristic tube lines. Tertian and Claisse (1982) have shown that the equation derived for the calculation of primary fluorescence of a specimen s in the context of *monochromatic* incident radiation λ in the absence of enhancement (Eqn 7.4)

$$I_{is\lambda} = \frac{Q_{i\lambda}C_i}{\mu_s^*} = \frac{Q_{i\lambda}C_i}{\mu_s' + \mu_s''} \tag{7.24}$$

may be retained in the case of a polychromatic source if it is assumed that there is a monochromatic equivalent wavelength λ_e that is representative of the whole polychromatic distribution, i.e. $\lambda = \lambda_e$ in the above expression. Noting that the effective μ for scattering is equal to $\mu_{s\lambda}$ csc $\psi' + \mu_{s\lambda}$ csc ψ'', it was also shown that an equation similar to Eqn (2.25) can be used to calculate scattered intensities:

$$I_{inc} = a + \frac{b}{\mu_{s\lambda_e}} \tag{7.25}$$

where I_{inc} is the intensity of the Compton scatter by the specimen of a characteristic line of the target element and a and b are constants; the constant a is usually small and often negligible (Figure 7.2).

As shown by Hower (1959), given that mass attenuation coefficients are virtually proportional between any two successive absorption edges, it then follows that

$$\mu_{s\lambda_e} \propto \mu_{s\lambda_i} \propto \mu_{s\lambda_b} \propto \mu_{s\lambda_{inc}} \qquad (7.26)$$

where λ_e is the estimated equivalent wavelength, λ_i is the wavelength of analyte i, λ_b is any appropriate wavelength, and λ_{inc} is the wavelength of the Compton (incoherent) scatter peak. Neglecting the small intercept a in Eqn (7.25) leads to

$$I_{is} = \frac{Q_1 C_i}{\mu_{s\lambda_i}} \qquad (7.27)$$

$$I_{inc} = \frac{Q_2}{\mu_{s\lambda_i}} \qquad (7.28)$$

where Q_1 and Q_2 are proportionality constants. Combining Eqn (7.27) and Eqn (7.28) results in

$$\frac{I_{is}}{I_{inc}} = Q_3 C_i \qquad (7.29)$$

where Q_3 is a new proportionality constant, indicating that the ratio I_{is}/I_{inc} is independent of specimen composition within the analytical context stated previously. This leads to the analytical formalism

$$C_{iu} = \left(C_i \frac{I_{inc}}{I_i} \right)_r \times \left(\frac{I_i}{I_{inc}} \right)_u \qquad (7.30)$$

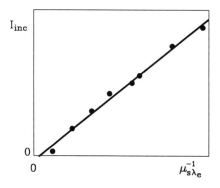

Figure 7.2 Correlation between incoherent scatter and the mass absorption of materials

It therefore follows from Eqns (7.24)–(7.29) that the following linear relationships illustrate the versatility of the incoherent scatter method:
(a) the linear relation between I_b and μ_s^{-1}, as shown in Figure 7.3,

$$\mu_s = a_0 + a_1 \frac{1}{I_{inc}} = b_0 + b_1 \frac{1}{I_b} \tag{7.31}$$

which provides further confirmation of the observation of Feather and Willis (1976) that not only I_{inc} measured at the Compton peak but also any measured background free of characteristic peaks may be used to evaluate μ_s;
(b) the linear relationship between I_{inc} and I_b (Figure 7.4), following the view of Feather and Willis (1976) that Compton 'peaks' may be visualized as background regions of elevated intensity, with the implication that a shorter counting time is needed to obtain the same precision as an intensity measured at a background position;
(c) the linear relationship between I_{b_1} and I_{b_2} (Figure 7.5).

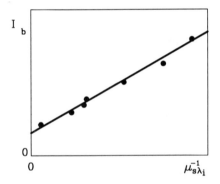

Figure 7.3 Correlation between background intensity and the mass absorption of materials

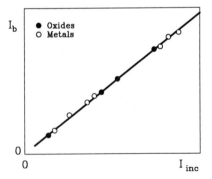

Figure 7.4 Linear correlation between background and incoherent peak intensity

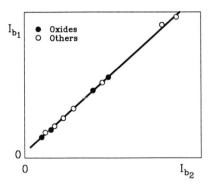

Figure 7.5   Linear correlation between background intensities

7.3.1.1   Numerical example

The following numerical example is an indirect application of the Compton scatter method in that it is based on calculated backgrounds obtained from theoretically calculated mass attenuation coefficients instead of measured Compton scatter intensities. This approach is taken in order to provide an example of the small but inherent errors introduced when the Compton method is applied in the context of a polychromatic incident source. Thus, to the extent that Compton scatter is proportional to μ_s^{-1}, data in Table 7.3 provide a valid demonstration of this method. These data were computed from fundamental parameters expressions for the following context: incident excitation source: Rh target tube operated at 45 kV, angle of incidence = 63°, and angle of emergence = 33°.

(a) The Zn $K-L_{2,3}$ line intensities listed in Table 7.3(a) are obtained from fundamental parameters computations wherein the proportionality constant g_i is arbitrarily set at 1000, i.e.

$$I_{Zn} = g_i \sum_{\lambda} \frac{I_\lambda C_{Zn}\mu_{Zn\lambda}}{\mu_s^*} \qquad (7.32)$$

e.g. for a specimen of 1000 ppm Zn in Mg, $I_{Zn} = 3050$ nominal c/s; for a specimen of 3000 ppm Zn in Fe, $I_{Zn} = 1151$ nominal c/s.

(b) Background intensity values at $\mu_{s(\lambda = 0.89 \text{ Å})}$ and $\mu_{s(\lambda = 1.09 \text{ Å})}$ near the $Rh_{K-L_{2,3(\lambda = 0.64 \text{ Å})}}$ Compton peak are used to compute background theoretical intensities from the relationship

$$I = \frac{Q}{\mu_{s\lambda}} \qquad (7.33)$$

Table 7.3 Theoretical data used to illustrate the Compton (incoherent) scatter method. Context: analyte Zn in Mg–Zn, Ti–Zn and Fe–Zn systems[a]

(a) Intensities calculated from fundamental theory (polychromatic excitation)

			I_{Zn} (nominal)			
λ	Binary	Zn (ppm):	1000	2000	3000	4000
Zn $K-L_{2,3}$	Mg–Zn		3050	6086	9110	12120
	Ti–Zn		589	1179	1769	2360
	Fe–Zn		383	767	1151	1536
(Rh_{inc}	Mg–Zn		305.8	302.1	298.5	295.9
0.64 Å)	Ti–Zn		56.78	56.69	(56.63)	56.56
	Fe–Zn		36.66	36.64	36.63	36.62
0.89 Å	Mg–Zn		116.6	115.3	114.0	112.9
	Ti–Zn		22.11	22.08	(22.06)	22.03
	Fe–Zn		14.32	14.32	14.31	14.31
1.09 Å	Mg–Zn		66.23	65.53	64.85	64.18
	Ti–Zn		12.71	12.69	(12.68)	12.67
	Fe–Zn		8.25	8.25	8.25	8.24

(b) Calculated concentrations based on binary

		Zn (ppm)			
		1000	2000	3000	4000
(Rh_{inc}	Mg–Zn	958	1935	2931	3934
0.64 Å)	Ti–Zn	996	1997	(3000)	4007
	Fe–Zn	1003	2010	3018	4028
0.89 Å	Mg–Zn	997	1998	2990	4016
	Ti–Zn	997	1998	(3000)	4008
	Fe–Zn	1001	2004	3009	4016
1.09 Å	Mg–Zn	990	1997	3021	4061
	Ti–Zn	997	1998	(3000)	4005
	Fe–Zn	998	1999	3000	4008

[a] Values in parentheses refer to data used for calibration.

where the proportionality constant is arbitrarily set at 1000. For example, in the case of a specimen of 3000 ppm Zn in Fe and $\lambda = 0.89$ Å,

$$\mu_{Zn} = 100.75 \text{ cm}^2 \text{g}^{-1}$$

$$\mu_{Fe} = 69.78 \text{ cm}^2 \text{g}^{-1}$$

$$\mu_s = (0.003 \times 100.75) + (0.997 \times 69.78)$$

$$= 69.87 \text{ cm}^2 \text{g}^{-1}$$

and

$$I = \frac{1000}{69.87} = 14.31 \text{ nominal } c/s$$

(c) Calculation of concentrations. If the data relating to 3000 ppm Zn in Ti are selected as the reference, then the calculation of C_{Zn} is obtained from Eqn (7.30) if I_{inc} is measured, or alternatively

$$C_{iu} = \left(C_i \, \frac{I_b}{I_i} \right)_r \times \left(\frac{I_i}{I_b} \right)_u \tag{7.34}$$

where I_b can be any interference free background, e.g. 0.89 Å or 1.09 Å, or alternatively, where I_b can be the background at the 2θ of the characteristic line of the analyte; i.e.

$$I_b = I_{\text{calculated } \lambda_i} \tag{7.35}$$

where λ_i can be any background for any given analyte. Thus, in the case of a specimen of 1000 ppm Zn in Mg, using Rh_{inc}

$$C_{Zn} = 3000 \, \frac{56.63}{1769} \times \frac{3050}{305.8}$$

$$= 96.037 \times 9.974 = 958 \text{ ppm}$$

and in the case of a specimen of 4000 ppm Zn:

$$C_{Zn} = 96.037 \times \frac{1536}{36.62} = 4028 \text{ ppm}$$

If intensities for 1.09 Å are used, in the case of 1000 ppm Zn in the system Zn–Mg

$$C_{Zn} = 3000 \, \frac{12.68}{1769} \times \frac{3050}{66.23}$$

$$= 21.504 \times 46.052 = 990 \text{ ppm}$$

and for 4000 ppm Zn in Zn–Fe

$$C_{Zn} = 21.504 \times \frac{1536}{8.24} = 4008 \text{ ppm}$$

Thus, following Feather and Willis (1976), Compton 'peaks' may be visualized as background regions of elevated intensity, with the implication that a shorter counting time is needed to obtain the same precision as for an intensity measured at a background position. However, if the Compton peak is subject to spectral interference from elements in a given specimen, then a more convenient background region may be used instead.

7.4   THE INTERNAL STANDARD METHOD

The use of an internal standard is a common practice in instrumental methods of analysis. In essence the concept is fairly simple: a known amount of element j not originally present is added to the sample, thus yielding a specimen of known concentration of j. In X-ray fluorescence spectrometry, this consists in adding a known (usually constant) proportion of a compound containing an element (the internal standard) having fluorescence properties similar to those of the analyte. The measured intensity emitted by element j is then a function of the matrix effect for the internal standard and, by implication, of analyte i.

The theoretical interpretation of the internal standard method is as follows. Consider a sample to which has been added a known amount of element j not originally present in that sample. Applying the simple relation of Eqn (3.46) to the resulting specimen z (sample + internal standard), the concentration of analyte i may be expressed as

$$C_{iu} = K_i I_{iz} M_{iz} \frac{W_z}{W_u} \qquad (7.36)$$

where W_u and W_z are the weights before and after the addition of j. Similarly, for element j added as an internal standard to the unknown

$$C_{jz} = K_j I_{jz} M_{jz} \qquad (7.37)$$

Ratioing Eqn (7.36) to Eqn (7.37) and solving for C_{iu} yields

$$C_{iu} = C_{jz} \frac{K_i}{K_j} \times \frac{I_{iz}}{I_{jz}} \times \frac{M_{iz}}{M_{jz}} \times \frac{W_z}{W_u} \qquad (7.38)$$

Given that by definition:

(a) element j is so chosen that it undergoes absorption and enhancement very similar to that of analyte i, the terms M_{iz} and M_{jz} are then essentially equal and cancel out;

(b) the ratio of the two proportionality constants K_i and K_j is a constant;

(c) element j is usually added at a fixed concentration, so that C_j, W_u and W_z are constants;

then Eqn (7.38) reduces to

$$C_{iu} = K \times \frac{I_{iz}}{I_{jz}} \qquad (7.39)$$

where K is a new proportionality constant.

Although simple in concept, the internal standard method presents problems in practical applications. The main problem is the selection of an internal standard. As well as being absent from the original sample, the analyte and the internal standard must be affected very similarly by all the elements of the specimen. This implies that the emitted characteristic radiations have similar wavelengths and preferably belong to the same series, i.e. both K or both L lines. It is imperative that there be no absorption edge of a major or minor constituent between the two emission lines. This condition imposes severe limitations when dealing with multielement samples that contain a number of elements present as minor constituents.

Although ideally suited for liquid samples, the internal standard method is not applicable to alloys unless these are first dissolved, a practice which is not normally contemplated in X-ray fluorescence. In the middle ground are powdered samples. Powders are amenable to the internal standard method in that sample and standard can be mixed, but every effort must be made to restrict the application to samples of the same type, i.e. having similar granulometry and mineralogy. These limitations may be eliminated if the sample plus the internal standard are subsequently mixed with a flux material, fused and cast as a solid glass disc.

7.4.1 Numerical example

The intensity data in Table 7. 4 are nominal intensities generated from fundamental parameters computations for the following analytical context: incident excitation source: W target tube operated at 45 kV, angle of incidence $= 63°$, angle of emergence $= 33°$.

The application of the internal standard method using data in Table 7.4 is as follows. For purposes of generating theoretical intensity data, niobium was chosen as the analyte and molybdenum as the internal standard; 7.4(a) is representative of four samples submitted for analysis; 7.4(b) represents two specimens that are prepared to establish the relative intensity of the analyte and the internal standard. In this case equal concentrations of niobium and molybdenum are added to the elements Mg and Ti. Intensities are computed from fundamental parameters expressions as described in Chapter 3. Table 7.4(c) lists the theoretical intensities computed from fundamental parameters for the four samples *after* the addition of 1% of Mo to each; for example 0.05 g of Mo added to 5.0 g of sample. It is obvious that the four Nb intensities are not linearly related to their respective Nb concentrations. Analysis proceeds along the following lines:

(a) the K constant in Eqn (7.39) is, from reference specimen 1,

$$1.00 = K \frac{4858}{4833}, \text{ thus } K = 0.995$$

similarly from reference specimen 2, $K = 0.999$, giving an average value $K = 0.997$;

Table 7.4 Theoretical data used to illustrate the internal standard method. Context: analyte Nb in the Si–Nb, Fe–Nb and Zn–Nb systems. Internal standard is Mo

(a) Samples treated as unknowns

Sample	Nb (%)	Matrix (%)
Sp1	0.70	99.3 Si
Sp2	0.70	99.3 Fe
Sp3	1.40	98.6 Fe
Sp4	0.70	99.3 Zn

(b) Data used for calibration

	Composition			Intensities	
	Nb (%)	Mo (%)	Matrix (%)	Nb	Mo
Ref 1	1.00	1.00	98.00 Mg	4858	4833
Ref 2	1.00	1.00	98.00 Ti	1010	1009

(c) Data used for calculating Nb concentrations

	$I_{Nb(theo)}$	$I_{Mo(theo)}$	Nb (calc %)	Nb (true %)
Sp1 + Mo	2260	3219		
Sp1			0.700	0.70
Sp2 + Mo	457	651		
Sp2			0.700	0.70
Sp3 + Mo	914	652		
Sp3			1.397	1.40
Sp4 + Mo	316	451		
Sp4			0.699	0.70

(b) Nb (%) is calculated from Eqn (7.39), e.g. in the case of samples 1 and 3

$$C_{Nb} = 0.997 \times \frac{2260}{3219} = 0.700$$

and

$$C_{Nb} = 0.997 \times \frac{914}{652} = 1.398$$

Although elements were used in the above examples, the process would more likely be carried out in oxide systems or in aqueous systems, e.g. 0.05 ml added to 5.0 ml when dealing with solutions.

7.5* THE STANDARD ADDITION METHOD

The standard addition method, also referred to as the addition or spiking method, is a variant of the internal standard method. As stated in the previous section, one of the inherent drawbacks of the internal standard method is finding an element that satisfies the criterion that *the analyte and the internal standard element must be affected in a very similar way by all the elements of the specimen*. Therefore an obvious solution is: *why not select the analyte as the internal standard*. Just as measuring the intensity I_j due to the presence of a known concentration of the internal standard C_j may be used to quantify matrix effects in a specimen, measuring ΔI_i, the increase in intensity from a known added concentration ΔC_i, may be used to quantify matrix effects.

The principle of the standard addition (or analyte addition) method is fairly simple. Consider two specimens: the original sample ($C_s = 1.0$) in which analyte i is present at concentration C_i, and a portion of the original sample to which has been added a known concentration ΔC_i of analyte i. The ideal criteria governing the addition are that ΔC_i be approximately equal to C_i, and that the standard addition be a very small fraction of the original sample. Net intensities are measured on both specimens, from which intensities it is possible to calculate ΔI_i, namely

$$\Delta I_i = I_{i,\text{sample + addition}} - I_{i,\text{original sample}}$$

$$= I_{iu+} - I_{iu} \tag{7.40}$$

The concentrations of analyte i in the original and spiked samples are then expressed in the form of Eqn (3.46):

$$C_{iu} = K_i I_{iu} M_{iu} \tag{7.41}$$

$$C_{iu+} = K_i I_{iu} M_{iu+} \tag{7.42}$$

Given that the method is based on the premise that the weight fraction of reference material added (W_r) is a very small fraction of the sample weight (W_u), the terms M_{iu} and M_{iu+} may for all practical purposes be considered as being equal; hence their ratio is essentially equal to unity. Thus Eqn (7.39) in this case reduces to

$$C_{iu} = \frac{\Delta C_i}{\Delta I_i} \times I_{iu} \tag{7.43}$$

In practice, two factors must be taken into consideration when applying the standard addition method, namely:

- although the fraction of reference material added may be small compared to the weight of sample, their ratio should be taken into account;
- the reference material added may not be the pure analyte.

If these two factors are taken into account, the concentration of analyte i in the unknown sample is given by

$$C_{iu} = C_{ir} \frac{(I_{iu}/I_{iu+})\,W_r}{1 - [(I_{iu}/I_{iu+})\,W_u]} \tag{7.44}$$

where C_{ir} is the concentration in the reference material and W_r and W_u are the proportions of the reference material and sample, respectively. Alternatively, according to Bertin (1975)

$$C_{iu} = C_{ir} \frac{I_{iu}/I_{iu+}}{1 + (W_u/W_r)\,[1 - (I_{iu}/I_{iu+})]} \tag{7.45}$$

7.5.1 Numerical example

The analytical context for Table 7.5 is identical to that for Table 7.4 as far as excitation and instrument geometry are concerned. Analyte i = Ni data relate to matrix effects due to Fe and Ti, both absorbers, whereas analyte i = Fe data relate to absorption and enhancement by Ni and Zr. The analyte concentrations are 1% in four hypothetical unknowns, and to each are made three different standard additions in the ratio 0.01:1.0, standard to unknown; e.g. $W_r = 0.1$ g added to $W_u = 10$ g of sample. The compositions of the three reference materials are: C_{ir1}, 50% analyte i in Si; C_{ir2}, 75% analyte i in Si; C_{ir3}, pure analyte i.

Table 7.5(a) lists the nominal intensities generated from fundamental parameters expressions for the compositions shown, and Table 7.5 (b) lists the analyte concentrations calculated using Eqns (7.44) and (7.45) and calculated as follows.

(a) The weight *fraction* required in Eqn (7.44) is given by $W_r = 0.1/10.1 = 0.009901$, and $W_u = 10.0/10.1 = 0.9901$.

(b) The concentration of nickel in the binary Ni–Fe based on the addition of C_{ir1} is given by, in the case of Eqn (7.44),

$$C_{Ni} = 50 \frac{(653/977) \times 0.009901}{1 - [(653/977) \times 0.9901]} = 0.978\%$$

and, in the case of Eqn (7.45),

$$C_{Ni} = 50 \frac{(653/977)}{1 + (10/0.1)\,[1 - (653/977)]} = 0.978\%$$

(c) The concentration of Fe in the binary Fe–Ni based on the addition of C_{ir2} is given by, using Eqn (7.45),

$$C_{Fe} = 75 \frac{3329/5686}{1 + (10/0.1)\,[1 - (3329/5686)]} = 1.03\%$$

Table 7.5 Theoretical data used to illustrate the standard addition method. Context: analyte 1% Ni in Fe–Ni and Ti–Ni systems; analyte 1% Fe in Fe–Ni and Fe–Zr systems

(a) Intensities calculated from fundamental theory (polychromatic excitation)

Specimen	ΔC_i (%)	I_i (nominal)			
		i = Ni		i = Fe	
		j = Fe	j = Ti	j = Ni	j = Zr
unk		653	1001	3329	1135
unk +1	0.50	977	1495	4891	1693
unk +2	0.75	1140	1743	5686	1794
unk +3	1.00	1301	1990	6472	2254

(b) Calculated concentrations

Based on	C_i (calc %)			
	i = Ni		i = Fe	
	j = Fe	j = Ti	j = Ni	j = Zr
unk +1 data	0.978	0.983	1.03	0.987
unk +2 data	0.983	0.989	1.03	0.991
unk +3 data	0.988	0.992	1.04	0.994

The differences between the calculated analyte concentrations and the concentrations known to be present arise because a key premise of the standard addition method is violated, namely that a linear intensity–concentration relationship is assumed. This is generally true for very low concentrations, i.e. less than 1%, but in a case of strong enhancement, as exemplified by analyte Fe in the system Fe–Ni, this assumption is not quite valid over the 0–2% Fe range. Thus, theory predicts a possible relative error of the order of 3% in this particular context.

7.6* THE HIGH DILUTION–HEAVY ABSORBER METHOD

In principle, the high dilution method aims at minimizing matrix effects by diluting all samples, unknowns and reference materials, to such a degree that the terms $M_{i(u+d)}$ and $M_{i(r+d)}$ are both essentially equal to M_{id}, where the subscript d represents the diluent. In other words, the high dilution method is equivalent to the analytical context of determining low elemental concentrations in specimens that are very similar in composition, namely the diluent. The method may be visualized in terms of the general equation Eqn (7.2) expressed in the formalism

$$C_{iu} = \frac{C_{ir}}{I_{ir}M_{i(r+d)}} \; I_{iu}M_{i(u+d)} \tag{7.46}$$

Thus the ultimate specimen preparation procedure depends on what maximum departure from unity of the ratio $(M_{i(u+d)}/M_{i(r+d)})$ is acceptable. This in turn depends on the nature of the samples, the magnitude of the matrix effects and the desired precision and accuracy of the analytical results. The diluted specimens may be visualized as

$$C_{iu} = K_i \times I_{iu} \times \left[1 + \sum_j m_{ij}C_j + m_{id}C_d \right]_u \tag{7.47}$$

and

$$C_{ir} = K_i \times I_{ir} \times \left[1 + \sum_j m_{ij}C_j + m_{id}C_d \right]_r \tag{7.48}$$

where, by definition, $C_i + C_j + \cdots + C_d = 1.0$. The goal is to have the product $m_{id}C_d \gg \sum_j m_{ij}C_j$, either by selecting a diluent such that $m_{id} \gg m_{ij}$ coefficients (heavy absorber) or by selecting a dilution factor such that $C_d \gg \sum C_j$ (high dilution). In either case Eqn (7.46) reduces to

$$C_{iu} = \left(\frac{C_i}{I_{i(r+d)}} \right) I_{i(u+d)} \tag{7.49}$$

7.6.1 Numerical example

Data for Table 7.6 were computed from fundamental parameters expressions for the context: incident excitation source: W target tube operated at 45 kV, angle of incidence $= 63°$, angle of emergence $= 33°$; analytes Ni and Fe in the binary Fe–Ni systems, elements Si and Cr as diluents, dilution ratios (sample:diluent) 1:9 and 1:99, which is representative of absorption and enhancement effects and dilution with both weak and strong absorbers. Data for $C_i = 50\%$ are selected as the reference materials, the others are treated as unknowns.

Table 7.6 Theoretical data used to illustrate the high dilution, heavy absorber method. Context: analytes Ni and Fe, diluents Si and Cr, dilution ratios 1:9 and 1:99

(a) Intensities calculated from fundamental theory (polychromatic excitation)

S no.	$C_i(\%)$	undiluted	I_i (nominal) dilution 1:9 in Si	dilution 1:9 in Cr	dilution 1:99 in Si	dilution 1:99 in Cr
For i = Ni; j = Fe; $C_i + C_j = 100\%$						
Ref.	50.0	4689	1313	403.0	160.3	39.8
S1	60.0	6188	1616	487.3	192.9	47.7
S2	40.0	3442	1025	320.0	127.9	31.8
S3	30.0	2386	750.7	238.2	95.6	23.8
S4	20.0	1480	488.9	157.6	63.5	15.9
S5	10.0	691.5	239.0	78.2	31.7	7.9
For i = Fe; j = Ni; $C_i + C_j = 100\%$						
Ref.	50.0	10306	1168	306.5	116.3	28.5
S6	60.0	11792	1389	366.8	139.6	34.2
S7	40.0	3736	944.3	245.9	93.2	22.8
S8	30.0	7037	715.8	185.0	70.0	17.1
S9	20.0	5131	482.5	123.7	46.7	11.4
S10	10.0	2883	244.1	62.0	23.4	5.7

(b) Calculated concentrations

S no.	$C_i(\%)$	undiluted	C_i (calc %) dilution 1:9 in Si	dilution 1:9 in Cr	dilution 1:99 in Si	dilution 1:99 in Cr
For i = Ni; j = Fe						
S1	80.0	66.0	61.5	60.5	60.2	59.9
S2	40.0	36.7	39.0	39.7	39.9	39.9
S3	30.0	25.4	28.6	29.6	29.8	29.9
S4	20.0	15.8	18.6	19.6	19.8	20.0
S5	10.0	7.4	9.1	9.7	9.9	9.9
For i = Fe; j = Ni						
S6	60.0	57.2	59.5	59.8	60.0	60.0
S7	40.0	42.4	40.4	40.1	40.0	40.0
S8	30.0	34.1	30.6	30.2	30.1	30.0
S9	20.0	24.9	20.7	20.2	20.1	20.0
S10	10.0	14.0	10.4	10.1	10.1	10.0

The application of the high dilution method using data in Table 7.6 is interpreted as follows:

(a) calculating C_{Ni} and C_{Fe} from the undiluted intensity data indicates the strong matrix effects; e.g. in the case of samples S3 and S8, both containing 30% of analyte:

$$C_{Ni} = \frac{50}{4689} \times 2386 = 25.4\%$$

$$C_{Fe} = \frac{50}{10\ 306} \times 7037 = 34.1\%$$

(b) low dilution (1:9) with a weak absorber (Si) does not compensate for matrix effects in this context; e.g. for the same two samples S3 and S8:

$$C_{Ni} = \frac{50}{1313} \times 750.7 = 28.6\%$$

$$C_{Fe} = \frac{50}{1168} \times 715.7 = 30.6\%$$

The benefits of stronger absorption and higher dilution are illustrated in the three rightmost columns in Table 7.6(b).

Although dilution methods are generally limited to powdered or liquid samples in practice, the choice of elements rather than oxides to generate intensity data does not negate the conclusions reached regarding this method.

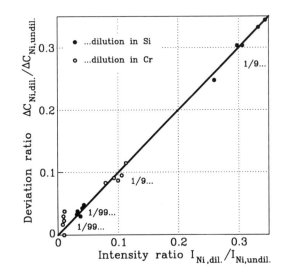

Figure 7.6 Plot of deviation ΔC_{Ni} (as a fraction of ΔC_{Ni} before dilution) versus measured intensity I_{Ni} (as a fraction of I_{Ni} before dilution)

The contexts were selected so as to maximize matrix effects and thus present a severe test of the method's applicability. The same proportionality constant g_i (Eqn 2.1) was used for all intensity calculations; hence the values listed are representative of the magnitude of the loss due to dilution.

As expected from Eqns (7.48) and (7.49), it does not matter much whether the correction for matrix effect is obtained from high dilution with a light absorber or from low dilution with a heavy absorber. This is shown in Figure 7.6, where the fraction of the error ΔC_i in the original specimen is plotted as a function of the fraction of the original fluorescence intensity after dilution. The four sets of points fall on the same line at 45°, indicating that a given reduction of matrix effect is obtained at a given cost of X-ray intensities. The cost may be high; for example, the intensity of the Fe or Ni $K-L_{2,3}$ lines in a binary specimen is likely to decrease at least tenfold before an accuracy of $\approx 1\%$ relative is reached.

7.7* MATRIX MATCHING METHODS

The term matrix matching is used here to designate the simplest approach for converting intensities to concentrations; one must have prior knowledge that: (a) the unknowns are all of one type; (b) type reference materials are on hand. The approach is based on the premise that, over short concentration ranges, M_{ir} and M_{iu} are likely to be similar and remain constant for materials of the type, so that they cancel out. Thus, plots of concentration versus intensity yield curves that can be defined by fairly simple mathematical expressions. In terms of the general relationship applicable to both unknown and reference material

$$C_i = K_i I_i M_i \qquad (7.50)$$

A plot of C_i versus I_i is linear only to the extent that the product K_i times M_i remains approximately constant. This condition is nearly met in low C_i concentration ranges. If the appropriate corrections are made in calculating net peak intensities, then the intercept of the regression equation should be negligible, and the analytical expression takes the form

$$C_i = K_i I_i \qquad (7.51)$$

where K_i is the slope of a linear least squares fit. Since residual backgrounds may be present, or if backgrounds have not been subtracted, a better fit should be obtained if the regression is not constrained through zero; i.e.

$$C_i = a + b I_i \qquad (7.52)$$

Given the fact that M_i is not likely to remain constant over wider concentration ranges of the analyte, a better fit may be obtained by resorting to non-linear models involving three coefficients; e.g.

$$C_i = a + bI_i + cI_i^2 \tag{7.53}$$

$$C_i = (a + bI_i)/(1 + cI_i) \tag{7.54}$$

$$C_i = a + b[1 - \exp(-cI_{iu})] \tag{7.55}$$

However, analysts must be aware that these approaches to matrix matching are entirely empirical and require not only extensive suites of reference materials but a judicious choice of appropriate reference materials, even in well defined and restricted analytical contexts.

7.7.1 Numerical example

Data in Table 7.7 are used to illustrate the process an analyst might follow in the application of the matrix matching method. Let us consider that the data in Table 7.7(a) represent seven reference materials numbered 1 to 7 that are used for calibration, and that the intensity data in Table 7.7(b) are to be converted to concentrations. A quick test to verify whether any one reference

Table 7.7 Data used to illustrate the matrix matching method

S no.	C_i (%)	I_i	C_i (calc %)[a]	Diff.	C_i (calc %)[b]	Diff.
(a) Specimens used for calibration.						
S1	3.95	1166	3.96	0.01	3.96	0.01
S2	6.03	1859	5.60	−0.43	5.98	−0.05
S3	5.00	1501	4.75	−0.25	4.96	−0.04
S4	11.96	4718	12.37	0.41	11.92	−0.04
S5	1.97	562	2.53	0.56	2.00	0.03
S6	11.03	4134	10.99	−0.04	11.03	0.00
S7	9.98	3601	9.73	−0.25	10.06	0.08
(b) Specimens treated as unknowns.						
S8	3.02	854	3.22	0.20	2.97	−0.05
S9	9.01	3112	8.57	−0.44	9.06	0.05
S10	6.89	2245	6.51	−0.38	7.01	0.12
S11	1.01	287	1.88	0.87	1.06	0.05
S12	10.92	4140	11.00	0.08	11.04	0.12
S13	8.15	2661	7.50	−0.65	8.04	−0.11

[a] C_i (calc %) $= a + bI$; $a = 1.1955$, $b = 0.002369$
    Diff. $= C_i$ (calc %) $- C_i$
[b] C_i (calc %) $= a + bI + cI^2$; $a = 0.0334$, $b = 0.003641$, $c = -2.37 \times 10^{-7}$
    Diff. $= C_i$ (calc %) $- C_i$

material may be chosen for calibration consists in dividing C_{ir} by I_{ir} for the seven reference materials. The fact that these values (i.e., K_i in Eqn (7.51)) vary from 0.0025 to 0.0035 indicates that a more sophisticated approach is definitely required. Calculating K_i using a 'least squares fit' regression constrained through the origin to be used in Eqn (7.51) yields

$$K_i = \frac{\sum C_{ir}^2}{\sum (C_{ir}I_{ir})} = 0.002741 \tag{7.56}$$

and does not fare much better. This is evident if the intensity data for the reference materials are treated as unknowns, which results in unacceptably large differences in the calculated concentrations compared with the known values.

The next step consists in assessing whether Eqn (7.52) is more appropriate to represent the intensity–concentration relationship for the seven reference materials. A least squares fit NOT constrained through the origin yields the values $a = 1.1955$ and $b = 0.002369$. To get an indication of whether these values yield a valid calibration, the seven intensities S1 to S7 are used to recalculate C_i. The results are listed in the column headed C_i (calc %)[a], and indicate that the differences are still too large when compared with the known concentrations. The only conclusion possible is that this approach does not provide a usable calibration.

Having concluded that a linear relationship is not appropriate in this context, the next step consists in evaluating a second order (quadratic) least squares fit, i.e. Eqn (7.53). A regression analysis of the seven reference materials yields the following values: $a = 0.0334$, $b = 0.003641$ and $c = -2.37 \times 10^{-7}$. As shown in the column headed C_i (calc %)[b], this represents a marked improvement over the previous attempts at calibration. This is reflected in Table 7.7(b), where the unknowns are computed using the linear with-intercept model and the second order model, respectively.

7.8 DISCUSSION

There is no doubt that the methods labelled 'global' in this presentation have little more in common than the fact that in all cases the matrix effects are compensated for without knowledge of sample composition or instrumental parameters. As a group, global methods are somewhat complementary in that some are best used for the determination of major constituents, whereas others are best suited for analysis at the trace level. Some require that additions be made to the samples and references; others are applicable to the samples 'as received'. These methods are therefore particularly useful in cases where the concentrations of one or two elements are to be determined in a few samples of truly unknown composition. Thus they are *specific* in the sense that they are generally applied in unique analytical contexts.

It is highly desirable that analysts heed the recommendations of Bertin (1975, Chap. 14) and Nielson (1979), namely that any quantitative analysis of a new sample type should be preceded by a qualitative analysis (2θ scan), and one should bear in mind that global methods are usually underlain by approximations based on assumptions that may limit their application in specific cases. In addition to the above precautionary measures, global methods should always be checked by analyzing suitable standard reference materials prior to applying them to unknowns.

7.9 SUMMARY

The following expressions are found to be applicable for 'global' correction for interelement effects, provided that a number of premises are met, namely in well defined analytical contexts:

The double dilution method:

$$C_{iu} = \left(\frac{C_i}{D_i}\right)_r D_{iu}; \qquad D_i = \frac{I_{i1} \times I_{i2}}{I_{i1} - I_{i2}}$$

The Compton (incoherent) scatter method:

$$C_{iu} = \left(C_i \frac{I_{inc}}{I_i}\right)_r \left(\frac{I_i}{I_{inc}}\right)_u$$

The internal standard method:

$$C_{iu} = K \frac{I_{iz}}{I_{jz}}$$

The standard addition method:

$$C_{iu} = C_{ir} \frac{I_{iu}/I_{iu+}}{1 + (W_u/W_r)\,[\,1 - (I_{iu}/I_{iu+})\,]}$$

The high dilution, dilution with heavy absorber method:

$$C_{iu} = \left(\frac{C_i}{I_{i(r+d)}}\right) I_{i(u+d)}$$

Matrix matching:

C_{iu} = empirical expressions as a function of I_{iu}.

8 SPECIAL ANALYTICAL CONTEXTS

The authors are well aware of the problem in terminology when designating certain analytical contexts as *special*. There is no doubt that what may be considered a special or unusual analytical request in one laboratory may be current practice in another. The aim of this chapter is therefore to examine a number of contexts where one or more criteria that are considered inherent to the ideal fundamental approach context are not met. To state a few examples: limited amount of sample available, low Z analytes or matrices, thin films etc.

8.1 GENERAL CONSIDERATIONS

In broad terms, special situations may be considered as arising if an additional step or steps have to be taken into consideration with regard to an established procedure or with regard to the 'ideal' context, i.e. dealing with thick, homogeneous, flat, stable (etc.) specimens. In some cases this may involve mechanical changes in spectrometer components, such as masking the irradiated area, and will only be addressed minimally. In other cases the problem may be resolved at the level of specimen preparation or by application of adequate mathematical corrections.

Consider the case involving the determination of low Z ($Z < 10$) analytes. The fact that fundamental parameters are less well known for low Z elements, that emitted intensities are very much dependent on the quality of specimen surface, and that scatter is much more prominent undoubtedly constitute a major problem. All these factors must be taken into consideration when contemplating this analytical context. Similarly, the determination of heavier elements in low Z matrices introduces the problem of accentuated matrix effects owing to the low absorption by the major constituents compared with the higher Z analytes.

In addition to expressing concentrations and influence coefficients in terms of oxides rather than elements, fused discs present a special analytical context for a number of reasons. Fusion automatically involves dilution of the sample and may involve loss of volatile constituents during the fusion process or, conversely, involve a 'gain on ignition' due to oxidation during the fusion process. If the flux to sample ratio is constant, there is a tendency for analysts to use modified coefficients in the sense that the term quantifying the matrix effect for the flux is taken in factor and the remaining influence coefficients are modified accordingly. It is also common practice to further modify the influence coefficients so that the correction for matrix effects is expressed in terms of concentration of oxides in the original sample rather than as a function of the concentrations in the fused disc specimen irradiated.

Consider also the case when analysts are called upon to carry out analysis on powdered materials, in which case particle size effects (see Section 9.3.2) come into play. In some instances the quantity of material submitted may be quite limited; hence the intensity measurements may have to be carried out on specimens that do not meet the criterion of 'infinitely thick' in respect to characteristic fluorescence emissions. In other cases, the samples submitted for analysis may be in the form of one or more thin layered films, usually deposited on a substrate. Algorithms have been developed for the determination of both composition and thickness of layered films by X-ray fluorescence techniques; hence the technique is often the method of choice in this important field of high technology.

8.2* LOW Z ANALYTES AND MATRICES

The gradual improvements in X-ray fluorescence instrumentation have led analysts to use this technique for the determination of low Z elements. This presents quite different problems to the analytical context in which higher Z elements are determined in low Z matrices, and therefore the two cases will be treated separately. Subjectively, in both cases low Z will refer to $Z < 10$, i.e. the analytes B, C, N, O and F, and low Z matrices will be representative of organic matter, plant materials, coals, petroleum products, etc.

8.2.1 Low Z analytes

There are a number of factors that come into play as the atomic number of the analytes decrease. There is no demarcation line denoting what constitutes a low Z analyte; it is rather the gradual decrease in intensity due to the fundamental parameters involved that underlies classifying low Z analytes as 'special'. The decrease in intensity can be ascribed to the following:

● conventional X-ray tubes are inefficient as an effective source of energy for

the excitation of low *Z* elements: the long wavelengths are absorbed by the
tube window;
- the excitation factors p_{λ_i} (the product of the probability of the ionization
  of an atom, the fluorescence yield and the probability of emission of the
  characteristic lines) are minimal (see Figure 8.1) and not well known (Broll
  et al., 1992);
- mass attenuation coefficients are quite large (see Table 8.1) and not well
  known for low *Z* elements (see Table 8.2);
- both coherent and incoherent scatter increase and should be taken into
  consideration (Campbell and De Forge, 1989);
- the critical or effective thickness decreases; i.e., the measured characteristic radiations emanate from a very thin surface layer, hence the more
  stringent requirement for a representative and better than average 'mirror
  like' surface;
- the need for a high vacuum path.

8.2.2 Low *Z* matrices

The problems associated with the determination of $Z > 10$ elements in low *Z*
matrices, typified by the determination of iron in carbonaceous material, stem
from the low absorption of the matrix, as illustrated in Figures 8.2 and 8.3.
The analytical context is a tungsten tube target (45 kV) incident source and
$63°/33°$ instrumental geometry. The lower curve in Figure 8.2 represents the
case where $R_i = C_i$, i.e. I_{Fe} is equal to the product $I_{(Fe)}C_i$, whereas the upper

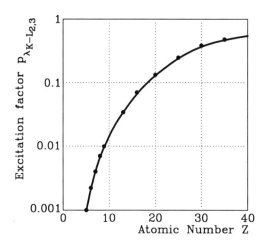

Figure 8.1 Plot of the fraction of the absorbed incident radiation that leads to the
fluorescence emission of a characteristic line p_{λ_i} (excitation factor) versus atomic
number *Z*

Table 8.1 Mass attenuation coefficients $(cm^2 g^{-1})$ of low Z elements for $K-L_{2,3}$ lines of elements $Z = 5$ to 9^a

		B	C	N	O	F
	E (eV)	183	277	392	525	677
Absorber	λ (Å)	67.6	44.7	31.6	23.6	18.3
H (1)		1723	474	157	61	26
He (2)		10677	3190	1136	466	210
Li (3)		32746	10420	3917	1682	789
Be (4)		69937	23698	9372	4195	2039
B (5)		2861?	39834	16610	7754	3902
C (6)		5945	2147	23586	11575	6079
N (7)		10118	3757	1553	16433	8823
O (8)		15774	5998	2529	1181	11928
F (9)		21803	8465	3635	1720	876

^{a} Data (some are rounded off) from Heinrich (1986).

Table 8.2 Mass attenuation coefficients $(cm^2 g^{-1})$ of low Z elements as a function of the $K-L_{2,3}$ lines of selected elements^a

		Mg	Si	Ca	Cr	Fe	Ni
	Analyte keV	1.254	1.740	3.690	5.411	6.399	7.471
Absorber	(Å)	9.887	7.127	3.360	2.291	1.937	1.659
H (a)		3.3	1.1	0.1	0.0	0.0	0.0
(b)		14.22	5.3	0.9	0.6	0.5	0.5
He (a)		28.4	9.5	0.8	0.2	0.1	0.1
(b)		27.32	10.1	1.0	0.5	0.4	0.32
Li (a)		115.2	39.8	3.3	0.9	0.5	0.3
(b)		107.2	39.8	4.1	1.3	0.8	0.58
Be (a)		322.0	114.8	10.1	2.9	1.7	1.0
(b)		282.7	104.9	10.7	3.4	2.0	1.27
B (a)		666.1	246.4	23.3	6.9	4.1	2.5
(b)		599.9	222.5	22.8	7.2	4.3	2.7
C (a)		1147	445.4	46.5	14.5	3.7	5.4
(b)		1153	471.0	46.5	14.2	8.5	5.2
N (a)		1766	705.8	76.4	20.0	14.4	9.0
(b)		1717	701.4	74.8	23.1	13.8	8.6
O (a)		2512	1032	116.5	36.9	22.3	13.9
(b)		2424	990.4	113.0	35.3	21.2	13.2
F (a)		3189	1344	158.6	50.8	30.7	19.3
(b)		3287	1343	162.5	51.2	30.8	19.3

^{a} (a) indicates data (rounded off) from Heinrich (1986, p. 72); (b) data (rounded off) from Leroux and Thinh (1977).

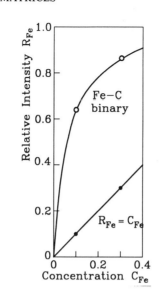

Figure 8.2 Plot of R_{Fe} versus C_{Fe} for binary system Fe–C; carbon is a weak absorber of iron K radiations

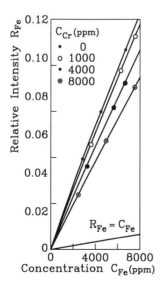

Figure 8.3 Illustration of the strong absorption effect of chromium in a light element matrix: analyte Fe, absorber Cr in a carbon matrix

curve is that of the plot R_{Fe} versus C_{Fe} for the Fe–C system. As a result of the low absorption of both the incident and characteristic iron radiations by carbon, 10% and 30% concentrations of iron have apparent concentrations of 64 and 87% iron, respectively. Figure 8.3, on the other hand, illustrates the drastic matrix effect encountered in the analysis of 98% plus carbon matrices for analyte iron (range 0–8000 ppm) when chromium is also present in the 0–8000 ppm range. In actual fact, the curves in Figures 8.2 and 8.3 are representative of any analyte in the presence of both lower and higher absorbing materials, which translates into negative and positive influence coefficients; in this case m_{FeC} and m_{FeCr} are roughly equal to -0.95 and 2.1, respectively.

A second factor that must be closely monitored is specimen thickness. J. P. Willis (1991) pointed out that, because of the low absorption of low Z matrices, the critical thickness (Eqn (8.5)) in respect to X-ray fluorescence emissions, e.g., analyte Mo $K-L_{2,3}$ in low ash coals, is of the order of 5 cm, whereas that for Rh $K-L_{2,3}$ is almost 10 cm. Willis goes on to warn analysts that, since it is impossible in most spectrometers to load samples thicker than ≈ 4 cm, it is also impossible to achieve infinite thickness in this context. Willis also points out that, for all spectrometers, the problem may be not only lack of infinite thickness but lack of infinite width; i.e., analysts using Mo or Rh target tubes must take considerably care in evaluating possible infinite thickness and infinite width problems in specimens of low mass absorption coefficient.

8.3   LIMITED SAMPLE QUANTITY

In keeping with the general trend concerning terminology in this chapter, the definition of what constitutes a limited quantity is subjective. It may range from the more common case of not receiving quite enough material to prepare specimens by an established procedure to receiving so little material that analysis by X-ray fluorescence is nearly impossible because of the difficulty in obtaining a single usable intensity measurement.

In some cases, the problem of limited sample quantity may be solved by the simple process of masking; i.e., the established procedure is followed to prepare smaller sized specimens, which are then irradiated using holders having smaller diameter masks. This involves recalibration under the same conditions and with longer counting times if the same precision is to be maintained. Another possibility is to dilute the samples so that there is enough material to proceed with a specimen preparation procedure already in place. In either case, the problem is surmounted only in part.

If handling very small quantities is the norm, recent developments in total reflection X-ray fluorescence spectrometry (TRXRF) provide a powerful tool for quantitative microanalysis. Prange and Schwenke (1992) examine a number of sample preparation procedures which have been adapted for TRXRF and have stood the test of practical application. Potential applications and special features of the technique are summarized as follows:

- quantitative microanalyses of samples in the 10–100 μg range;
- forensic applications;
- microinclusions;
- minute amounts of radioactive solutions;
- multielement capability down to parts per trillion;
- environmental samples like rainwater, seawater etc.

8.4* THIN SPECIMENS

As shown in Section 2.3, unless specimens meet the criteria of being 'infinitely thick' in respect to X-ray fluorescence emissions, the emitted intensity of a characteristic line will be a function of the specimen thickness. For a monochromatic incident excitation source of wavelength λ, the fluorescence intensity I_t emitted by a specimen of finite thickness t is deduced from Eqn (2.21), leading to

$$I_t = C_i \frac{Q_{i\lambda}}{\mu_s^*} (1 - e^{-k}) \tag{8.1}$$

where

$$Q_{i\lambda} = q p_{\lambda_i} I_\lambda \mu_{i\lambda} \tag{8.2}$$

$$k = \rho \mu_s^* t \tag{8.3}$$

A specimen is considered as infinitely thick in respect to X-ray fluorescence emissions when t is sufficiently large to make the exponential term in the above equation negligible compared with unity. The intensity I_∞ is then given by

$$I_\infty = C_i \frac{Q_{i\lambda}}{\mu_s^*} \tag{8.4}$$

Conventionally that condition is met if the ratio I_t/I_∞ is greater than 0.999, i.e. the measured intensity is at least 99.9% of the intensity that would be emitted if the specimen were much thicker. This is commensurate with k in Eqn (8.3) being equal to or greater than 6.91, hence

$$t_{\text{critical}} = \frac{6.91}{\rho \mu_s^*} \tag{8.5}$$

It follows from rationing Eqns (8.1) and (8.4) that

$$(I_t/I_\infty) = 1 - e^{-k} \tag{8.6}$$

For sufficiently small values of k, $1 - e^{-k}$ is approximately equal to k; hence, substitution in the above leads to

$$(I_t/I_\infty)_{k \ll 1} = k \tag{8.7}$$

Given that $k = \rho \mu_s^* t$ and that ρ and μ_s^* are constants for a given specimen under fixed operating conditions, the values listed in Table 8.3 or the plot (Figure 8.4) of the ratio I_t/I_∞ as a function of k (hence indirectly as a function of t) clearly demarcate the three regions observed by Liebhafsky and Zemany (1956):

(1) the linear region, i.e. thin specimens (Eqn 8.7); e.g. $(I_t/I_\infty)_{k \ll 1} = k$

Table 8.3 Intensity ratio I_t/I_∞ as a function of $k = \rho \mu_s^* t$

k	I_t/I_∞	Rel. error	Region
0.001	0.00100	0.0	
0.01	0.00995	0.5	
0.02	0.01980	1.0	(1)
0.05	0.04877	2.4	
0.10	0.09516	4.8	
0.20	0.1813		
0.50	0.3935		
0.6931	0.5000		(2)
1.0	0.6321		
2.0	0.8647		
5.0	0.99326		
6.91	0.99900		(3)
10.0	0.99996		

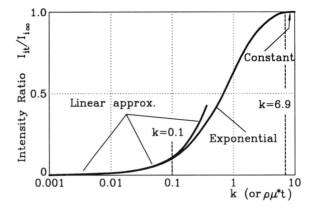

Figure 8.4 Intensity ratio $I_{it}/I_{i\infty}$ as a function of $k = \rho \mu^* t$

(2) the exponential region, i.e. intermediate thickness (Eqn 8.6); $(I_t/I_\infty) = 1 - e^{-k}$

(3) the constant intensity region, i.e. thick specimens; I_t/I_∞ is essentially equal to unity (Eqn 8.6), e.g. $(I_t/I_\infty)_{k > 6.91}$.

8.5* THIN FILMS

X-ray fluorescence has found wide application in the analysis of thin films. In many instances thin films consist of coatings deposited on substrates, and therefore the compositions of both film and substrate are not true unknowns but generally materials of a type, so that their approximate compositions are known. Initially, empirical methods were used for the simultaneous determination of the thickness and composition based on well characterized sets of similar reference materials. Subsequently, methods were developed based on the fundamental parameters approach, allowing the determination of thickness and composition of single and multi-layer thin films, in some instances via calibrations using thick reference materials. For example, calibration curves I_t versus t may be generated by computing theoretical values of I_t for a given material by varying the thickness in the exponent k in Eqn (8.6)

$$I_t = I_\infty(1 - e^{-k}) \tag{8.8}$$

8.5.1 Determination of film thickness

Let us consider the simple case of a thin coating of pure element i on a substrate of pure element j . For monochromatic incident wavelength λ, an expression for thickness t of layer i may be derived from Eqn (8.8) transformed to

$$e^{-k} = 1 - \frac{I_{ic}}{I_{(i)}} \tag{8.9}$$

where c stands for coating. Taking the logarithm of each side

$$k = -\ln\left(1 - \frac{I_{ic}}{I_{(i)}}\right) \tag{8.10}$$

Combining the above with Eqn (8.3) and solving for t yields

$$t = -\frac{1}{\rho\mu_i^*} \ln\left(1 - \frac{I_{ic}}{I_{(i)}}\right) \tag{8.11}$$

where

$$\mu_i^* = \mu_{i\lambda} \csc \psi' + \mu_{i\lambda_i} \csc \psi'' \tag{8.12}$$

In contrast to the above method, in which the film thickness is determined by measurement of a characteristic line of element i, the thickness of the thin film may also be determined using a characteristic line of the substrate element j. The context may then be visualized as a coated specimen that undergoes partial absorption of the incident radiation and of the fluorescent radiation characteristic of element j. As shown by Zemany and Liebhafsky (1956), the relationship between the intensities emitted by element j from the coated specimen and the uncoated substrate in this case results in the expression

$$\frac{I_{jc}}{I_{(j)}} = \exp(-\rho\mu^*_{i\lambda_j}t) \tag{8.13}$$

where I_{jc} is the intensity of the characteristic line of element j for the coated specimen and $I_{(j)}$ is the intensity for a bulk specimen of pure element j, and

$$\mu^*_{i\lambda_j} = \mu_{i\lambda} \csc \psi' + \mu_{i\lambda_j} \csc \psi'' \tag{8.14}$$

The thickness t (cm) is then given by

$$t = -\frac{1}{\rho\mu^*_{i\lambda_j}} \ln\left(\frac{I_{jc}}{I_{(j)}}\right) \tag{8.15}$$

and the coating weight W (mg cm^{-2}) by

$$W = -\frac{1}{\mu^*_{i\lambda_j}} \ln\left(\frac{I_{jc}}{I_{(j)}}\right) \tag{8.16}$$

8.5.1.1 Numerical examples

The thickness t in the above expressions is expressed in cm units. Analysts generally prefer to work in units of weight per unit area, i.e. the product thickness times density. Commonly, coating weight W is expressed in mg cm^{-2} units; hence

$$W \text{ (mg cm}^{-2}) = t \text{ (cm)} \times \rho \text{ (g cm}^{-3}) \times 1000 \tag{8.17}$$

Data in Table 8.4 relate to coatings of gold on a copper substrate. In the present context, the incident excitation source wavelength is 0.615 Å (Rb $K-L_{2,3}$) and the angles of incidence and emergence are each equal to $45°$ (csc $45° = 1.4142$). Given μ_{Au} for λ (Rb $K-L_{2,3}$) = 77.08 cm^2 g^{-1} and μ_{Au} for λ (Au $L_3-M_{4,5}$) = 125.24 cm^2 g^{-1}, the effective mass attenuation coefficient of the coating is given by Eqn (8.12)

$$\mu^*_{Au} = (77.08 \times 1.4142) + (125.24 \times 1.4142)$$
$$= 286.12 \text{ cm}^2 \text{ g}^{-1}$$

The first four rows (Nos 1–4) represent the linear region, where there is a linear relationship between I_{Au} and the thickness of the Au film. The next four rows (Nos 5–8) are representative of gold coating standard reference materials (National Institute for Standards and Technology SRM 2318a), namely,

Table 8.4   Theoretical values of $k = \rho\mu_s^* t$ and intensity ratio $I_{Au,c}/I_{(Au)\infty}$ as a function of specimen thickness t, gold coating over copper substrate

No.	t (cm)	W (mg cm^{-2})	t (Å)	k	$I_{Au,c}/I_{(Au)\infty}$
1	1.0×10^{-8}	0.00019	1	5.551×10^{-5}	5.551×10^{-5}
2	1.0×10^{-7}	0.0019	10	5.551×10^{-4}	5.549×10^{-4}
3	1.0×10^{-6}	0.0194	100	0.00555	0.00553
4	1.0×10^{-5}	0.194	1000	0.0555	0.0540
5	8.0×10^{-5}	1.552	8000	0.4441	0.359
6	1.5×10^{-4}	2.910		0.8326	0.565
7	3.0×10^{-4}	5.820		1.6652	0.811
8	7.0×10^{-4}	13.58		3.8855	0.979
9	1.0×10^{-3}	19.4		5.551	0.996
10	1.24×10^{-3}	21.1		6.91	0.999
11	2.0×10^{-3}	38.8		11.1	0.9999+

nominal coatings 0.8, 1.5, 3 and 7 μm thick on a copper substrate, for which the average weights per unit area are determined by weight and area measurements. The thickness range covers the exponential region in Figure 8.4; hence these materials may be used for calibration for the measurement of weight per unit area of gold coatings of equivalent purity. The last three rows (Nos 9–11) relate to coatings that can be considered infinitely thick with respect to Au L_3–$M_{4,5}$ X-ray emissions.

Example 1

Let us consider that a pure gold bulk specimen will be used for calibration and that $I_{Au,c}$ is the intensity for the gold-coated specimen. The ratio $I_{Au,c}/I_{(Au)}$ is given by Eqns (8.3) and (8.6). For $t = 8 \times 10^{-5}$ cm (sample 5)

$$k = \rho\mu_{Au}^* t$$
$$= 19.4 \times 286.12 \times 8 \times 10^{-5}$$
$$= 0.4441 \qquad\qquad (8.18)$$

$$I_{Au,c}/I_{Au} = 1 - e^{-0.4441}$$
$$= 0.3586$$

and for sample 6, where $t = 1.5 \times 10^{-4}$ cm

$$I_{Au,c}/I_{(Au)} = 0.5651$$

The determination of the thickness of gold films in this context would then involve measuring $I_{Au,c}$ and $I_{(Au)}$ and computing the ratios $I_{Au,c}/I_{(Au)}$.

Consider the case where this ratio is 0.8108 (sample 7). The thickness is then given by Eqn (8.11):

$$t = -\frac{1}{\rho \mu_{Au\lambda}^{*}} \ln\left(1 - \frac{I_{Au,c}}{I_{(Au)}}\right)$$

$$= -\frac{1}{19.4 \times 286.12} \ln(0.1892)$$

$$= 3.0 \times 10^{-4} \text{ cm} \tag{8.19}$$

or the coating weight can be obtained directly from

$$W = \frac{1000}{\mu_{Au}^{*}} \ln\left(1 - \frac{I_{Au}}{I_{(Au)}}\right)$$

$$= -\frac{1000}{286.12} \ln(0.1892)$$

$$= 5.82 \text{ mg cm}^{-2} \tag{8.20}$$

Example 2

Table 8.5 relates to the same specimens except that the thickness of the gold films is determined by the 'absorption method', i.e. by measuring the intensity of the coated specimens and that of a pure bulk specimen of the substrate element, in the present context copper. Given that μ_{Au} for the λ (Cu $K-L_{2,3}$) = 203.64, the effective mass attenuation coefficient obtained from Eqn (8.14) is

$$\mu_{Au\lambda Cu}^{*} = (77.08 \times 1.4142) + (203.64 \times 1.4142)$$

$$= 396.98$$

Table 8.5   Theoretical values for $k = \rho \mu_{s}^{*} t$ and intensity ratio $I_{Cu}/I_{(Cu)\infty}$ as a function of specimen thickness t, gold coating over copper substrate

No.	t (cm)	W (mg cm^{-2})	t (Å)	k	$I_{Cu}/I_{(Cu)\infty}$
1	1.0×10^{-8}	0.00019	1	7.702×10^{-5}	0.9999+
2	1.0×10^{-7}	0.0019	10	7.702×10^{-4}	0.9992
3	1.0×10^{-6}	0.0194	100	0.00770	0.992
4	1.0×10^{-5}	0.194	1000	0.0770	0.926
5	8.0×10^{-5}	1.552	8000	0.616	0.540
6	1.5×10^{-4}	2.910		1.155	0.315
7	3.0×10^{-4}	5.820		2.310	0.0993
8	7.0×10^{-4}	13.58		5.391	0.0046
9	1.0×10^{-3}	19.4		7.702	0.00045
10	1.24×10^{-3}	24.1		9.550	0.0000
11	2.0×10^{-3}	38.8		15.4	0.0000

Substitution of the appropriate values in Eqn (8.13) yields in the case of sample 5, where $t = 8.0 \times 10^{-5}$ cm

$$\frac{I_{Cu,c}}{I_{(Cu)}} = \exp(-19.4 \times 396.98 \times 0.00008)$$

$$= 0.540$$

On the other hand, if the ratio of the emitted Cu intensity for the coated substrate to that of the uncoated substrate is known, i.e. measured experimentally, then the thickness of the coating is given by Eqn (8.15):

$$t = -\frac{1}{\rho \mu_{i\lambda_j}^{\star}} \ln\left(\frac{I_{jc}}{I_{(j)}}\right) \tag{8.21}$$

Substitution in the context $I_{Cu,c}/I_{(Cu)} = 0.0993$ (sample 7) yields

$$t = -\frac{\ln 0.0993}{19.4 \times 396.98}$$

$$= 3.0 \times 10^{-4} \text{ cm}$$

and, for W, the coating weight

$$W = -\frac{\ln 0.0993}{396.98} \times 1000$$

$$= 5.82 \text{ mg cm}^{-2}$$

The theoretical data in rows 5–8 in Tables 8.4 and 8.5 are in good agreement with experimental data reported by J. E. Willis (1988).

8.5.2 Multi-layer thin films

The above numerical examples relating to the determination of thickness or coating weight are representative of the simplest context, namely a thin layer of a pure element on a pure element substrate. They can also apply to coatings and substrates when the measured element is not present in both. Practical applications may require the determination not only of the thickness but also the composition of multi-layered substrates. The process is then inherently more complex, given that the following criteria have to be taken into consideration (Mantler, 1984):

- all secondary enhancement effects from elements within the same layer as well as in the other layers and the substrate;
- the contribution of backscattered X-ray tube radiation from substrates;
- the need to use iterative procedures; in principle all concentrations and thicknesses can be calculated from experimental count rates provided that no measured element is present in more than one layer;

● the maximum number of unknowns is restricted by the errors in the intensities and the resulting numerical instability of the system of equations.

8.6 PARTICULATES

That the fluorescence intensities from non-homogeneous and homogeneous specimens of the same composition may be quite different is well known. Thus, if samples consisting of small, separate particles are used directly as specimens for X-ray fluorescence analysis, one must consider *heterogeneity effects*. Following Tertian and Claisse (1982), the term heterogeneity effect is retained rather than particle size effect because it is more encompassing. Claisse (1957) and Claisse and Samson (1962) proposed a semi-quantitative interpretation of the origin of heterogeneity effects based on microabsorption; i.e., if a specimen is made up of two or more phases, and if the size of the heterogeneities is large compared with the penetration depth of the X-rays, then most of the fluorescence process occurs in single particles. This being the case, the fluorescence intensity depends on the composition of the phase that contains the analyte and is almost independent of the composition of the other phases. If the size of the particles is decreased, a greater number of photons have a fraction of their path lengths in more than one particle and the emitted intensity then depends more and more on the overall composition of the specimen (see Section 9.3).

8.7 OXIDE SYSTEMS

Although it is entirely feasible to consider oxygen (or other elements) as simply another constituent in samples such as rocks, soils, cements, etc., analysts have generally shown a preference for expressing concentrations, intensities, influence coefficients etc. in terms of oxides (or more complex compounds). Consider for example the system i–j–o, where o represents the element oxygen and C_{io}, I_{io} designate oxides. From Eqn (1.27)

$$\mu_{io}^* = C_i \mu_i^* + C_o \mu_o^* \tag{8.22}$$

$$\mu_{jo}^* = C_j \mu_j^* + C_o \mu_o^* \tag{8.23}$$

where both $C_i + C_o$ and $C_j + C_o = 1$. The intensity for the binary $C_{io} + C_{jo} = 1$ is given by

$$I_{io} = \frac{I_\lambda C_{io} \mu_{i\lambda}}{C_{io} \mu_{io}^* + C_{jo} \mu_{jo}^*} \tag{8.24}$$

and for the pure oxide io

$$I_{(io)} = \frac{I_\lambda \mu_{i\lambda}}{\mu_{io}^*} \qquad (8.25)$$

Ratioing and solving for I_{io} leads to

$$I_{io} = \frac{I_{(io)} C_{io} \mu_{io}^*}{C_{io} \mu_{io}^* + C_{jo} \mu_{jo}^*} \qquad (8.26)$$

Substituting $1 - C_{jo}$ for C_{io} in the denominator of the above equation and solving for C_{io} yields

$$C_{io} = \frac{I_{io}}{I_{(io)}} \left[1 + C_{jo} \left(\frac{\mu_{jo}^* - \mu_{io}^*}{\mu_{io}^*} \right) \right] \qquad (8.27)$$

defining

$$a_{ijo} = \left(\frac{\mu_{jo}^* - \mu_{io}^*}{\mu_{io}^*} \right) \qquad (8.28)$$

and

$$R_{io} = \frac{I_{io}}{I_{(io)}} \qquad (8.29)$$

It is customary to drop the subscript o when it is clearly evident that one is dealing with an oxide context and to revert to the conventional formalism:

$$C_i = R_i \left[1 + \sum_j a_{ij} C_j \right] \qquad (8.30)$$

8.7.1 Numerical example

The following numerical example relates to a three-oxide mixture: 20% of oxide i, the analyte consisting of 50% i and 50% oxygen; 40% of oxide j, consisting of 70% j and 30% oxygen; and 40% of oxide k, consisting of 60% k and 40% oxygen. Let us say that $\mu_i = 100$, μ_j 250, $\mu_k = 400$ and $\mu_o = 50 \text{ cm}^2 \text{g}^{-1}$. For pure oxide io, the mass attenuation coefficient is

$$\mu_{io} = (0.50 \times 100) + (0.50 \times 50)$$
$$= 75 \text{ cm}^2 \text{g}^{-1}$$

Similarly, for oxides j and k, μ_{jo} and μ_{ko} are equal to 190 and 260 $\text{cm}^2 \text{g}^{-1}$, respectively. The mass attenuation coefficient for the mixture expressed in terms of oxides is

$$\mu_{mix} = (0.20 \times 75) + (0.40 \times 190) + (0.40 \times 260)$$
$$= 195 \text{ cm}^2 \text{g}^{-1}$$

indicating that the mixture is $195/75 = 2.6$ times more absorbing than the pure oxide i. Assuming that enhancement is negligible, influence coefficients can be computed directly from oxide mass attenuation coefficients, namely

$$m_{ij} = a_{ij} = \frac{\mu_{jo} - \mu_{io}}{\mu_{io}}$$

$$= \frac{190 - 75}{75} = 1.5333 \qquad (8.31)$$

$$a_{ik} = \frac{260 - 75}{75} = 2.4667$$

The verification that oxide influence coefficients are applicable for the specimen proceeds as follows.

(a) Calculate R_i for pure oxide i and the specimen in terms of the elemental influence coefficients a_{ij}, a_{ik} and a_{io}

$$a_{ij} = \frac{250 - 100}{100} = 1.50$$

$$a_{ik} = \frac{400 - 100}{100} = 3.00$$

$$a_{io} = \frac{50 - 100}{100} = -0.50$$

For the mixture with elemental concentrations $C_i = 0.10$, $C_j = 0.28$, $C_k = 0.24$ and $C_{oxygen} = 0.38$

$$R_i = \frac{0.1}{1 + (1.50 \times 0.28) + (3.0 \times 0.24) + (-0.50 \times 0.38)}$$

$$= 0.05128$$

and for the pure oxide

$$R_i = \frac{0.5}{1 + (-0.50 \times 0.5)} = 0.66667$$

(b) Given that by definition R_{io} for the pure oxide io is equal to unity, it therefore follows that R_{io} for the specimen is given by

$$R_{io} = \frac{0.05128}{0.66667} = 0.07692$$

Substitution in Eqn (8.30), in which the subscripts o have been dropped even though the model refers to an oxide context, yields

$$C_i = 0.07692 [1 + (1.5333 \times 0.40) + (2 \cdot 4667 \times 0.40)]$$
$$= 0.20$$

Although this treatment was examined in terms of absorption effects only, the concept is equally applicable for fundamental influence coefficients m_{ijo}. The same properties apply to m_{ij} and m_{ijo} coefficients; they are variables as a function of specimen composition. If m_{ijo} coefficients are treated as constants in a well defined analytical context, it is preferable to use the symbol α_{ij} to indicate that this involves an approximation, i.e.

$$C_i = R_i \left[1 + \sum_j \alpha_{ij} C_j \right] \tag{8.32}$$

rather than Eqn (8.30).

8.8* FUSED DISC SPECIMENS

The practice of eliminating particle size and mineralogical effects by preparing solid solutions (fused discs) when dealing with powdered rocks, cements, etc., is well established, and is treated in the next chapter. Fused discs present a clear-cut case of the specimen (the material submitted for irradiation) and the sample (the material submitted for analysis) being quite different. The fusion process automatically introduces a diluting effect and subjects the sample to high temperatures; hence the following must be taken into consideration:

- the different values for $C_i, C_j, \ldots$ in the sample and in the specimen;
- the matrix effect of the flux;
- the loss on fusion LOF if volatile materials are present;
- the gain on fusion GOF if materials in a reduced state are present and are oxidized during the fusion process.

8.8.1 Influence coefficients modified for flux

Consider the relationship between the analytical expression Eqn (8.32) correcting for matrix effects in terms of oxides in the original sample and in the fused disc of that sample. For a multielement sample

$$C_i = R_i [1 + \alpha_{ij} C_j + \cdots + \alpha_{in} C_n]$$
$$= R_i \left[1 + \sum_j \alpha_{ij} C_j \right] \tag{8.33}$$

where $C_i, C_j, \ldots$ are concentrations in the original sample, $\alpha_{ij}, \alpha_{ik}, \ldots$ are oxide influence coefficients, R_i is the intensity *relative to the pure oxide*, and

$$C_i + \sum_j C_j = 1.0 \tag{8.34}$$

For the fused disc of that sample containing a concentration C_f of flux and $(1 - C_f)$ of the sample, i.e.

$$C_{sample} + C_{flux} = 1.0 \tag{8.35}$$

Eqn (8.33) becomes

$$(1 - C_f)C_i = R_i\left[1 + \sum_j \alpha_{ij}C_j(1 - C_f) + \alpha_{if}C_f\right] \tag{8.36}$$

where R_i now relates to intensities measured for fused discs relative to the pure analyte, i.e.

$$R_i = \frac{C_i(1 - C_f)}{1 + \sum_j \alpha_{ij}C_j(1 - C_f) + \alpha_{if}C_f} \tag{8.37}$$

and

$$R_{i,fd} = \frac{I_{i,fd}}{I_{(i),fd}} \tag{8.38}$$

The matrix coefficients α_{ij} are basically the same in Eqns (8.32) and (8.36) but, being a function of composition, they would take on slightly different values.

It is generally common practice when preparing fused discs to use fixed weights of sample and flux. It can be shown that if the weight ratio of sample to flux is in the range 0.2 or less, it follows that in this context the α coefficients may be considered as constants. Inasmuch as the product $\alpha_{if}C_f$ is constant in Eqn (8.36) it may be taken out of the correction term. Rearranging terms and solving for C_i in the sample:

$$(1 - C_f)C_i = R_i(1 + \alpha_{if}C_f)\left[1 + \sum_j \frac{\alpha_{ij}}{1 + \alpha_{if}C_f} C_j(1 - C_f)\right] \tag{8.39}$$

$$C_i = R_i \frac{1 + \alpha_{if}C_f}{1 - C_f}\left[1 + \sum_j \frac{1 - C_f}{1 + \alpha_{if}C_f} \alpha_{ij}C_j\right] \tag{8.40}$$

or

$$C_i = R_i \frac{1}{f}\left[1 + \sum_j f\alpha_{ij}C_j\right] \tag{8.41}$$

where

$$f = \frac{1 - C_f}{1 + \alpha_{if}C_f} \tag{8.42}$$

is the flux factor and is always < 1.

Thus the conventional formalism of Eqn (8.32) can be retained to yield C_i in the original sample from fused disc intensities relative to the fused disc intensity of pure analyte i, with

$$R_{i,fd} = R_i \, \frac{1}{f} \tag{8.43}$$

and

$$\alpha_{ij,fd} = f\alpha_{ij} \tag{8.44}$$

When there is no ambiguity that one is dealing with an oxide context and fused discs, the subscripts fd are usually dropped and Eqn (8.43) is simplified to the conventional form

$$C_i = R_i \left[1 + \sum_j \alpha_{ij}C_j \right] \tag{8.45}$$

where C_i relates to the concentration of oxide i, R_i relates to the intensity $I_{i,fd}$ relative to $I_{(i),fd}$, and a_{ij} relates to oxide influence coefficients modified for a fused disc context.

Thus, Eqn (8.45) satisfies analysts who generally prefer to visualize the correction model in terms of the concentrations of the oxides *in the sample* rather than concentrations in the fused disc specimen. Summarizing, dilution or fusion of a sample with a flux does not change the conventional formulation; it merely modifies both the matrix effect coefficients and the relative intensity as a function of the flux factor. The reader is reminded that, although the influence coefficients are generally treated as constants when the concentration of the flux in the fused disc is roughly 80% plus, for lower flux concentrations and better accuracy the influence coefficients $\alpha_{ij}, \alpha_{ik} \ldots \alpha_{if}$ may be defined in COLA formalism (Section 5.7.1), i.e.

$$\alpha_{ij} = \alpha_{ij,hyp} + \sum_k \alpha_{ijk}C_k \tag{8.46}$$

8.8.1.1 *Numerical example*

Data in Table 8.6 are used to illustrate the computation of modified influence coefficients for the correction of matrix effects in fused disc specimens. The data relate to three samples: a pure oxide and two binary oxides. All three are fused using a mixture made up of 1.5 g of sample and 6.0 g of flux. The following computations are carried out.

Table 8.6 Theoretical data used to illustrate the concept of modified influence coefficients in the context of fused disc specimens

(a) Context

$$W_s = 1.5 \text{ g} \qquad W_f = 6.0 \text{ g}$$
$$C_s = 0.2 \qquad C_f = 0.8$$
$$\alpha_{ij} = 1.4 \qquad \alpha_{if} = -0.75$$

(b) Samples

Sample	C_i	C_j	R_i	$R_{i,fd}$
1	1.00	0.00	0.5	1.0
2	0.20	0.80	0.06410	0.12820
3	0.30	0.70	0.10067	0.20134
4	0.60	0.40	0.23438	0.46875
5	0.80	0.20	0.35088	0.70175

(a) The flux factor given by Eqn (8.42) is

$$f = \frac{1 - C_f}{1 + \alpha_{if}C_f}$$

$$f = \frac{1 - 0.8}{1 + (-0.75 \times 0.8)} = 0.5$$

(b) The intensity relative to that of the pure oxide for fused discs is obtained by Eqn (8.37); for disc 1

$$R_i = \frac{1.0 \times (1 - 0.8)}{1 + (-0.75 \times 0.8)} = 0.5$$

and for disc 3

$$R_i = \frac{0.3 \times (1 - 0.8)}{1 + [1.4 \times 0.7 \times (1 - 0.8) + (-0.75 \times 0.8)]}$$

$$= 0.10067$$

(c) The relative intensity is given by Eqn (8.43); in the case of sample 3

$$R_{i,fd} = 0.10067 \times \frac{1}{0.5}$$

$$= 0.20134$$

(d) The oxide coefficient modified for the fused disc $\alpha_{ij,fd}$ is given by Eqn (8.44)

$$\alpha_{ij,fd} = 0.5 \times 1.4$$
$$= 0.7$$

(e) Substitution in Eqn (8.45) yields in the case of sample 3

$$C_i = 0.20134[1 + (0.7 \times 0.7)]$$
$$= 0.3$$

and in the case of sample 4

$$C_i = 0.46875[1 + (0.7 \times 0.4)]$$
$$= 0.6$$

As expected, the relative intensity for disc 1 is unity, i.e.

$$R_{i,fd} = 0.5 \times \frac{1}{0.5}$$
$$= 1$$
$$= R_{(i),fd}$$

The change of fluorescence intensity due to dilution can be calculated by Eqn (8.37); i.e. R_i for disc 1 is 0.5, a loss of 50% relative, but in the case of disc 2 the loss is only 34% after a 5-fold dilution in both cases, as shown in Figure 8.5.

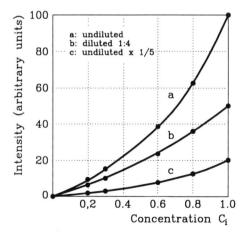

Figure 8.5 Illustration of the moderate decrease in intensity for dilution with low absorbers. Line b represents emitted intensity after a five-fold dilution; line c is given by undiluted intensity divided by five

8.8.2 Correction for loss on fusion (LOF)

The advantage of using a fusion procedure to eliminate heterogeneity effects may be offset by a problem which, unless it is addressed, may lead to erroneous results. The high temperatures required to melt the fluxing compound in order to dissolve samples and produce fused discs can lead to the loss of some components during the fusion process. Carbonates, hydrated minerals, sulphides, organic matter, etc., which as a group constitute what may be termed 'volatile', are likely to produce CO_2, H_2O, etc., i.e. compounds that are lost and thus not present in the fused disc. Analysts have a number of options for dealing with 'volatiles'.

(a) A simple procedure consists in determining the loss on ignition prior to fusion. The ignited residue is then treated in the normal manner and the results obtained are converted to concentrations in the original samples.

(b) A second procedure consists in igniting a normal weight of sample and mixing the residue with a quantity of flux that maintains the sample to flux ratio constant. The analysis then proceeds normally followed by conversion of the results to that in the original sample. Consider for example samples A and B, each containing equal concentrations of i and j. Sample A has no LOF, and if 1.5 g of A is fused after the addition of 6.0 g of flux (1.5×4), the concentrations of C_i and C_j in the fused disk are equal to 10%. Sample B contains 10% of volatiles; hence $C_i = C_j = 0.45$. If 1.5 g of sample B is ignited, the weight after ignition is 1.35 g. Now, if 5.4 g of flux (1.35×4) is used instead, the concentrations of C_i and C_j in fused disc B are still equal to 10%. This will result in $C_i = C_j = 0.5$, which must then be corrected for the presence of 10% of volatiles; i.e. the concentrations in B are given by $C_i = 0.5 \times 0.9 = 0.45 = C_j$.

(c) A third option consists in applying a correction based on the concentration of the volatiles C_v where $C_v = C_{LOF}$; i.e. the term $\alpha_{iLOF}C_{LOF}$ is included in the conventional formalism based on oxides

$$C_i = R_i \left[1 + \sum_j \alpha_{ij}C_j + \alpha_{iLOF}C_{LOF} \right] \qquad (8.47)$$

Tertian (1975) and Tertian and Claisse (1982) visualize the correction for LOF by considering the volatile constituents expelled during fusion as ghost components at the original concentration but with no absorption, i.e. μ_{LOF} is zero. The practical significance is that the analytical sum remains unchanged; for example, in a three element specimen

$$C_i + C_j + (C_k - C_v) + C_{LOF} = 1 \qquad (8.48)$$

Given that the mass attenuation coefficient μ_{LOF} is equal to zero, the influence coefficient α_{iLOF} can be determined from the expression

$$\alpha_{iLOF} = \frac{0.0 - \mu_i}{\mu_i}$$

$$= -1.0 \qquad\qquad (8.49)$$

and the loss on fusion coefficient modified for the flux $\alpha_{iLOF,fd}$ is given by Eqn (8.44):

$$\alpha_{iLOF,fd} = f \times (-1.0) \qquad\qquad (8.50)$$

From a different point of view, de Jongh (1979) identified LOF with an element of atomic number zero, which constitutes an equivalent approach. The advantage, however, is that the de Jongh algorithm allows the analyst to select any one component for elimination in the correction process. Hence the recommendation by de Jongh that the 'volatile' be selected for elimination because its concentration is generally unknown unless determined by an alternative method.

It is interesting to note that the computation of $\alpha_{iLOF,fd}$ need not involve knowledge of the composition of the volatile materials. The fact that $\alpha_{iLOF,fd}$ can be computed from the knowledge of α_{if} and C_f, namely

$$\alpha_{iLOF,fd} = -\frac{1 - C_f}{1 + \alpha_{if}C_f} \qquad\qquad (8.51)$$

led Lachance (1979) to visualize the α_{iLOF} coefficient as 'an α_{if} coefficient in disguise', i.e. relating mainly to a correction for the excess of flux in fused discs if components are volatilized during fusion.

8.8.2.1 Numerical example

The analytical context of Table 8.6 is retained in Table 8.7, i.e. the α_{ij}, α_{if}, C_s, C_f, W_s and W_f values are identical in both tables. Sample 1 relates to the pure oxide i, hence the weights involved are 1.5 g of oxide i and 6.0 g of flux. There being no volatiles, the weight of the fused disc is 7.5 g. Sample 2 relates to a binary, oxides i and j again with no volatiles present. Both sample 3 and sample 4 contain 20% of oxide i as in sample 2 but have 10 and 30% volatiles, respectively. Thus the volatile fraction lost during fusion leads to $W_{disc} < 7.5$ g. The last three samples also contain volatiles at the expense of both C_i and C_j. Because sample 2 relates to oxides i and j, and no volatile constituents, the fused disc weighs 7.5 g and $\alpha_{ij,mod}$ is equal to 0.7, as shown previously in relation to Table 8.6.

Table 8.7 Theoretical data used to illustrate the application of a correction for constituents that volatilize during fusion in the context of fused disc specimens

(a) Context

$$W_s = 1.5 \text{ g} \qquad W_f = 6.0 \text{ g}$$
$$C_s = 0.2 \qquad C_f = 0.8$$
$$\alpha_{ij} = 1.4 \qquad \alpha_{if} = -0.75$$

(b) Samples

Sample	C_i	C_j	C_{LOF}	W_{disc} (g)	$R_{i,fd}$
1	1.00	0.00	0.00	7.50	1.00
2	0.20	0.80	0.00	7.50	0.12821
3	0.20	0.70	0.10	7.35	0.13889
4	0.20	0.50	0.30	7.05	0.16667
5	0.18	0.72	0.10	7.35	0.12380
6	0.16	0.64	0.20	7.20	0.11869
7	0.14	0.56	0.30	7.05	0.11269

Computations are carried out in the following sequence:

(a) R_i for disc 2 is calculated by substitution in Eqn (8.37)

$$R_i = \frac{0.2 \times (1 - 0.8)}{1 + (1.4 \times 0.8 \times (1 - 0.8) + (-0.75 \times 0.8)}$$

$$= 0.064103$$

(b) $R_{i,fd}$ is calculated using Eqn (8.43); the value of f is identical to the computation relating to Table 8.6, i.e. $f = 0.5$:

$$R_{i,fd} = 0.064103 \times \frac{1}{0.5}$$

$$= 0.12821$$

(c) The calculation of R_i for samples 3–7, where the weight of the fused disc is no longer equal to 7.5 g owing to loss on ignition, requires the calculation of C_i, C_j and C_f in the specimen, i.e. the fused disc; thus, in the case of disc 3

$$C_{i,fd} = \frac{C_i \times W_s}{W_{disc}} = \frac{0.2 \times 1.5}{7.35} = 0.040816$$

Similar calculations give

$$C_{j,fd} = 0.14286 \qquad \text{and} \qquad C_{f,fd} = 0.81633$$

Solving for R_i for disc 3

$$R_i = \frac{0.040816}{1 + (1.4 \times 0.14286) + (-0.75 \times 0.81633)}$$

$$= 0.069444$$

and substituting in Eqn (8.43)

$$R_{i,fd} = 0.069444 \times \frac{1}{0.5}$$

$$= 0.13889$$

(d) The coefficient $\alpha_{iLOF,fd}$ is obtained from Eqn (8.50) or Eqn (8.51)

$$\alpha_{iLOF,fd} = 0.5 \times (-1.0)$$

$$= -0.5$$

or

$$\alpha_{iLOF,fd} = -\frac{(1 - 0.8)}{1 + (-0.75 \times 0.8)}$$

$$= -0.5$$

(e) Solving Eqn (8.45) for C_i using modified influence coefficients for matrix element j and loss on ignition in the case of discs 3 and 6

$$C_i = 0.13889[1 + (0.7 \times 0.7) + (-0.5 \times 0.1)]$$

$$= 0.20$$

and

$$C_i = 0.11869[1 + (0.7 \times 0.64) + (-0.5 \times 0.2)]$$

$$= 0.16$$

respectively. In the case of disc 3, if the correction for loss on ignition is omitted, the result is then

$$C_i = 0.13889[1.49]$$

$$= 0.2069$$

Clearly, the correction for LOF cannot be ignored.

8.8.3 Correction for gain on fusion (GOF)

Under quite different circumstances from the above, in which fusion involved loss of volatiles, there may be a gain on ignition during the fusion process owing to the presence of elements at lower oxidation state. Typical examples are the presence of ferrous iron or zinc sulphide in geological samples. The iron will be oxidized during fusion, not only because of the high temperature but also because oxidizing agents are often (indeed, must be) added if the fusion is carried out using platinum crucibles. The presence of sulphide sulphur is more of a problem, given that oxidation may be only partial, and therefore some sulphur may volatilize during fusion; but in the presence of strong oxidizing agents loss of sulphur may be prevented.

It therefore follows that the gain in weight and the nature of the oxidized product must be known in order to make a correction based on theoretical considerations. The gain in weight may be obtained from 'weight before fusion' and 'weight after fusion' measurements or from the stoichiometry of the oxidation products as part of the iteration process. For example, a value for α_{iGOF} may be computed theoretically if it is assumed that ferrous iron is oxidized to the ferric state, i.e. a gain in weight due to oxygen. The inclusion of a correction for GOF results in an expression similar to Eqn (8.47), namely

$$C_i = R_i \left[1 + \sum_j \alpha_{ij}C_j + \alpha_{iGOF}C_{GOF} \right] \tag{8.52}$$

The expression defining α_{iGOF} is derived as follows: recalling Eqn (2.28) and dispensing with the subscript λ,

$$I_i = I_{(i)} \frac{C_i\mu_i^*}{\mu_s^*} \tag{8.53}$$

leads to

$$\frac{I_i}{I_{(i)}} = \frac{C_i\mu_i^*}{\mu_s^*} \tag{8.54}$$

where, if the weight gain is taken into consideration

$$\mu_s^* = \frac{C_i\mu_i^* + C_j\mu_j^* + \cdots + C_{GOF}\mu_{GOF}^*}{1 + C_{GOF}} \tag{8.55}$$

$$C_{GOF} = W_{GOF}/W_s \tag{8.56}$$

Multiplying both sides of the equal sign in Eqn (8.55) by $(1 + C_{GOF})$ leads to

$$\mu_s^*(1 + C_{GOF}) = C_i\mu_i^* + C_j\mu_j^* + \cdots + C_{GOF}\mu_{GOF}^* \tag{8.57}$$

Substituting $1 - C_j - C_k - \cdots$ for C_i (N.B. C_{GOF} is not included) leads to

$$\mu_s^*(1 + C_{GOF}) = [\mu_i^* + C_j(\mu_j^* - \mu_i^*) + \cdots + C_{GOF}\mu_{GOF}^*] \tag{8.58}$$

Substituting the right hand side of the above equation in the denominator of Eqn (8.54) and dividing numerator and denominator by μ_i^* leads to

$$\frac{I_i}{I_{(i)}} = \frac{C_i}{1 + \alpha_{ij}C_j + \cdots + \alpha_{iGOF}C_{GOF}} \tag{8.59}$$

Solving for C_i yields the conventional formalism

$$C_i = R_i\left[1 + \sum_j \alpha_{ij}C_j + \alpha_{iGOF}C_{GOF}\right] \tag{8.60}$$

where

$$\alpha_{ij} = \frac{\mu_j^* - \mu_i^*}{\mu_i^*} \tag{8.61}$$

$$\alpha_{iGOF} = \frac{\mu_{GOF}^*}{\mu_i^*} \tag{8.62}$$

and after modification for the fused disc context

$$\alpha_{ij,fd} = \alpha_{ij}f \tag{8.63}$$

$$\alpha_{iGOF,fd} = \frac{\mu_{GOF}^*}{\mu_i^*}f \tag{8.64}$$

In the context chosen in Table 8.8 to illustrate corrections for GOF (context identical to that in Tables 8.6 and 8.7), the mass attenuation coefficients have been assigned the values $\mu_i^* = 120 \text{ cm}^2\text{g}^{-1}$ and $\mu_{GOF}^* = 24 \text{ cm}^2\text{g}^{-1}$.

8.8.3.1 *Numerical example*

Table 8.8 lists compositions of seven binary samples; the pure analyte along with three samples having no GOF, and three that are representative of samples that would undergo a gain in weight during the fusion process. Also listed are the weight of the resulting fused discs W_{disc} and computed values for $R_{i,fd}$. It is noted that sample 1 is identical to samples 1 in Tables 8.6 and 8.7 and that sample 2 is identical to sample 5 in Table 8.6. Computations are carried out in the following sequence, e.g. for disc 3.

(a) The concentrations, C_i, C_j, C_f and C_{GOF} in the fused disc are given by:

$C_i = (0.8 \times 1.5)/7.62 = 0.15748$

$C_j = (0.2 \times 1.5)/7.62 = 0.03937$

$C_f = 6.0/7.62 = 0.7874$

$C_{GOF} = 0.12/7.62 = 0.01575$

Table 8.8 Theoretical data used to illustrate the application of a correction for gain in weight during fusion in the context of fused disc specimens

(a) Context

$$W_s = 1.5 \text{ g} \qquad W_f = 6.0 \text{ g}$$
$$C_s = 0.2 \qquad C_f = 0.8$$
$$\alpha_{ij} = 1.4 \qquad \alpha_{if} = -0.75 \qquad \alpha_{iGOF} = -0.80$$

(b) Samples

Sample	C_i	C_j	W_{GOF} (g)	W_{disc} (g)	$R_{i,fd}$
1	1.00	0.00	0.00	7.50	1.00
2	0.80	0.20	0.00	7.50	0.70175
3	0.80	0.20	0.12	7.62	0.69686
4	0.80	0.20	0.36	7.86	0.68725
5	0.70	0.30	0.0	7.50	0.57851
6	0.70	0.30	0.105	7.605	0.57511
7	0.50	0.50	0.00	7.50	0.37037
8	0.50	0.50	0.075	7.575	0.36900

(b) R_i is given by

$$R_i = \frac{0.15748}{1 + (1.4 \times 0.03937) + (-0.75 \times 0.7874) + (-0.80 \times 0.01575)}$$

(c) $R_{i,fd}$ is obtained from Eqn (8.43)

$$R_{i,fd} = 0.34843 \times \frac{1}{0.5} = 0.69686$$

(d) The value $\alpha_{ij,fd} = 1.4 \times 0.5 = 0.7$ is retained.

(e) $\alpha_{iGOF,fd}$ is given by Eqn (8.64)

$$\alpha_{iGOF,fd} = \frac{24}{120} \times 0.5$$

$$= 0.1$$

(f) C_{GOF} is given by Eqn (8.56)

$$C_{GOF} = \frac{0.12}{1.5} = 0.08$$

Substitution in Eqn (8.52) yields

$$C_i = 0.69686 [1 + (0.7 \times 0.2) + (0.1 \times 0.08)]$$
$$= 0.80$$

and in the case of discs 7 and 8

$$C_i = 0.37037 [1 + (0.7 \times 0.5)]$$
$$= 0.50$$

$$C_i = 0.36900 [1 + (0.7 \times 0.5) + (0.1 \times 0.05)]$$
$$= 0.50$$

respectively.

8.8.4 General theory of LOF and GOF

The mere knowledge of the total weight change after fusion is not sufficient to calculate compositions in the original sample when LOF and GOF *are present at the same time*. However, corrections are possible provided that the chemical transformation that occurs in each component is known. Consider a specimen in which no LOF and no GOF is observed. Recalling Eqn (8.53) for a monochromatic excitation source,

$$I_i = I_{(i)} C_i \frac{\mu_i^*}{\mu_s^*} \tag{8.65}$$

leads to the intensity ratio, i.e. the relative intensity $R_{i,fd}$; in the case of fused discs,

$$R_{i,fd} = \frac{I_{,fused\ disc}}{I_{(i),fused\ disc}} \tag{8.66}$$

$$= \frac{C_i \mu_i^*}{C_i \mu_i^* + C_j \mu_j^* + C_k \mu_k^* - C_L \mu_L^* + C_n \mu_n^* + C_G \mu_G^* + C_f \mu_f^*} \tag{8.67}$$

where the subscripts L and G refer to LOF and GOF of individual compounds, respectively. All concentrations refer to the specimen (i.e. the fused disc). A correction for the sum of concentrations $(1 + C_G - C_L)$ not being unity equally affects the numerator and denominator and cancels out. Let us assume that compound k has a LOF (e.g. $CaCO_3$, which is transformed to CaO and CO_2) and that compound n has a GOF (e.g. ZnS, which is transformed to $ZnSO_4$). Combining the terms $C_k \mu_k$ and $C_l \mu_L$ and taking C_k in factor leads to $(\mu_k^* - (C_L/C_k)\mu_L^*)$ where

$$\frac{C_L}{C_k} = \frac{Mol.\ wt.\ LOF}{Mol.\ wt.\ compound\ before\ LOF} \tag{8.68}$$

By the same token, combining terms $C_n\mu_n$ and $C_G\mu_G$ and taking C_n in factor leads to the term $(\mu_n^* - (C_G/C_n)\mu_G^*)$ where

$$\frac{C_G}{C_n} = \frac{\text{Mol. wt. GOF}}{\text{Mol. wt. compound before GOF}} \tag{8.69}$$

Defining

$$\mu_{kLOF}^* = \mu_k^* - \frac{C_L}{C_k}\,\mu_L^* \tag{8.70}$$

and

$$\mu_{nGOF}^* = \mu_n^* + \frac{C_G}{C_n}\,\mu_G^* \tag{8.71}$$

leads to

$$R_{i,fd} = \frac{C_i^* \mu_i}{C_i\mu_i^* + C_j\mu_j^* + C_k\mu_{kLOF}^* + C_n\mu_{nGOF}^* + C_f\mu_f^*} \tag{8.72}$$

Expressed in the conventional influence coefficient formalism

$$R_{i,fd} = \frac{C_i}{1 + C_j\alpha_{ij} + C_k\alpha_{ik,mod} + C_n\alpha_{in,mod} + C_f\alpha_{if}} \tag{8.73}$$

where concentrations refer to the fused disc, or in Eqn (8.45) formalism

$$C_i = R_i\left[1 + \sum_j \alpha_{ij,fd}C_j\right] \tag{8.74}$$

where concentrations are those of the compounds in the original sample. In the above context, compound (oxide) j is inert, compound k ($CaCO_3$) has a LOF, and compound n (ZnS) has a GOF. Thus

$$R_{i,fd} = R_1\frac{1}{f} \tag{8.75}$$

$$\alpha_{ij,fd} = \frac{\mu_j^* - \mu_i^*}{\mu_i^*}\times f_f \tag{8.76}$$

$$\alpha_{ik,mod} = \frac{\mu_{kLOF}^* - \mu_i^*}{\mu_i^*}\times f_f \tag{8.77}$$

$$\alpha_{in,mod} = \frac{\mu_{nGOF}^* - \mu_i^*}{\mu_i^*}\times f_f \tag{8.78}$$

8.8.4.1 Numerical example

The overall analytical context in Table 8.9 is representative of the following:

- i is an inert oxide, analyte i;
- j is an inert oxide, matrix element j;
- k is a compound that has a LOF equal to 44%, e.g. $CaCO_3$;
- n is a compound that has a GOF equal to 66%, e.g. ZnS.

Assuming nominal values of μ for the chemical elements, the effective mass attenuation for compounds can be computed directly from $\Sigma C_i \mu_i$; thus

$$\mu_{CaCO_3} = (0.40 \times 170) + (0.12 \times 10) + (0.48 \times 20)$$
$$= 78.8 \text{ cm}^2 \text{g}^{-1}$$

$$\mu_{LOF} = \mu_{CO_2} = (0.273 \times 10) + (0.727 \times 20)$$
$$= 17.3 \text{ cm}^2 \text{g}^{-1}$$

$$\mu_n = \mu_{ZnS} = (0.67 \times 180) + (0.33 \times 80)$$
$$= 147.0 \text{ cm}^2 \text{g}^{-1}$$

$$\mu_{CaO} = (0.714 \times 170) + (0.286 \times 20)$$
$$= 127.1 \text{ cm}^2 \text{g}^{-1}$$

The α coefficients applicable directly to the fused disc specimen are given by the usual formalism

$$\alpha_{ij} = \frac{288 - 120}{120} = 1.4$$

$$\alpha_{iCaO} = \frac{127.1 - 120}{120} = 0.05917$$

$$\alpha_{in} = \frac{147.0 - 120}{120} = 0.225$$

$$\alpha_{iGOF} = \frac{20 - 120}{120} = -0.83333; \qquad \text{GOF is oxygen}$$

$$\alpha_{if} = \frac{33.6 - 120}{120} = -0.72$$

In the case of sample 7, the LOF is given by

$$W_{LOF} = C_k \times W_s \times 0.44$$
$$= 0.3 \times 1.5 \times 0.44 = 0.198 \text{ g}$$

Table 8.9 Data used to illustrate the correction for LOF and GOF when the stoichiometries of LOF and GOF are known

(a)

	Elements			Compounds	
	μ (cm^{2}g^{-1})	Atomic wt.		μ (cm^{2}g^{-1})	Mol. wt.
C	10	12	k = CaCO$_3$	78.8	100
O	20	16	CO$_2$	17.3	44
S	80	32	n = ZnS	147.0	97
Ca	170	40	CaO	127.1	56
Zn	180	65			

(b) $W_s = 1.5$ g $W_f = 6.0$ g $C_s = 0.2$ $C_f = 0.8$

$\mu_i = 120$ cm^{2}g^{-1} $\mu_j = 288$ cm^{2}g^{-1} $\mu_k = 78.8$ cm^{2}g^{-1}

$\mu_n = 147.0$ cm^{2}g^{-1} $\mu_L = 17.3$ cm^{2}g^{-1} $\mu_G = 20$ cm^{2}g^{-1}

$\mu_f = 33.6$ cm^{2}g^{-1}

(c)

			Samples			Fused disc	
	C_i	C_j	C_k	C_n		W_{disc}	$R_{i,fd}$
1	0.2	0.0	0.0	0.0		7.5	1.0
2	0.2	0.8	0.0	0.0		7.5	0.13086
3	0.2	0.0	0.8	0.0		6.972	0.23627
4	0.2	0.0	0.0	0.8		8.292	0.17756
5	0.2	0.2	0.4	0.2		7.434	0.18401
6	0.2	0.2	0.2	0.4		7.764	0.17289
7	0.1	0.2	0.3	0.4		7.698	0.08791

The GOF is given by

$$W_{GOF} = C_n \times W_s \times 0.66$$
$$= 0.4 \times 1.5 \times 0.66 = 0.396 \text{ g}$$

The weight of the fused disc is given by

$$W_{disc} = W_s + W_f - W_{LOF} + W_{GOF}$$
$$= 1.5 + 6.0 - 0.198 + 0.396$$
$$= 7.698 \text{ g}$$

$$C_{i,fd} = (0.1 \times 1.5)/7.698 = 0.01949$$

$$C_{j,fd} = (0.2 \times 1.5)/7.698 = 0.03897$$

$$C_{CaO,fd} = (0.3 \times 0.56 \times 1.5)/7.698 = 0.03274$$

$$C_{ZnS,fd} = (0.4 \times 1.5)/7.698 = 0.07794$$

$$C_{GOF,fd} = 0.396/7.698 = 0.05144$$

$$C_{f,fd} = 6.0/7.698 = 0.77942$$

Hence the sum of concentrations in the disc is equal to

$$0.01949 + 0.03897 + 0.03274 + 0.07794 + 0.05144 + 0.77942 = 1$$

The flux factor is given by

$$f = \frac{1 - 0.8}{1 + (-0.72 \times 0.8)}$$

$$= 0.4717$$

The relative intensity $I_{i,fd}/I_{(i),fd}$ is given by

$$R_{i,fd} = \frac{C_{i,fd}}{1 + \sum \alpha_{ij} C_{j,fd}} \frac{1}{f}$$

$$= \frac{0.01949}{0.46999} \times \frac{1}{0.4717}$$

$$= 0.08791$$

The μ_k coefficient modified for LOF is given by

$$\mu_{k LOF} = \mu_{CaCO_3,LOF} = 78.8 - (17.3 \times 0.44) = 71.2 \text{ cm}^2 \text{g}^{-1}$$

The μ_n coefficient modified for GOF is given by

$$\mu_{n GOF} = \mu_{ZnS,GOF} = 147.0 + (20 \times 0.66) = 160.2 \text{ cm}^2 \text{g}^{-1}$$

The $\alpha_{ij,fd}$, $\alpha_{ik,fd}$ and $\alpha_{in,fd}$ applicable in the conventional formalism

$$C_i = R_i \left[1 + \sum_j \alpha_{ij,fd} C_j \right] \tag{8.79}$$

are given by

$$\alpha_{ij,fd} = 1.4 \times 0.4717 = 0.6604$$

$$\alpha_{ik,fd} = \frac{71.2 - 120}{120} \times 0.4717 = -0.1918$$

$$\alpha_{in,fd} = \frac{160.2 - 120}{120} \times 0.4717 = 0.1580$$

The above calculations need be made once only for a given analytical context; thus, the corrections for matrix effects in the case of samples 5 and 7 take the form

$$C_i = 0.1840 [1 + (0.6604 \times 0.2) + (-0.1918 \times 0.4) + (0.158 \times 0.2)]$$

$$= 0.2000$$

and

$$C_i = 0.0879\,[1 + (0.6604 \times 0.2) + (-0.1918 \times 0.3) + (0.158 \times 0.4)]$$
$$= 0.1000$$

In summary, a condensed yet comprehensive expression for converting intensities to concentration for the fusion method has been developed. It is complementary to the option of working exclusively in the context of fundamental influence coefficients and the fused disc specimen, i.e. $R_{i,fd}$ and coefficients $m_{ij,fd}, m_{ik,fd}, \ldots$, concentrations $C_{i,disc}, C_{j,disc}, \ldots$ etc. On the other hand, it is possible to compute modified influence coefficients for compounds that generate either a LOF or a GOF provided that the stoichiometry between the compound in the sample and in the fused disc is known. The symbol α is retained to indicate that a polychromatic source introduces approximations (albeit likely to be very slight) in the proposed model.

8.9 LIQUIDS

In addition to having to determine element concentration in samples submitted in liquid form, analysts will on occasion have recourse to dissolution of solid materials in order to take advantage of the many alternatives for subsequent treatment that are not possible in the case of solids. However, there is no doubt that dealing with liquid specimens may present problems of its own unless special precautions are taken. This is clearly brought out by Bertin (1975), who lists 20 principal advantages and 17 disadvantages of specimen presentation in liquid form. Among the advantages listed are:

- dissolution is particularly advantageous when the same material is received in various forms such as bar, foil, drillings etc.;
- analytical techniques, e.g. dilution, standard addition etc. that are particularly useful for one-of-a-kind analyses are applicable to almost any type of liquids;
- standards are easily prepared by dissolution of soluble compounds at convenient concentration;
- liquids are better adapted to on-line process control.

However, there are a number of disadvantages associated with liquid specimens:

- solution techniques are usually not as rapid and convenient as direct excitation of the original material;
- evaporation may occur during irradiation, with consequent change in concentration; spectrometer components may be subject to corrosive vapours;
- liquid cells are generally somewhat more difficult to handle;

- problems attend the use of liquid specimens in a helium path for low Z analytes.

For a comprehensive treatment of liquid specimen techniques and of the various liquid specimen cells that have been proposed for handling liquids, the reader is referred to Bertin (1975), Chapter 17.

8.10 CONCLUSIONS

There is no doubt that the topics examined in this chapter are diverse, but their commonality is the limitation posed by certain analytical contexts. In some instances the problems are more or less surmountable, but at the price of taking very special precautions (establishing the sample type prior to proceeding with the analysis) and following a rigid protocol for specimen preparation, so that differences between specimen and references (powdered materials) used for calibration are negligible. Analysts should be cautious when relying on matching powdered samples and reference materials lest differences between the two creep in. This may lead to gross errors, but the fact that the measurements are highly reproducible tends to give a false impression of 'accuracy'.

Heterogeneity problems may be surmounted by dissolution procedures (involving time, cost, etc.), yielding either aqueous or solid solutions. Although both are quite versatile, the former may introduce problems in handling (special specimen holders, corrosion, etc.), whereas the latter may introduce the need for corrections to compensate for the loss of volatiles or gain in weight due to oxidation during the fusion process. The need for applying loss on ignition or gain on ignition corrections may be circumvented by pre-igniting all samples before fusion and determining the loss/gain in weight in order to report concentrations on the 'as received' basis. In some cases ignition may lead to sintering or to hygroscopic products (e.g. $CaCO_3$ converted to CaO), thereby introducing handling problems. If the stoichiometry of the compounds before and after fusion is known, it is then possible to compute influence coefficients that incorporate and thus compensate for LOF and GOF.

9 ANALYTICAL STRATEGIES

The aim of this chapter is to examine a number of factors that must be considered in the development of analytical procedures in X-ray fluorescence spectrometry. The plural 'strategies' is used expressly to emphasize the fact that there is generally more than one procedure to achieve the desired result, the consistent output of compositional data that meet an acceptable level of precision and accuracy. When one considers the wide-ranging applications of the X-ray fluorescence technique, it is readily understood that only general guidelines can be given for the development of analytical procedures. The chapter is envisaged in the framework in which concentration C is a function of the four factors K, I, M and S, and examines procedural steps in method selection; i.e. it offers a guide to the often heard reflection of one who is newly initiated to the X-ray fluorescence technique: 'but in my particular case, how do I put all this together?'

9.1 GENERAL CONSIDERATIONS

Although not endless, the number of permutations and combinations of options that analysts may be faced with in developing an analytical scheme are many. Let us consider some widely diversified contexts:

- a laboratory may be called upon to provide compositional data in a research environment, where the first step would be to carry out a qualitative analysis, but another laboratory may deal with type materials, i.e. manufactured products of roughly known composition;
- a laboratory may be considering the acquisition of a new spectrometer, in which case the advantages and disadvantages of both wavelength and energy dispersion will need examination *in the context of that laboratory's present and/or future requirements*;

- a laboratory may be considering the application of newly developed instrumental technology, in which case the analyst will be breaking new ground; the laboratory may be considering the implementation of a comprehensive analytical scheme for a new material or simply wishing to upgrade any one of concentration range, precision, etc. of a method in current use, and so on.

In general terms, analytical strategy may be envisaged as dealing with each of the factors K, I, M and S in the following progression:

Approach−Method−Procedure−Protocol

where the term approach is used to designate the general or overall process selected; e.g. analysts may select the fusion approach to eliminate particle size effects, heterogeneity effects etc. As there is more than one method for preparing fused specimens, one analyst may opt for carrying out the fusion using graphite crucibles and a muffle furnace, whereas another may opt for any one of using Pt−Au crucibles and gas burners or an induction furnace. Any given method may lead to a number of different procedures, depending upon the sample weight, the sample to flux ratio, the duration of the melting cycle, the casting and cooling processes, etc. Protocol designates those rarer cases where a number of laboratories may wish to produce interchangeable analytical data, e.g. various plants of a parent company. By means of round-robin tests, the laboratories involved evaluate a given procedure in great detail and reach agreement on a common detailed procedure to which all laboratories are strictly committed.

Analysts may also find it beneficial to recognize and evaluate each of the above in turn so as to be aware of their attributes or characteristics such as, say, in a given calibration process:

- Premise(s):
  what is initially taken for granted; e.g. many reference materials are on hand, reference materials are appropriate, etc.;
- Concept:
  what basic principle is involved; e.g. in the case of a statistical treatment of experimental data, what is the number of reference materials that is adequate;
- Mathematical expression:
  selection of a first or second order model for least squares regression;
- Implication:
  regression techniques are empirical in nature and fairly simple, but computer facilities and appropriate software are required;
- Application:
  given the above, the method is limited to a well defined analytical context.

The above outline is quite subjective and presented merely as a guide to what analysts may have to take into consideration; in many instances fewer factors need be considered.

The advances in technology that have made it possible to operate spectrometers unattended on a continuous basis, with the resulting need to treat very large amounts of data on-line, require the analyst to be informed in a number of related fields, e.g. statistics, computer technology, software etc. A common requirement in quantitative analysis is the need to determine 'values of central tendency', i.e. to select a single value that best represents a number of replicate measurements, and to assign a confidence level to this value. Another is the need for instrumental calibration, which generally involves some form of least squares fitting of a mathematical relationship between two or more variables that are functions of C_i and I_i. The broad field of regression analysis offers a variety of methods for these purposes.

Last but not least, the need also exists to be aware of and refer to the fairly extensive literature on all aspects of quantitative analysis in general and X-ray fluorescence spectrometry in particular. This includes textbooks, which systematically cover the theoretical aspects of X-ray physics relating to X-ray fluorescence, instrumentation and practical applications that have withstood the test of time. On the other hand, recent issues of international and national journals, proceedings of conferences, symposia etc., are invaluable to keep abreast of the *state of the art* and of new orientations.

9.2 ANALYTICAL CONTEXT

First and foremost is the need for a clear and detailed visualization, in mind but preferably set down 'on paper', of the overall context. Samples will be submitted for analysis, and the goal is to arrive at a consensus on the quality of the analytical data reported so that there must be some assurance that the data are not misleading to the submitter. This applies regardless of whether the X-ray laboratory is part of a research or an industrial environment, where the former involves being presented with 'new' materials on a daily basis and the latter involves analyses relating to samples from ongoing production of a 'known' material. If truth be told, however, most analysts in industry find themselves somewhere in a middle position of having to analyse 'a few' new materials on occasion. Thus, the constraints on options available to the analyst are influenced by:

- type of samples submitted;
- whether comprehensive or partial analysis is required;
- analytes and their expected range of concentration;
- expected precision, accuracy, productivity etc.;

- whether cost per result reported is a major concern;
- the resources available, instrumental and personnel.

9.2.1 Nature of samples submitted

The term sample is used to designate the material submitted to the laboratory for analysis and should be representative of the bulk from which it was obtained, although this is not generally the analyst's responsibility. The term specimen is used to designate the material introduced into the spectrometer for irradiation. In some cases sample and specimen may be one and the same (alloys, thin films, liquids), but more generally some transformation of the sample (particulates such as powdered geological materials) is necessary to ensure that the specimens meet the required criteria of homogeneity, surface finish and thickness.

Firstly, one must consider the stability of the materials submitted, which may be affected by such factors as solids picking up moisture or precipitates forming in liquids if allowed to stand for any length of time. Next to be considered are all the possibilities for contamination, which may vary depending on the complexity of the specimen preparation processes. For example, in some cases two different specimen preparation schemes may be needed to circumvent the inherent contamination during the grinding process. This is definitely required if there is danger of contamination of analyte elements occurring at low trace levels. The same care must be taken concerning contamination by materials added to the sample (binders, diluents etc.). Simply stated, the analyst's watchword must be: 'know thy samples, know thy specimens'.

9.2.2 Instrumental resources

In the same vein, one might add 'know thy spectrometer', so as to be able to use it to its full potential in any given analytical context. This is more demanding in the case of wavelength dispersion because more components are involved than in energy dispersion, hence there are more options such as choice of X-ray tube target, operating voltage and current, collimators, analyzing crystals and detectors. In either case, a listing of all characteristic emissions of the tube target element should be made and verified against a scan of a low Z material of high purity in order to identify any peaks that may be due to impurities in the anode material, tube window etc. This scan should be repeated periodically, as peaks due to impurities may vary in intensity as a function of time.

9.2.3 Sample composition

It would be foolhardy for an analyst to proceed with the development of analytical procedures without a general knowledge of what elements are present and in roughly what concentrations. This information is crucial if one is to:

- avoid contamination during specimen preparation;
- make a valid selection of peak and background 2θ angles;
- determine whether spectral interference corrections are needed;
- proceed to the calibration process, etc.

Thus, if presented with a sample of an unknown material for analysis, the analyst has no other recourse but to carry out a comprehensive qualitative analysis, which will provide a rough classification of what elements are present in major, minor or trace concentrations. On the other hand, if the sample can be classified with assurance as to type, then that knowledge is usually sufficient to dispense with a qualitative analysis.

9.2.4 Precision and accuracy

Precision and accuracy are terms used to express the degree of agreement between two values, and are often confused. In quantitative compositional analysis, *precision* refers to the difference between an individual measurement (e.g. I_i) or determination (e.g. C_i) and what is regarded as the *best known value*, that is, the average value of a large number of independent replicate measurements (arithmetic mean) or determinations of I_i or C_i, respectively. In other words, precision refers to degree of scatter about an average. In the context of a normal or Gaussian distribution and a relatively small number of replicates (around 10), precision may be expressed in terms of the standard deviation σ, a measure of dispersion about the arithmetic mean:

$$\sigma = \left(\frac{\sum d^2}{n-1}\right)^{1/2} \qquad (9.1)$$

where d is the difference between a given measurement x_n and the arithmetic mean $\bar{x}$ of n replicate measurements of x, i.e.

$$\bar{x} = \frac{\sum x}{n} \qquad (9.2)$$

$$d = x - \bar{x} \qquad (9.3)$$

For example, the standard deviation may be used to provide a measure of confidence of a measured intensity I_i; this is usually expressed by the probability that a single measurement I_i is within $\pm\sigma$ of the value obtained if

10 (or so) I_i measurements were made and averaged. From statistical theory, it is expected that a single measurement I_i will lie within the ranges

$I_i \pm \sigma$ 68.3% of the time
$I_i \pm 2\sigma$ 95.4% of the time
$I_i \pm 3\sigma$ 99.7% of the time

which are also called confidence limits, as shown in Figure 9.1. The confidence level for the precision of a single determination C_i can be obtained by carrying out a number of replicate determinations of C_i and applying the above mathematical process.

The mathematical process for assessing the *accuracy* of C_i is identical to the above except for the important distinction that d now expresses the difference between the value C_i as determined and the *true* value of C_i. It therefore follows that accuracy can only be assessed using a number of reference materials of known composition, i.e. preferably materials where the *true* concentrations are assumed to be the average of C_i values in close agreement obtained from two or more unrelated analytical techniques.

Once the analytical context and the goals have been concisely stated, the next step is to develop procedures aimed at achieving these goals, bearing in mind that there is no simple scheme or set of methods for achieving all of them. It therefore follows that analysts are more often than not faced with having to make a set of compromises. Briefly stated, the analyst must consider collectively the four factors that have a direct bearing on the precision and

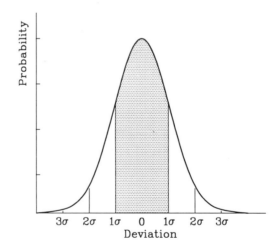

Figure 9.1 Area of the probability curve included within the range plus and minus one standard deviation

accuracy of C_i, namely K, I, M and S. These will now be taken in the normal operational sequence:

- S specimen preparation;
- I intensity measurements;
- K calibration;
- M correction for matrix effects.

9.3* SPECIMEN PREPARATION

Broadly speaking, the goal is to prepare specimens such that all the physical properties that influence the emitted intensity of characteristic lines, apart from composition, have been minimized to the point of being negligible. For reasons of expediency, it may be acceptable in some analytical contexts to develop a specimen preparation procedure wherein the goal is not to minimize the effect of physical properties as such, but simply to match the physical properties of sample and reference materials so that the effects cancel out. Ideally, specimen preparation should be such that the measured intensity of spectral lines will not be influenced by the physical properties of the specimen. Indeed, the effects of surface roughness, inhomogeneity, particle size and shape and distribution, mineralogy, method of casting, degradation of surfaces due to oxidation or hydration etc. on emitted intensity are well documented. In practice, specimen preparation is generally envisaged within the framework of the three categories: bulk solids, powders and liquids.

9.3.1 Bulk solids—alloys, glasses, plastics, ceramics

In addition to being representative of the sample, a specimen should ideally meet the criteria of being homogeneous, flat and infinitely thick in respect of emitted intensities, and in practice the preparation process should be relatively simple, rapid and highly reproducible. The initial step is the preparation of a rough specimen of appropriate size (generally circular in shape) by chill-casting or milling of an ingot, bar or stock and slicing. Rarely is either of the flat surfaces satisfactory for direct irradiation; hence one side is selected for further treatment. This consists of milling in the case of the usual metallic alloys and grinding in the case of hard alloys and brittle materials such as ceramics; i.e. where surface roughness is approximately 40 μm. This is followed by polishing with progressively finer abrasives until the surface is free from striations. The surface finish is of prime importance because striations give rise to the so-called shielding effect, leading to lower emitted intensities. Although spinning the specimen does smooth out this effect to some degree, specimens and standard reference materials should be similar.

The combination of long wavelength analytes in a heavy matrix, e.g. Mg and Al in Fe, requires polishing down to the order of 1 μm.

Two problems may arise during a mechanical polishing process: (a) smearing of a softer component over the specimen surface (electrolytic polishing is a non-mechanical alternative but unfortunately is time-consuming); (b) contamination depending on the abrasives used, e.g. Al from Al_2O_3 or Si from SiC.

9.3.2 Pressed powder pellets

Although, for simplicity, powder samples are sometimes used directly as specimens in X-ray fluorescence spectrometry, analysts should be aware that this technique may not always yield sufficiently reproducible intensities owing to segregation of various sized particles and degree of compactness at the surface. These effects can be greatly minimized by pelletizing or briquetting with or without the addition of a binder. The following criteria must be taken into consideration in the preparation of pressed pellets.

(a) As shown by Blanquet (1986) in the context of monochromatic excitation, negligible enhancement, perfectly flat specimens and the fluorescent element being present in one phase only, for a given overall specimen composition the heterogeneity effect is a function of the size and concentration of heterogeneities (usually particles) and the effective mass attenuation coefficient of the fluorescent and non-fluorescent compounds (mineralogical effect).

(b) For a given compound in a given mixture, the heterogeneity factor H

$$H = \frac{I_{\text{equivalent particle diameter}}}{I_{\text{particle diameter}=0}} \qquad (9.4)$$

as a function of particle size is given in Figure 9.2 (Claisse and Samson, 1962) for analyte Sr from $SrCl_2$ mixed with $CaCO_3$ or with KBr and incident radiation 0.71 Å. The overall effect may be described as follows:

(i) within the first plateau region (small particles), the fluorescence intensity is essentially the same as for a homogeneous specimen ($H \approx 1$);

(ii) a transition zone wherein intensity decreases or increases as the particle size increases;

(iii) a second plateau (large particles) which tends towards

$$H = \frac{\mu_m^* \rho_m}{\mu_f^* \rho_f} \qquad (9.5)$$

(where f = fluorescent compound, m = total matrix including fluorescent compound, ρ = density) when the particle size of the fluorescent compound increases. Given that the transition zone spans a large range of particle size, it is often difficult to obtain a high degree of precision if intensities are measured on powdered specimens;

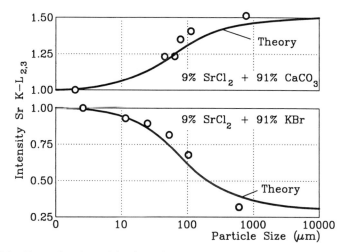

Figure 9.2 Example of particle size effect on Sr fluorescence

(c) Even when particle size and the overall specimen composition are constant, variations in fluorescence intensities can be predicted if the analyte occurs in two minerals (mineralogical effect). Typical examples would be analyte copper occurring as the mineral chalcocite (CuS) or chalcopyrite ($CuFeS_2$) or analyte zinc occurring as in end members of the mineral sphalerite ((ZnFe)S), where the iron content may reach a maximum of 18%. An example of mineralogical effect is shown in Figure 9.3. On the second plateau, the particle size is so large that the excitation process is totally inside single particles of the mineral and, consequently, the emitted intensity is a function of the total absorption in that mineral only. Thus, if the analyte is present in more than one compound, its measured fluorescence intensity is of little value because it now depends on the composition of those compounds, which are usually in unknown proportions.

9.3.3 Fused discs

Fusion is an extremely effective technique which transforms compounds into solid solutions (glasses), thereby eliminating all vestige of the particle size and mineralogy of the original material. Typically, this is achieved by:

- heating a mixture of sample and flux at high temperature;
- maintaining the high temperature until the flux melts and dissolves the sample;
- agitation during the dissolution process to obtain a homogeneous liquid;
- pouring the melt into a mould which, on controlled cooling, produces an amorphous solid, the specimen.

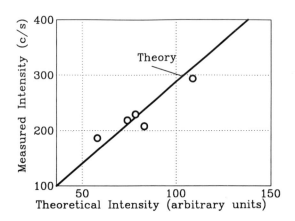

Figure 9.3 Mineralogical effect for mixtures of 1% Sr (as various salts) in calcite. Excitation radiation 0.71 Å

More specifically, the analyst has to take into consideration a number of factors that influence the consistent production of high quality fused disc specimens, e.g.:

(a) the granulometry of the sample;
(b) the composition and physical properties of the flux and additives;
(c) the sample to flux ratio;
(d) the choice of crucibles and moulds;
(e) the heating cycle;
(f) agitation during the fusion process;
(g) casting and cooling conditions.

(a) To ensure that the portion of sample taken is representative, and to facilitate dissolution, the material to be analysed should be finely ground. Grinding the material to -200 mesh (75 μm) is generally acceptable. Finer grinds will reduce fusion time but will compromise the chemical determination of FeO (if required) through likely oxidation during the grinding process.

(b) For oxide materials, lithium borates have found wide acceptance, namely lithium metaborate ($LiBO_2$, m.p. 850°C), which reacts as lithium oxide in the presence of an acidic oxide, e.g. SiO_2, and lithium tetraborate ($Li_2B_4O_7$, m.p. 920°C), which reacts with basic oxides, e.g. K_2O, CaO etc. As a universal flux, analysts have resorted to various mixtures of the two lithium salts, in some cases adding Li_2CO_3 to increase the basicity or LiF if a more acidic melt is preferable. Sodium tetraborate (provided sodium is not being determined) is sometimes used for its low melting point (741°C) in spite of its hygroscopic property. Sodium hexametaphosphate and potassium pyrosulphate have found limited application. Fluxes should preferably be granular (0.3–0.5 mm diameter); because of their higher density, they occupy less space (smaller fusion crucibles), and as very fine powders fluxes have a greater affinity for absorbing humidity from the atmosphere to a variable extent. The selected fluxes should be of high purity and each new batch should be checked for the presence of elements being determined at trace levels.

One or more compounds are sometimes added to the flux for specific purposes, namely:

- heavy absorbers (e.g. La_2O_3, BaO_2) are added to decrease matrix effects at the cost of decrease in emitted intensity;
- oxidizers are added to prevent damage to platinum crucibles when organic material or partially oxidized elements are present;
- internal standards may be added if required by the analytical method;
- non-wetting or releasing agents (iodide or bromide salts) may be added in order to obtain a more complete transfer of the melt to the mould and to facilitate the removal of the glass disc from the mould.

(c) The choice of sample to flux ratio depends on the goal in mind:

- sample to flux ratios $\approx 1:100$ result in line intensities proportional to concentration; however, the peak to background ratio is low and the limit of detection relatively high, thus such a ratio is unsuited to the determination of trace constituents;
- ratios $\approx 1:10$ yield higher intensities but generally require correction for matrix effects;
- 'minimum flux', e.g. 1:2 ratio, is used if low trace levels are to be determined; in these cases correction for matrix effects is necessary (Eastell and Willis, 1990).

(d) Fusions have been carried out in platinum-rich alloy (Pt–5% Au) or graphite crucibles, and Pt–5% Au is by far the preferred material for moulds. A mirror-polish crucible or mould is generally not wetted by the molten glass, thus minimizing sticking to crucible and mould. Care must be taken with platinum ware to maintain oxidizing conditions during the fusion process.

(e) The heating cycle should incorporate a low temperature pre-heat (200–300°C, addition of a nitrate, no agitation) in order to oxidize materials prior to fusion or to liberate the volatiles gradually. The temperature is then increased (to ≈ 1100°C, with agitation) and maintained until dissolution is complete.

(f) Agitation during fusion speeds up the dissolution process and is necessary to achieve complete homogenization of the melt. Preferably the agitation should produce friction between adjacent layers in the melt, i.e. by the creation of convection currents.

(g) When dissolution is complete, the liquid melt is simply poured into a mould hot enough to avoid sudden crystallization. Desirable and undesirable conditions for cooling are illustrated in Figure 9.4.

Glass is a supercooled liquid, and hence is an unstable material that tends to crystallize. Above the melting point, the glass is liquid and is obviously amorphous. Just below the melting point, the tendency to crystallize is small; as the temperature decreases, the tendency to crystallize increases but the increasing viscosity of the material slows down the transformation. In simple terms, successful glass discs are produced when the cooling curve avoids crossing the crystallization zone as in curve 2.

9.3.4 Liquids

Provided that proper liquid specimen holders are at hand, liquids offer a number of advantages in specific analytical contexts. Liquid samples may in any event be submitted for analysis; additionally, the dissolution of a material in acid media, e.g. an alloy submitted as drillings, wire or foil, is certainly an option if the following points are considered:

- sample and standard reference material need not be of the same form;
- liquids are homogeneous;
- they are easily amenable to standard addition methods;
- synthetic standard references based on pure elements or oxides are easily prepared;
- dilution in aqueous media does not reduce intensities by a factor equivalent to the dilution factor;
- solution is available for analysis by alternative methods;

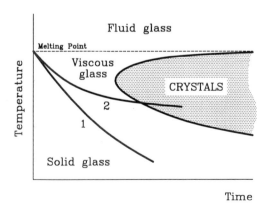

Figure 9.4 Typical cooling curves of molten glass in the preparation of fused discs. Curve 2 results in crystallized and cracked discs

- possibility of pre-concentration by chemical methods;
- possibility of absorbing liquids in various media (paper, cellulose powder, inert oxide) if handling liquids poses problems.

9.4* INTENSITY MEASUREMENTS

Having the assurance that specimens can be prepared with a high level of reproducibility, the next step aims at establishing an intensity measurement scheme that will lead to the consistent output of *valid net intensities*. The word valid is used to describe data that are not only *reproducible* but also reliable in the sense of being *factual*, i.e. not biased. As intensities registered at peak 2θ angles are functions not only of concentrations but also of background, counting loss, poor resolution, drift etc., it is imperative that a counting strategy take due consideration of all the processes involved. Although treated in a given order, the following factors are interrelated and must ultimately be considered as a group.

9.4.1 Precision, counting statistics

Given both instrumental stability and the proper selection of operating conditions, precision is firstly a function of counting time. In contrast to many analytical techniques, the quality of an X-ray fluorescence intensity measurement is a function of the total number of counts registered. Similarly to emissions from radioactive sources, the emission of photons is a random process; hence, if a constant flux of X-ray photons is measured repeatedly for a fixed time period, the total numbers N of counts registered are not identical. However, if this measurement is made a number of times under identical operating conditions, the distribution is approximately Gaussian (Figure 9.1). Then, by definition, the standard deviation is simply

$$\sigma = \sqrt{N} \tag{9.6}$$

and the relative standard deviation, or coefficient of variation, is given by

$$\varepsilon = 1/\sqrt{N} \tag{9.7}$$

Thus, as shown in Figure 9.1, the probability of obtaining a given value N from a single measurement can be calculated; i.e. N can be expected to be within the range

$N \pm \sigma$ 68.3% of the time

$N \pm 2\sigma$ 95.4% of the time

$N \pm 3\sigma$ 99.7% of the time

Thus, the risk that N will differ by more than $\pm 3\sigma$ from the average value if 10 or so measurements have been made is only 0.3%, which expresses the degree of confidence one may have in the measured I_i value. It must be noted that these confidence limits represent the minimum error, even if measured with flawless instrumentation. It must also be noted that the above relates to replicate readings on a single specimen. The standard deviation calculated for the same number of replicate measurements of different specimens of the same sample would be somewhat greater. Finally, replicate measurements of replicate specimens for a number of samples provides a measure of the uncertainties inherent in the specimen preparation procedure under consideration.

Analysis ultimately requires as a minimum the subtraction of background intensities and a ratioing of intensities or its equivalent in order to convert net intensities to concentration, i.e. an expression of the type

$$C_{iu} = \left(\frac{C_i}{I_{i,\text{peak}} - I_{\text{bkg}}}\right)_r \times (I_{i,\text{peak}} - I_{\text{bkg}})_u \tag{9.8}$$

and the two multiplicants must be taken into consideration when evaluating precision. The rule of variance implies that the statistical error on a difference, in this case $I_{i,\text{net}}$, is given by the standard deviation on the net intensity σ_{net}, where

$$\sigma_{\text{net}} = (\sigma_{\text{peak}}^2 + \sigma_{\text{bkg}}^2)^{1/2} \tag{9.9}$$

and the statistical uncertainty on a ratio is given by

$$\sigma_{\text{ratio}} = \frac{I_{i,\text{peak}}}{I_{\text{bkg}}} \left[\left(\frac{\sigma_{\text{peak}}}{I_{i,\text{peak}}}\right)^2 + \left(\frac{\sigma_{\text{bkg}}}{I_{\text{bkg}}}\right)^2\right]^{1/2} \tag{9.10}$$

A more comprehensive treatment of intensity measurement strategy may be found in Tertian and Claisse (1982, Chapter 19) and in Jenkins et al. (1981, Chapters 5 and 11).

9.4.2 Excitation source selection

Given the choice of a wide variety of X-ray tubes, for maximum excitation efficiency one would select the tube with characteristic lines of the target element slightly shorter than the absorption edge of the analyte. This is only practical in the context of the determination of a few analytes at low trace levels; otherwise, the frequent change of X-ray tubes in order to obtain maximum intensities is too onerous. This problem can be circumvented by the use of secondary targets as the excitation source, but this approach is limited to energy dispersive spectrometers. Generally, a tube selection has been made when the spectrometer is acquired and analysts operate within this context. This being the case, the analyst then selects kV/mA settings that tend to maximize peak to background ratios, namely high kV and low current for the shorter wavelengths, gradually changing to low kV and high current as the wavelength of the characteristic lines gets longer. Beam filters may also be used to eliminate interference from the tube lines and to improve peak to background ratios, especially in low Z matrices.

On occasion, analysts can use tube voltage to eliminate undesired spectral lines. Consider for example the determination of tantalum at trace levels in the presence of high niobium concentrations. The second order Nb $K-L_3$ and $K-L_2$ peaks are in the same general 2θ region as the Ta L_3-M_5 and L_3-M_4 peaks. However the critical excitation voltages Nb K and Ta L_3 are 19 and 9.9 keV, respectively. Therefore, if an operating voltage of 18 kV is selected, the Ta L lines are emitted (albeit at lower intensities) but the Nb lines are no longer present.

9.4.3 Line selection

The selection of characteristic lines is fairly straightforward in that the choices are quite limited. Based on the premise that K lines are more intense than L lines, which are in turn more intense than M lines, the $K-L_{2,3}$ line is generally the initial choice for analytes having atomic numbers lower than 50 or so, and the $L_3-M_{4,5}$ line for elements with Z higher than 50. An exception is the selection of $K-M_3$ L_2-M_4 or L_3-N_5 lines in order to lower count rates or to avoid or minimize spectral interferences. The latter implies that intensities of characteristic lines may have to be measured for elements not requested in order to make overlap corrections. The general rule is that it is preferable to select a line free from spectral interference, even if less intense, and to compensate by counting for a longer time. Possible line interference is easily checked from tabulations of sequential 2θ values for a given analyzing crystal, as shown in Table 9.1 for analyte Zn using LiF $2d = 4.028$ Å. Some lines are not emitted if the X-ray tube voltage does not exceed the critical excitation voltage of the interfering element or if pulse discrimination is used to minimize interference due to higher order reflections.

9.4.4 Collimation

The prime function of collimators is to limit the angular divergence of the incident excitation source so that an essentially parallel beam arrives at the analyzing crystal. In practice, the degree of collimation is used to balance sensitivity and resolution. The choice of a coarser collimator results in higher count rates but broader peaks, whereas the choice of a fine collimator results in much sharper peaks and lower intensities. The ultimate choice between coarse, intermediate and fine collimation is made in conjunction with other components, namely, available analysing crystals and pulse discrimination.

Table 9.1 Verification of spectral interference. Analysing crystal: LiF $2d = 4.208$ Å

2θ	Element	Z	Line	n	$n\lambda$
41.47	Mo	42	$K-L_2$	2	1.426
41.53	Ce	58	$K-L_3$	4	1.428
41.53	Cd	48	$K-M_2$	3	1.428
41.68	Re	75	L_3-M_5	1	1.433
*41.74	Zn	30	$K-L_3$	1	1.435
*41.86	Zn	30	$K-L_2$	1	1.439
41.89	U	92	L_2-M_4	2	1.440
41.92	Lu	71	L_1-M_2	1	1.441
42.02	Re	75	L_3-M_4	1	1.444
42.05	Gd	64	$K-L_{2,3}$	5	1.445
37.12	W	74	L_2-M_4	1	1.282
37.18	Th	90	L_1-N_2	2	1.284
*37.18	Zn	30	$K-N_{2,3}$	1	1.284
37.21	Ta	73	L_3-N_5	1	1.285
37.24	Ru	44	$K-L_3$	2	1.286
37.36	W	74	L_3-N_1	1	1.290
37.48	Ru	44	$K-L_2$	2	1.294
*37.54	Zn	30	$K-M_3$	1	1.298
37.60	Hf	72	$L_3-O_{4,5}$	1	1.298
37.63	I	53	$K-L_3$	3	1.299
37.72	W	74	L_1-M_2	1	1.302
37.81	Sn	50	$K-M_3$	3	1.305

9.4.5 Crystal selection

Although d spacing is generally the prime criterion in the selection of the analysing crystals, a crystal's diffracted intensity, thermal expansion, stability and possibly cost may limit one's options. For example, LiF 200 is one of the best all-round crystals, having good sensitivity and resolution, and is widely used for elements of Z 19–92. LiF 220, on the other hand, may be used for elements of Z 22–92, but it is rated as only fair for sensitivity although high for resolution. Some crystals (e.g. Ge 111) have no second order diffraction, and hence are chosen to eliminate second order spectral interferences. PET 002 has high reflectivity but unfortunately tends to deteriorate on prolonged exposure to X-radiations, and has a very high thermal expansion coefficient; hence its use generally requires a temperature controlled environment. Bertin (1975, Appendix 10) lists information for 72 analysing crystals and multilayer films, the latter being the preferred choice for elements of low Z ($Z < 13$).

9.4.6 Detector selection

The two commonly used detectors are the scintillation detector and the gas flow proportional counter. The choice of detector is mainly based on the counting efficiency at the wavelength of the analyte line selected. The scintillation counter, being more efficient at shorter wavelengths, is generally used if λ_i is shorter than 1.3 Å. The flow proportional counter (90% argon, 10% methane) is more efficient at longer wavelengths and is generally used if λ_i is longer than 2.0 Å. Both detectors are used in tandem for the intermediate wavelength region $\approx 1.3–2.0$ Å when this option is available. If both detectors have similar counting efficiency, the one having the higher peak to background ratio is normally selected.

9.4.7 Correction for background

The subtraction of a valid background intensity from the measured intensity of a characteristic line, i.e. the intensity measured at a given 2θ angle, takes on more and more importance as the intensity of the peak decreases. Ultimately, the presence of an element at trace levels rests on the ability to detect a very small significant difference between peak and background intensities. Given the fact that it is not possible to measure the *exact* background under a peak directly, the background intensity is usually the resultant of interpolation between measurements at *interference free* positions close to the peak. As shown in Figure 9.5, three cases need be considered:

- the background is essentially constant in the vicinity of the peak, in which case a single measurement on either side is adequate;
- the background is sloping but not curved in the vicinity of the peak;
- the background is curved in the vicinity of the peak.

In the last two cases, the background intensity I_{bkg} (at the peak position of the characteristic line of an element) for a given specimen is computed using the expression

$$I_{bkg} = I_{b_1} - (I_{b_1} - I_{b_2})F_b \qquad (9.11)$$

where I_{b_1} is the background intensity at 2θ, position 1; I_{b_2} is the background intensity at 2θ, position 2; I_{bkg} is the background intensity at 2θ, and peak position F_b is given by

$$F_b = \left(\frac{I_{b_1} - I_{bkg}}{I_{b_1} - I_{b_2}}\right)_{blank} \qquad (9.12)$$

where the blank is a type of material similar to the specimens but free of the analyte. If this condition cannot be met, it may then be necessary to measure the background at three or more positions and to interpolate using a second order least squares fit.

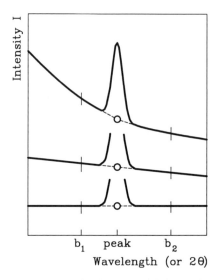

Figure 9.5   Intensity measurements for background correction

9.4.8   Correction for spectral interference

Intensity measurements must be corrected for peak overlap prior to calibration analysis. The application of a correction for spectral overlap is all the more important because it depends on specimen composition, which is not known *a priori* in the case of *true* unknowns. In such cases it is imperative that a qualitative analysis be carried out in order to confirm the absence of overlap

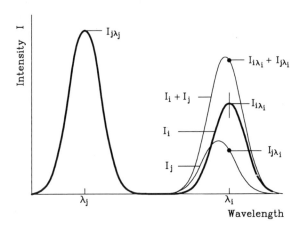

Figure 9.6   Intensity measurements for peak overlap correction

peaks for each analyte; otherwise there is the great danger of reporting a concentration for a given element that may not even be present. The approach usually selected is either compensation or pulse discrimination.

In all cases, the net intensity $I_{i,cor}$ of element i corrected for peak overlap due to element j (Figure 9.6) can be computed using the expression

$$I_{i,cor} = I_{i\lambda_i} - I_{j\lambda_i}$$
$$= I_{i\lambda_i} - O_j I_{j\lambda_j} \qquad (9.13)$$

where all intensities have been previously corrected for background. The factor O_j is obtained from measurements made on *interference references*, i.e. a blank material that does not contain analyte i to which have been added small amounts (e.g. ≈ 0.2, 0.4 or 0.6%) of the purest available compound of element j. Intensities are then measured at a characteristic line of element j free from spectral interference and at the characteristic line of element i that is overlapped by a different line of element j. The ratio of these intensities yields the constant O_j

$$O_j = \frac{I_{j\lambda_i}}{I_{j\lambda_j}} \qquad (9.14)$$

A more precise value of O_j may be obtained from a linear least squares fit through the origin as shown in Figure 9.7.

If the spectral overlap is due to second order peaks, then pulse discrimination is more likely to be the approach chosen. Let us consider the case examined in Section 9.4.2, i.e. the possible overlap of second order Nb $K-L_2$ and $K-L_3$ and Ta L lines. Although these lines occur in the same 2θ region, their pulse amplitudes are quite different. For example, the energies of the Nb $K-L_3$ and Ta L_3-M_5 lines are 16.8 and 8.1 keV, respectively; i.e., the Nb pulses have roughly twice the voltage of those due to Ta. Thus, if the detector gain is such that the Ta pulses are statistically distributed around 15 V, the Nb pulses would then be distributed around 31 V (often referred to as pulse

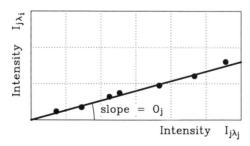

Figure 9.7 Typical spectral overlap correction. The $I_{j\lambda_i}$ versus $I_{j\lambda_j}$ curve relates to specimens free from element i

heights). Hence, by setting the base level at 12 V and the upper level at 18 V, the Nb second order radiations would no longer be registered, since they are well outside that 6 V window.

9.4.9 Correction for dead time

Dead time refers to the time interval during which a detector gives no response to an incident photon because the latter arrives too soon after a previous photon has triggered a pulse. At high counting rate, these 'lost' photons result in measured intensities that are lower than the *true* emitted intensity. The length of time during which the detector is insensitive and gives no response is referred to as the *dead time*. Given the dead time t_d in seconds, measured intensities I_{meas} are corrected for dead time to yield the true intensity I_{true} using the expression

$$I_{true} = \frac{I_{meas}}{1 - I_{meas} t_d} \qquad (9.15)$$

Correcting for dead time (if significant) is absolutely necessary if intensity data are subsequently processed using expressions derived from theoretical principles. Table 9.2 lists typical measured intensities for a dead time t_d equal to 5.0×10^{-7} seconds and provides a clear indication that a large error would result in such a context if dead time were not taken into consideration.

In practice, however, it is the dead time of the overall detection–readout components of the spectrometer that is the concern of the analyst. Thanks to advances in electronic technology, most spectrometers now provide automatic compensation for dead time such that the linear range is extended to one million counts per second.

If compensation for dead time is not provided automatically, linearity can be verified and t_d calculated based on the concepts that the emitted intensity

Table 9.2 Data used to illustrate detector dead time

No.	$I_{i,true}$	$I_{i,meas}$	Ratio	
1	200	200.0		
2	1000	999.5	2:1	4.9975
3	5000	4986	3:2	4.9885
4	10 000	9950		
5	25 000	24 691	5:3	4.9521
6	50 000	48 780	6:4	4.9025
7	100 000	95 238		
8	250 000	222 222	8:6	4.5556
9	500 000	400 000	9:7	4.2000

of a characteristic line is a linear function of tube current, and the ratio of the intensity emitted by two characteristic lines of an element is constant, e.g. $K-L_{2,3}:K-M_3$.

In the first instance, intensities are measured under operating conditions in which higher count rates are obtained by increasing the tube current, whereas in the second case measurements are made on specimens with increasing concentrations of a given element, e.g. increasing amounts of iron oxide in a silica matrix.

If I_{true} is known, then t_d is given by

$$t_d = \frac{I_{true} - I_{meas}}{I_{true} I_{meas}} \tag{9.16}$$

e.g. in the case of data point 6, Table 9.2

$$t_d = \frac{50000 - 48780}{50000 \times 48780} = 5.0 \times 10^{-7} \, \text{s}$$

On the other hand, t_d can be calculated from measured intensities of two characteristic lines, e.g. the $K-L_{2,3}$ and the $K-M_3$ lines of an element. A number of paired data in Table 9.2 (true intensities in the ratio 5:1 representing $K-L_{2,3}:K-M_3$) can serve to represent the decrease in the ratio of the intensities of two characteristic lines due to dead time, as shown under the heading 'Ratio'. In practice, the value of t_d is computed from the measured intensity data, namely

$$t_d = \frac{R_t - \dfrac{I_{K-L_{2,3},\text{meas}}}{I_{K-M_3,\text{meas}}}}{(R_t - 1)I_{K-L_{2,3},\text{meas}}} \tag{9.17}$$

where R_t is the ratio measured at low count rates. Taking R_t as equal to 5 and substituting data for Nos. 6 and 4 yields

$$t_d = \frac{5 - 4.9025}{(5 - 1) \times 48780} = 5.0 \times 10^{-7} \, \text{s}$$

9.5* CALIBRATION

At some stage, all instrumental techniques for quantitative analysis involve the determination of one or more factors that convert signal intensity to concentration; hence calibration involves the measurement of emitted intensities of characteristic lines for specimens of known composition. Although in many instrumental methods the calibration factor is given simply by dividing concentration by intensity, i.e. $K = C/I$, which implies a linear relationship between intensity and concentration, X-ray fluorescence analysts more generally have to work within the framework $C = KIM$, in which case the calibration constant K_i is given by

$$K_i = \left(\frac{C_i}{I_i M_i}\right)_r \tag{9.18}$$

where K_i may be regarded as a *quantification of sensitivity*, i.e. concentration per unit signal *corrected for matrix effects*. If the term M_i is not included in the denominator, then K_i is a variable, implying that sensitivity varies from specimen to specimen. From Eqn (9.18) it follows that, under fixed operating conditions, the smaller the value of the constant K_i the greater the sensitivity, that is, the greater likelihood of detecting lower concentrations. At this point it is important to differentiate between:

- 'intrinsic' sensitivity, which is a function of X-ray physics, namely, fluorescence yield, mass attenuation coefficients etc.; these are the factors that result in low Z elements having lower emitted intensity than high Z elements and account for L lines being less intense than K lines and M lines being less intense than L lines; and
- 'spectrometer' sensitivity, which in addition to the above is a function of spectrometer configuration and operating parameters.

It is the latter that analysts are primarily concerned with in day-to-day laboratory operations.

9.5.1 Definition

A survey of the literature clearly indicates that X-ray fluorescence analysts have used the term calibration to designate a number of related but quite different processes, for example:

(a) the periodic remeasurement of a given material to verify instrumental drift as a function of time;

(b) the determination of the regression coefficients a and b in expressions of the type

$$C_i = a + bI_i \tag{9.19}$$

where a is the intercept and b is the slope for the general equation of a straight line;

(c) the determination of empirical influence coefficients by regression of experimental data in expressions of the type

$$C_i = a + I_i(b + r_{ii}I_i + r_{ij}I_j + \cdots) \tag{9.20}$$

$$\left(\frac{C_i}{R_i} - 1\right) = C_j r_{ij} + C_k r_{ik} + \cdots \tag{9.21}$$

(d) the determination of the calibration (sensitivity) constant K_i wherein absorption and enhancement effects are taken into consideration based on theoretical principles.

9.5.2 Mathematical models

In the framework of $C = KIM$, one can visualize mathematical models that have been used for quantitative X-ray fluorescence analysis in three categories, namely:

- M is ignored; i.e. C is viewed as a function of I only;
- M is evaluated globally for each analyte in each specimen as a result of well defined specimen preparation methods (Chapter 7), i.e. calibration and analysis are carried out simultaneously; this topic is not considered in the following treatment;
- M is calculated; i.e. C is viewed as a function of I and M.

The fact that X-ray fluorescence spectrometers became commercially available prior to the comprehensive development of X-ray fluorescence theory led analysts to use analytical methods that ignored the factor M, as in (a)–(c) below. This is always a viable approach if its inherent limitations are taken into consideration, e.g. if limited to type materials, short concentration ranges, etc. In order to overcome these shortcomings, algorithms were developed in which the factor M, i.e. correction for matrix effects, is evaluated based on corrections that are either empirical (d) or theoretical (e and f) in nature. The following models, presented in order of increasing ability to be applied to more comprehensive analytical contexts, are used initially for calibration and subsequently for analysis:

(a) linear constrained through the origin;
(b) linear with intercept;
(c) non-linear with intercept;
(d) based on the concept of empirical influence coefficients;
(e) based on the concept of semi-empirical influence coefficients;
(f) based on fundamental theory.

(a) Based on the premises that *net* intensities should be nil if the analyte is not present in the specimen, and that *net* intensities are a linear function of concentrations (which implies that M_{iu} and M_{ir} are essentially equal), then a calibration factor can be calculated from data for a single reference material, namely

$$K_i = \left(\frac{C_i}{I_i}\right)_r \tag{9.22}$$

This is obviously questionable unless used strictly to provide an initial estimate of concentration. Given net intensity data for a number of type reference materials, coefficients a and b in Eqn (9.19) can be computed by least squares fitting. If concentration is plotted on the y axis and intensity on the x axis, and the calibration line goes through the origin, then coefficient a is equal to zero (Figure 9.8, line 1) and

$$b = K_i = \left(\frac{\sum C_i^2}{\sum C_i I_i} \right)_r \tag{9.23}$$

This approach at least provides a rough means of evaluating the validity of assuming a linear relationship, hence that K_i is essentially a constant. If C_{ir} values are recalculated by treating the I_{ir} data as unknowns and the results are compared to the known concentrations, one is then in a position to assess whether this model is acceptable in a given application (see column headed $C_i(1)$ in Table 9.3).

(b) If there is a possibility of a small residual background, and even more so if backgrounds are not subtracted, then it is by far preferable to use the general expression for a linear least squares fit with intercept, i.e. in which the coefficient a is not equated to zero (Figure 9.8, line 2); hence

$$C_i = a + bI_i \tag{9.24}$$

where

$$b = \frac{\sum C_i I_i - \dfrac{\sum C_i \sum I_i}{n}}{\sum I_i^2 - \dfrac{(\sum I_i)^2}{n}} \tag{9.25}$$

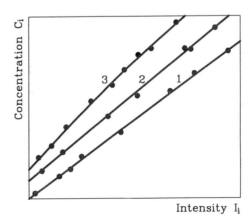

Figure 9.8   Matrix matching method, typical first and second order calibration curves

Table 9.3 Data used to illustrate calibration by linear and second order regression[a]

	$C_i(\%)$	$I_i(c/s)$	$C_i(1)$	$C_i(2)$	$C_i(3)$
Calibration					
R1	0.10	218	0.116	0.089	0.101
R2	0.20	412	0.219	0.197	0.200
R3	0.40	790	0.419	0.407	0.399
R4	0.50	972	0.516	0.508	0.499
R5	0.60	1152	0.612	0.608	0.600
R6	0.70	1328	0.705	0.706	0.701
R7	0.80	1501	0.797	0.802	0.802
R8	1.00	1828	0.971	0.984	0.999
Analysis					
U1	(0.30)	603	0.320	0.303	0.299
U2	(0.90)	1670	0.887	0.896	0.903

[a] $C_i(1) = 5.309 \times 10^{-4} \times I_i$;
$C_i(2) = -0.0325 + 5.560 \times 10^{-4} \times I_i$;
$C_i(3) = -0.0070 + 4.871 \times 10^{-4} \times I_i + 3.454 \times 10^{-8} \times I_i^2$

and

$$a = \frac{\sum C_i}{n} - b\frac{\sum I_i}{n}$$ (9.26)

(see Table 9.3, column headed $C_i(2)$). Equation (9.24) is based on regression of C on I, which assumes that intensities are not subject to errors but that concentrations are. If it is assumed that both concentration and intensity are subject to error, then slope b is given by

$$b = \frac{\left[\dfrac{\sum C_i^2 - \dfrac{(\sum C_i)^2}{n}}{n-1} \right]^{1/2}}{\left[\dfrac{\sum I_i^2 - \dfrac{(\sum I_i)^2}{n}}{n-1} \right]^{1/2}}$$ (9.27)

(c) The above models are based on the premise of a linear relationship between intensity and concentration, and involve the determination of one coefficient if the regression is constrained to pass through the origin and two coefficients if it is not. On the other hand, it may be obvious from a plot of C versus I that the intensity–concentration relationship is not a straight line but a curve (Figure 9.8, curve 3). This being the case, three models involving three coefficients can be considered:

$$C_i = a + bI_i + cI_i^2 \tag{9.28}$$

$$C_i = a + bI_i/(1 + cI_i) \tag{9.29}$$

$$C_i = a + b[1 - \exp(-cI_i)] \tag{9.30}$$

depending on whether one assumes a quadratic (Table 9.3, column headed $C_i(3)$), a hyperbolic or an exponential relationship, respectively. A common feature of Eqns (9.23)–(9.30) is that they are based strictly on finding a statistical correlation between C–I data and do not involve any X-ray fluorescence theory; i.e. identical models are used to process data without any knowledge of the analytical context.

(d) The common feature of the following models is the introduction of a matrix effect correction term based on the concept of influence coefficients obtained strictly by regression of experimental data (see Chapter 6).

$$C_i = a + I_i(b + I_i r_{ii} + I_j r_{ij} + \cdots) \tag{9.31}$$

$$I_i = C_i I_{(i)} - (I_i C_j)r_{ij} - (I_i C_k)r_{ik} - \cdots \tag{9.32}$$

$$\left(\frac{C_i}{R_i} - 1\right) = C_j r_{ij} + \frac{C_k}{1 + C_i}\, r_{ik} + \cdots \tag{9.33}$$

$$\left(\frac{C_i}{R_i} - 1\right) = C_j r_{ij} + C_k r_{ik} + \cdots \tag{9.34}$$

It is highly recommended that more than the minimum number of reference materials be used in order to compensate somewhat for uncertainties in the data, in which case the set of equations relating to Eqn (9.34) takes the form

$$\sum \left(\frac{C_i}{R_i} - 1\right)C_j = r_{ij} \sum C_j^2 + r_{ik} \sum C_j C_k + \cdots + r_{in} \sum C_j C_n$$

$$\sum \left(\frac{C_i}{R_i} - 1\right)C_k = r_{ij} \sum C_k C_j + r_{ik} \sum C_k^2 + \cdots + r_{in} \sum C_k C_n \ldots \tag{9.35}$$

$$\sum \left(\frac{C_i}{R_i} - 1\right)C_n = r_{ij} \sum C_n C_j + r_{ik} \sum C_n C_k + \cdots + r_{in} \sum C_n^2$$

(e) Rather than use empirical influence coefficients, the following calibration models are based on the concept of semi-empirical α influence coefficients (Chapter 5), in which case the term $K_i = C/IM$ is given by

$$K_i = \left(\frac{C_i}{I_i\left[1 + \sum_j \alpha_{ij}C_j\right]}\right)_r \approx 1/I_{(i)} \tag{9.36}$$

if α_{ij} is made equal to the theoretical influence coefficient m_{ijr} computed for the composition of a type material or an average composition and is treated as a constant, thus limiting the applicable concentration range. This range can be extended by using the Claisse–Quintin model

$$\alpha_{ij} = \left(\alpha_j + \alpha_{jj}C_j + \sum_k \alpha_{jk}C_k\right)_i \tag{9.37}$$

or extended even further by using the COLA model

$$\alpha_{ij} = \left[\alpha_{j_1} + \alpha_{j_2}\frac{C_M}{(1 - C_M)\alpha_{j_3}} + \sum_k \alpha_{jk}C_k\right]_i \tag{9.38}$$

$$= \alpha_{ij,hyp} + \sum_k \alpha_{ijk}C_k \tag{9.39}$$

or Tertian's unified model (Tertian, 1988)

$$C_i = \frac{R_i}{1 + \varepsilon_i}\left[1 + \sum_j \alpha_{ij,hyp}C_j\right] \tag{9.40}$$

In all four models the influence coefficients are linked to theoretical data generated by fundamental parameters expressions and are calculated *a priori*.

(f) The following expressions (Chapters 3 and 4) share the distinct feature that the calculated calibration constant is the reciprocal of the intensity emitted by a specimen of pure element i, i.e.

$$K_i = K_{(i)} = 1/I_{(i)} \tag{9.41}$$

$$= \left(\frac{P_i + S_i}{I_i P_{(i)}}\right)_r \tag{9.42}$$

$$= \left(\frac{C_i}{I_i[\ldots]}\right)_r \tag{9.43}$$

where the bracketted term $[\ldots]_r$ may be any fundamental influence coefficient algorithm, for example:

the Broll–Tertian (1983) model

$$[\ldots]_r = 1 + \sum_j \left\{a_{ijp} - e_{ijp}\frac{C_i}{R_i}\right\}_r C_{jr} \tag{9.44}$$

the Rousseau (1984a) model

$$[\ldots]_r = \left[\frac{1 + \sum_j a_{ijp}C_j}{1 + \sum_j e_{ijp}C_j}\right]_r \tag{9.45}$$

or the Lachance (1988) model

$$[\ldots]_r = \left[1 + \sum_j m_{ij}C_j\right]_r \qquad (9.46)$$

In the case of the de Jongh model

$$C_i = (a + bI_i)\left[1 + \sum_j m_{ijn}C_j\right] \qquad (9.47)$$

the influence coefficients are also computed from fundamental theory. In the case of the JIS model

$$C_i = (a + bI_i + cI_i^2)\left[1 + \sum_j \alpha_{ij}C_j\right] \qquad (9.48)$$

the influence coefficients are determined either from theory or by regression analysis, and the coefficients a, b and c are computed from regression analysis of experimental data.

9.6* THE ANALYTICAL PROCESS

Broadly speaking, analysis may be visualized as the antithesis or the opposite of calibration, i.e. the calibration process is simply reversed. If a satisfactory correlation between intensity and concentration has been established (whether or not it included a correction for matrix effects), the correlation should apply within the limits of the model to unknowns that have been prepared and treated identically. A first estimate of a satisfactory correlation simply involves treating the intensities obtained from the reference materials as unknown specimens and re-calculating C_i, which can then be compared with the known values. Needless to say, the differences should be minimal. Prior to proceeding with the analysis of unknowns, steps should be taken to:

- confirm the validity of the calibration;
- compensate for instrumental drift;
- monitor the quality of the compositional data reported in order to maintain the same level of confidence in their precision and accuracy.

9.6.1 Confirmation of the validity of a calibration

If the calibration approach used is empirical in nature, then its validity is very much a function of the number of reference materials, their composition and, concentration ranges etc. It is widely recognized that the coefficients simply represent the best set of numbers that fit the data regressed and often do not reflect what one would expect from X-ray fluorescence theory. This does not mean that the method should be rejected but that it should be used with caution. Ideally, a large number of reference materials should be used for calibration and an equally large number of reference materials *that were not used for calibration* should be used to confirm that acceptable analytical data are generated. Unfortunately, this is often a difficult task in practice. It is also widely recognized that extrapolation is not an option when using an empirical approach.

If the calibration model includes a correction for matrix effects, then it is somewhat easier to assess its validity in the sense that the purpose for including a matrix effect correction is to linearize data that were scattered, hence not linear in the first place. Whether or not linearization was successful can be estimated by comparing plots of C_{ir} versus I_{ir} with plots of C_{ir} versus $(I_i M_i)_r$, or checking the constancy of $C_{ir}/(I_i M_i)_r$ values.

9.6.2 Counting monitors

Except for the singular case where the calibration and analytical processes are carried out simultaneously, analysts are faced with the prospect of carrying out a calibration at time t_c and performing analyses at a later date, that is, at time t_a. Consider for example the context wherein the X-ray spectrometer is operational on a continuous basis. Over a period of time (days, weeks, ...) the gradual deterioration of spectrometer components with age results in a gradual decrease in 'spectrometer' sensitivity, so-called long term drift. This can be eliminated by frequent recalibration, but this can be time consuming in the context of a multielement analytical scheme. The problem can be circumvented by the use of counting monitors. The prerequisites of a counting monitor are presence of the analyte and stability, so that any change in the emitted intensity of a characteristic line is ascribed to a change in sensitivity rather than a deterioration of the monitor. The process may be visualized as normalizing 'time of analysis' (subscript t_a) intensity data to 'time of calibration' (subscript t_c) intensity data.

Normalization is based on the premise that, whereas the absolute net intensities of a characteristic line for two specimens may decrease as a function of time (fixed instrumental operation), their ratio should essentially remain constant. Thus, as part of the calibration process, I_i is measured repeatedly for the monitor and an average value $I_{i,\text{monitor at time of calibration}}$ (I_{i,mon,t_c}) is calculated. The analytical procedure is then segmented in time spans that involve re-reading the counting monitor, yielding $I_{i,\text{monitor at time of analysis}}$ (I_{i,mon,t_a}) values along with I_{iu,t_a} values. The intensities measured at time 't_a' are then normalized to time 't_c' using the expression

$$I_{iu,t_c} = I_{iu,t_a} \frac{I_{i,\text{mon},t_c}}{I_{i,\text{mon},t_a}} \qquad (9.49)$$

It is most important to remind analysts at this stage that although the abbreviated symbols I_{iu} and I_{ir} are used extensively, in the context of quantitative analysis it is *always* implied that these intensities are corrected for background, peak overlap, dead time, instrumental drift, etc.

9.6.2.1 Numerical example

The following treatment examines hypothetical data relating to dead time and instrumental drift corrections.

(a) Analytical context:

$$C_{ir} = 0.2, \; O_j = 0.1$$

$$M_{ir} = 1.1, \; M_{iu} = 1.25$$

dead time = 0.5 μs
uncertainty of the spectral overlap correction = $\pm 5\%$ and the 'true' intensities are

$$I_{ir} = 38\,400 \quad I_{iu} = 15\,000$$

$$I_{mon} = 30\,000 \quad I_j = 8000$$

$$I_{bkg} = 600$$

in which case $K_{(i)}$ is given by

$$K_{(i)} = \frac{0.2}{38\,400 \times 1.1}$$

$$= 4.734 \times 10^{-6}$$

and C_{iu} by

$$C_{iu} = 4.734 \times 10^{-6} \times 15\,000 \times 1.25$$

$$= 0.08878$$

(b) the 'true' emitted intensities at time of calibration are

$$I_{ir} = 38\,400 + 600 = 39\,000$$

$$I_{iu} = 15\,000 + 600 + (8000 \times 0.1) = 16\,400$$

$$I_{mon} = 30\,000 + 600 = 30\,600$$

$$I_j = 8000 + 600 = 8600$$

(c) given a dead time of 0.5 μs, the actual intensities measured are

$$I_{ir} = \frac{39\,000}{1 + (39\,000 \times 0.5 \times 10^{-6})} = 38\,254$$

$$I_{iu} = 16\,167$$

$$I_{mon} = 30\,139$$

$$I_j = 8563$$

$$I_{bkg} = 599.8$$

(d) at time of analysis, as a result of instrumental drift, all intensities are reduced by 5% and by dead time, thus

$$I_{iu} = \frac{(16\,400 \times 0.95 = 15\,580)}{1 + (15\,580 \times 0.5 \times 10^{-6})} = 15\,460$$

$$I_{mon} = 28\,654$$

$$I_j = 8137$$

$$I_{bkg} = 570$$

Consider the case in which a counting monitor is not used and the analyst is unaware that a dead time correction is needed. The net intensity I_{iu} is then computed as

$$I_{iu} = 15\,460 - 570 - [(8137 - 570) \times 0.1]$$
$$= 14\,133.3$$

and

$$C_{iu} = 4.734 \times 10^{-6} \times 14\,133.3 \times 1.25$$
$$= 0.0836, \text{ versus the 'true' value of } 0.08878$$

However, if a counting monitor is used and the analyst is aware that the dead time is 0.5 μs, then intensities are firstly corrected for dead time, e.g. in the case of the monitor

$$I_{mon} = \frac{28\,654}{1 - (28\,654 \times 0.5 \times 10^{-6})} = 29\,070$$

which is then corrected for background

$$I_{mon} = 29\,070 - 570 = 28\,500$$

i.e. the net intensity of the counting monitor has decreased to 28 500 from the value of 30 000 at the time of calibration. The drift factor is therefore equal to

$$28\,500/30\,000 = 0.95$$

Knowledge of this value makes it possible to convert all measured intensities to their equivalent at the 'time of calibration' and obtain the 'true' value for C_{iu}.

9.6.3 Quality control

The need for developing a quality assurance procedure, in order to verify that the precision and accuracy of reported compositional data is maintained over a somewhat extended time span, is unquestionable. Assuming that all four factors S, K, I and M have been adequately dealt with, it would be most unfortunate, when a procedure is first established, if the validity of analytical results were subsequently questioned because of ageing of instrumental components, new operators needing more direction or just plain complacency. The number of items that need to be monitored depends on the *cost* or the *price to pay* if a single or a few reported results turn out to be *wrong*, i.e. outside the expected accuracy. For example, the on-going analytical process may be visualized in terms of weekly cycles made up of daily subcycles; daily cycles made up of three or more subcycles, etc. Let us consider a very simple quality control scheme, namely a daily cycle consisting of three subcycles. The three criteria that will be monitored as a function of elapsed time from calibration are:

(a) the counting reference, which reflects instrument stability;
(b) the net intensities of replicate specimens, which reflect the precision of the K and I factors;
(c) the concentrations of replicate unknowns and/or reference materials.

(a) The counting monitor should be read systematically during the calibration process and the standard deviation computed. It is then the first specimen measured in the daily cycle and the ratios $I_{mon,t_a}/I_{mon,t_c}$ are printed for each analyte. If all the ratios are within the expected range (e.g. confidence limits such as $\pm 2\sigma$) the cycle proceeds; if not, then the analyst must intervene directly in order to rectify the problem. In actual practice one would also normally plot a moving average value (i.e. the average value for the first seven days, the average value day 2 to day 8, etc.), which reflects the trend of instrumental drift. A closer examination of these data may be helpful in indicating the source of the problem; e.g. anomalous data may all relate to the scintillation detector.

(b) Measuring I_{iu} for a newly prepared replicate specimen can provide confirmation that the specimen preparation and intensity measurement processes are still reliable. This step is invaluable for detecting new sources of contamination.

(c) Determining C_{iu} or C_{ir} for a newly prepared replicate specimen provides an assessment of the overall reproducibility of the analytical procedure inasmuch as all four factors K, I, M and S are involved. This is especially the case with theoretical correction methods for matrix effects wherein the factor M is a function of specimen composition.

Given that all three criteria have been found to be within the expected boundaries, analysis of unknown specimens can then proceed. The subcycle (change of operator etc.) may simply involve re-measuring the monitor to verify that all is in order. A more thorough verification short of a full calibration would generally be carried out following any repair, down period, etc.

9.6.4 Analytical precision and the detection limit

As stated at the beginning of this chapter, analytical methods are *almost* unique to each laboratory; i.e., there are bound to be slight differences in the procedures even when laboratories are dealing with identical materials. As long as methods meet the precision and accuracy requirements of the submitter they can be considered as adequate for that context. However, analysts are increasingly required to quantify the reliability of reported compositional data owing to the more stringent demands being made of the results. It is therefore imperative that analysts be fully aware of the small but likely significant discrepancies between theoretical principles and practical application. As stated in a report of the Analytical Methods Committee of The Royal Society of Chemistry (Thompson *et al.* 1987):

Analysts often refer to the performance characteristic (e.g. the precision) on an *analytical method*. This usage is misleading because 'the method' *per se* contributes only part of the total variation that is observed during its use. If the method is used at a number of separate locations and on different occasions, other factors contribute to (and usually dominate) the total variation.

The report seeks to clarify the concept of detection limit in a manner consistent with the IUPAC definition, to reconcile the varying usages of the term, and to maintain the general utility of detection limits by means of a *simple operational definition*.

The concept is examined in the context of within-laboratory and between-laboratory variations; thus, detection limits refer to 'practical conditions' and realistically define the smallest concentration of analyte that can be detected under appropriate conditions. It is proposed that this definition overcomes the criticism that the usual detection limit concept is unrealistic and well below what can be achieved in practice. This usage obviates the need for alternative 'practical limits' such as 'limits of determination'. The authors go on to examine 'instrumental detection limits': how they are unrealistically low, cannot be matched in practice, and are therefore misleading to the unwary.

In the final analysis it comes down to quantifying two quite different concepts, namely:

- the smallest concentration difference that can be reported, exemplified by the ability to detect small differences at high C_i concentrations;
- the smallest concentration that can be reported, usually referred to as the lower limit of detection.

Consider for example a specimen with 18.00% C_i. The question is, what is the likely spread in reported C_i if the same specimen is resubmitted as a blind duplicate over an extended time span, or conversely, what is the probability that a specimen of slightly different composition can be so identified. The problem is somewhat different if the same question is asked when C_i occurs at trace levels. In this context, Thompson et al. (1987) conclude that:

- despite the apparent simplicity of the detection limit concept, a degree of confusion surrounds its use and that of similar terms;
- the term 'detection limit' is clearly worth retaining even though it cannot be rigorously interpreted in terms of confidence intervals.

The premise is that, in practical analysis for elements at trace levels, precision (and ultimately, accuracy) in analytical systems is a function of the analyte concentration; i.e. the standard deviation increases as C_i increases. The relationship can often be expressed by the simple model

$$C_i = C_i \pm (a + bC_i) \tag{9.50}$$

where the coefficient a relates to uncertainties in computing $I_{i,net}$ and the coefficient b relates to uncertainties in the calibration and the correction (or lack thereof) for matrix effects. The coefficients a and b can be so chosen as to encompass 50%, 90% ... or 1σ, 2σ, 3σ (3σ recommended) of the data. With regard to precision, the data may relate to C_i values for duplicates, whereas with regard to accuracy the data would relate to C_i for reference materials (Figure 9.9). Let us consider, for example, a situation in which the concentration range is 2000 ppm and the coefficients a and b are equal to 2 and 0.02, respectively; hence the typical values are 20 ppm ± 2.4 ppm, 200 ppm ± 6 ppm and 2000 ppm ± 42 ppm. The detection limit is then a matter of interpretation; i.e. whether one considers a result with ±50%, 90% etc. error as quantitative and useful to the submitter. This is likely to depend on the context. For example, the result 5 ppm ± 2.1 ppm is of little value within a suite of concentrations ranging from 0 to 10 or 20 ppm, but it may be acceptable within a suite of concentrations ranging from 0 to 1500 ppm.

A more annoying problem arises in the presence of significant spectral overlap because of the inherent error associated with the correction due to the

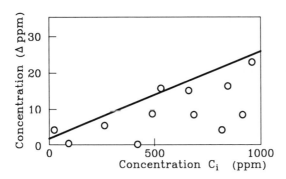

Figure 9.9 Schematic representation of the uncertainty of analytical data as expressed by Eqn (9.50)

presence of element j. The total uncertainty associated with the determination of C_i can then be expressed as

$$C_i = C_i \pm (a + bC_i + o_jC_j) \tag{9.51}$$

with the uncertainty of the overlap correction visualized as

$$\Delta C_i \propto \Delta I_i \propto I_{j\lambda_i} \propto I_{j\lambda_j}O_j \propto o_jC_j$$

Consider the data in Table 9.4 relating to the following context:

- coefficient a is equal to 2 ppm;
- coefficient b is equal to 2%;
- the factor o_j, i.e. the uncertainty associated with the spectral overlap correction due to element j, is equal to 0.5%.

Thus, in the presence of 2000 ppm C_j, the uncertainty introduced due to spectral overlap is given by $0.005 \times 2000 = \pm 10$ ppm C_i. If $C_i = 20$ ppm, this results in a total uncertainty of ± 12.4 ppm compared with ± 2.4 ppm if element j is absent. In this context, it is obvious that the spectral overlap correction governs the validity of results at low C_i concentrations.

9.6.5 Software

Advances in electronic technology have led to the availability of powerful microcomputers at prices that have certainly been a welcome benefit to X-ray fluorescence analysts. The state of the art is such that the computer not only provides for automatic operation but also allows the on-line processing of experimental data, right up to the printout of the report of analyses, for most if not all currently produced X-ray fluorescence spectrometers. This is certainly the case in energy dispersive instruments, where one cannot imagine

Table 9.4   Data illustrating the level of uncertainty at trace
levels of C_i in the presence of spectral overlap, Eqn (9.51)

Analytical context: $a = 2$ ppm; $b = 2\%$; $o_j = 0.5\%$

		Uncertainty ($\pm$ ppm) arising from			
C_i(ppm)	C_j(ppm)	a	b	$o_j C_j$	Total
20	0	2	0.4	0	2.4
20	60	2	0.4	0.3	2.7
20	200	2	0.4	1.0	3.4
20	1000	2	0.4	5.0	7.4
20	2000	2	0.4	10.0	12.4
400	0	2	8.0	0	10.0
400	200	2	8.0	1.0	11.0
400	1000	2	8.0	5.0	15.0
400	3000	2	8.0	15.0	25.0

manually handling the enormous amount of data produced. Thus the software
that is generally part and parcel of energy dispersive spectrometers normally
provides for:

- manual or automatic operation;
- selection of kV, mA, primary filter, secondary target excitation, counting
  time, characteristic line, channel width, peak identification etc.;
- smoothing the spectrum to reduce statistical fluctuations;
- deconvolution of spectra, i.e. subtraction of background, escape and sum
  peaks, and peak overlap;
- film thickness routines;
- a choice of empirical and theoretical models for calibration and
  subsequent analysis.

Similarly, in the case of modern automated wavelength dispersive spectrometers, it is taken for granted that a computer not only controls the insertion
of reference standards, specimens and counting monitors, but also sets the
operating parameters (kV, mA, insertion of filters, 2θ angles and counting
times for peaks and backgrounds) for every element whose concentration is to
be determined or is only needed for overlap corrections etc. The raw intensity
data can then be processed to yield net peak intensities including a correction
for drift. Net intensity files for standards are used for calibration and, subsequently, intensity files for unknowns are processed to yield concentrations.
In most cases a choice of empirical and theoretical models is offered for the
correction for matrix effects.

Software is generally supplied by the manufacturer, and in most cases offers a selection of options for data processing. If the 'software package' is available only in compiled form (the usual case), it is very difficult if not impossible to modify the software so as to incorporate desirable features that are not provided. In the use of X-ray fluorescence software, analysts may give priority to simplicity or to versatility, depending on the analytical context. This has become a relatively minor concern, given the memory and speed of execution of modern computers. Unfortunately, if the software is versatile to the point of being able to deal with most analytical contexts, the analyst faced with applying non-routine procedures will then be requited to make a number of judicious selections from a host of features provided, entailing an attendant loss of simplicity. Yet this is by far preferable for the experienced analyst. It is advantageous if the software offers the option of bypassing one or more sub-routines if the analytical context is only that of quality control. It would appear that in such cases 'the goal is *that the operator need press no more than three buttons*'. In the main, software simply provides analysts with a greater facility to specify how intensity data will be acquired and processed for calibration and analysis. Software cannot *per se* guarantee the quality of results unless it relates to a protocol application; i.e. all the details of the procedure to be followed are fixed and must be followed in minute detail. Thus, although software should make the analyst's task of applying analytical methods easier, it does not preclude the need to be knowledgeable of the principles of X-ray fluorescence and to keep abreast of current developments.

9.7 SUMMARY

At this stage there can be no doubt that there are many options that have to be exercised in the development of a successful analytical procedure. The overall process may be envisaged in the framework that the accuracy of C_i determined by X-ray fluorescence analysis is a resultant of the interaction between four factors, namely:

S, the physical nature of the specimen;
I, the validity of the net intensities of characteristic lines;
M, fundamental and instrumental parameters;
K, the calibration constant.

Regardless of the physical nature of the samples submitted for analysis, the specimen preparation procedure *must* be such that, as a minimum, it is reproducible, i.e. well within the precision expected for the concentrations ultimately reported. Ideally all specimens should meet the criteria of being flat, homogeneous and infinitely thick. Barring this, analysts must have some assurance that reference materials and unknowns have very similar physical properties so that their effects cancel out or are compensated for differences mathematically. Should the specimens not meet the criterion of infinite thickness, then an appropriate correction must be applied. Care must be taken to ensure that contamination during the specimen preparation process is negligible or else compensated for. The ultimate criterion is assurance that any further work on any given intensity data will be meaningful.

The production of valid net intensity data on an ongoing basis requires close attention, especially when dealing with true unknowns. In the worst case scenario, the intensity of a characteristic line emitted at a given 2θ angle will have to be corrected for:

- a sloping curved background;
- overlap by one or more characteristic lines of other elements present in the specimen or emanating from the tube target;
- dead time;
- instrumental drift.

The ultimate criterion is that the values assigned to I_i, I_j, ... relate *only* to the emitted characteristic radiations of elements i, j, ... and are independent of when they were measured, i.e. are corrected for instrumental drift.

The factors K and M are closely related in that the factor M which quantifies the matrix effect will directly or indirectly affect K values: directly in the case where the model is based on $C = KIM$ and indirectly if calibration and analysis are based on $C = KI$. The latter implies that the values M_{iu} and M_{ir} are similar to the point that a matrix effect correction can be ignored. These include methods covered in Chapter 7 such as matrix matching, standard addition, high dilution, double dilution etc.; for example

$$C_{iu} = \left(\frac{C_i}{R_i}\right)_r I_{iu}$$

or

$$C_{iu} = \left(\frac{\Delta C_i}{\Delta I_i}\right)_u I_{iu}$$

both of which imply a linear relationship between intensity and concentration (true in some limited analytical contexts).

In some cases it is possible to select an empirical approach based on least squares fitting of experimental data. These may take the form

$$C_{iu} = a + bI_i$$

which implies a linear relation; or if curvature is suspected

$$C_{iu} = a + bI_{iu} + cI_{iu}^2$$

In order to extend the concentration range, one may apply theoretically calculated corrections for matrix effects, namely

$$C_{iu} = \left(\frac{C_i}{I_i M_i}\right)_r I_{iu} M_{iu}$$

$$= K_i I_{iu}[\ldots]_u$$

where $[\ldots]$ is any given fundamental influence coefficient model.

Conversely, concentration may be calculated directly from classical fundamental parameters formalism, i.e.

$$C_{iu} = \frac{R_{iu(expr)}}{R_{iu(theo)}} C_{iu(estimate)}$$

where

$$R_{iu(expr)} = \left(\frac{P_i + S_i}{I_i P_{(i)}}\right)_r I_{iu}$$

and

$$R_{iu(theo)} = \left(\frac{P_i + S_i}{P_{(i)}}\right)_{u(estimate)}$$

All theoretical approaches are based on the iterative process by which one makes successively better estimates of concentrations until the ratio $R_{iu(expr)}/R_{iu(theo)}$ is essentially equal to unity. The fact that X-ray fluorescence is the technique of choice in such a wide variety of materials is clearly attested by the many examples of successful applications outlined in Part II.

PART II
APPLICATION

Part II aims at presenting a comprehensive and systematic suite of numerical examples relating to topics covered in Part I. The material is generally presented in the same sequence as the theory was developed in Part I and, whenever possible, involves carrying out theoretical calculations for which experimental data are available in the literature. This provides an objective assessment of both the exactness of the theoretical expressions and the validity of the experimental data, i.e. a window on the extent to which theory is applicable in practice. Although it may seem strange to set limits on the application of theoretically derived expressions, it must be remembered that theoretical developments are based on what might be described as *ideal situations*, which may not prevail in practice. One of the key points to be brought out is that, even in the most simple analytical contexts (mono-chromatic incident X-radiations, binary systems etc.), theoretical calculations do not lead to absolute values. The reason is that the accuracy with which the input data, i.e. the fundamental parameters, are known is limited, there being an uncertainty associated with each parameter used. Thus their values differ slightly depending on their source.

The greater problem, however, is likely to rest on the experimental measurement process. Experienced analysts are well aware that, because of time, cost and a host of other constraints, not all specimens may be perfectly homogeneous, instrumental operations may not always be carried out under idealized conditions, etc. The anomalies show up when measurements are made on two or more reference materials. Ideally, given the measured intensity of a characteristic line of an element in a well defined analytical context, it should be possible to *predict* the measured intensity of the same characteristic line in any other specimen of known composition. There are ample examples in the literature where this has been achieved within the reproducibility expected. It could hardly be otherwise. Were it not so, i.e. if it could be proven beyond all doubt that confirmed experimental data are not in agreement with what is predictable from theoretical principles, then the whole foundation of quantitative X-ray fluorescence analysis, nay of X-ray physics, would be undermined. But what of those occasions when the anomalies are somewhat greater than expected? A case in point would be the application of theoretical corrections to experimental data relating to intensities measured on pressed powder specimens of rock samples. According to Jenkins (1979), the fact that particle size and mineralogical effects are also present are much more likely to account for anomalies than the theoretical principles of X-ray fluorescence emission.

In the main, Part II presents a comprehensive integrated treatment of applications of the theory developed in Part I. The Cr–Fe–Ni system is generally used for continuity and in order to have access to experimental data that have appeared in the literature, thereby providing confirmation by independent sources. Based on actual tables of fundamental parameters or algorithms that have been proposed to facilitate generating them for data processing, numerical examples are given that lead to further tabulations. This process is continued throughout, thus providing readers with the equivalent of a set of questions and answers; i.e. calculations similar to the examples shown can be carried out, if desired, and the results compared to the data in the appropriate tables. At any time, the reader may refer back to Part I for the theoretical aspect of any given expression in Part II; e.g. theoretical considerations relating to the application of a given expression in Part II.1 will be found in Part I, Chapter 1.

II.1 X-RAY PHYSICS

II.1.4 CRITICAL EXCITATION (BINDING) ENERGIES

The energy binding electrons to the nucleus of atoms is a fundamental entity that is at the heart of quantitative X-ray fluorescence calculations. Binding energies are obtained from critical evaluations of very exacting experimental measurements of the minimum energy required to cause the emission of the different series of characteristic lines. The energy level data in Table II.1.1 may be used as the source material to compute the data in Tables II.1.3–II.1.5.

In order to provide greater facility for computer data processing, algorithms have been developed that generate critical excitation energies as a function of the atomic number of the element under consideration. For example, Poehn *et al.* (1985) proposed the following expression:

$$E_{crit.} \text{ (keV unit)} = a + bZ + cZ^2 + dZ^3 \tag{II.1.1}$$

i.e. in the case of the K energy level of element i

$$E_{K,i} = a + bZ_i + cZ_i^2 + dZ_i^3 \tag{II.1.2}$$

Table II.1.2 lists the values for the coefficients a, b, c and d in Eqn (II.1.2) that may be used to calculate E_K, E_{L_1}, E_{L_2} and E_{L_3}. Thus, the critical excitation energy $E_{K,Fe}$ necessary to eject Fe ($Z = 26$) K level electrons is given by

$$
\begin{aligned}
a &= -1.304 \times 10^{-1} & &= -0.1304 \\
+ bZ & \quad -2.633 \times 10^{-3} \times 26 & &-0.685 \\
+ cZ^2 & \quad +9.718 \times 10^{-3} \times 26^2 & &+6.569 \\
+ dZ^3 & \quad +4.144 \times 10^{-5} \times 26^3 & &\underline{+0.728} \\
& & &= \quad 7.098 \text{ keV}
\end{aligned}
$$

versus 7.112 keV, the value listed in Table II.1.1.

Table II.1.1   Critical (binding) energies (keV)[a]

	E_K	E_{L1}	E_{L2}	E_{L3}	E_{M3}	E_{M4}	E_{M5}	E_{N5}
Si	1.839	0.149	0.099	0.099				
Cr	5.989	0.695	0.584	0.575	0.043	0.002	0.002	
Fe	7.112	0.846	0.721	0.708	0.054	0.004	0.004	
Ni	8.333	1.008	0.872	0.855	0.068	0.004	0.004	
Mo	20.000	2.866	2.625	2.520	0.392	0.230	0.227	0.002
W	69.525	12.100	11.544	10.207	2.281	1.872	1.809	0.245

[a] Data (rounded off) from Bearden and Burr (1967).

Table II.1.2   Fit parameters for critical (binding) energies, Eqn (II.1.1)[a]

Level	E_K	E_{L_1}	E_{L_2}	E_{L_3}
Z	11–63	28–83	30–83	30–83
a	-1.304×10^{-1}	-4.506×10^{-1}	-6.018×10^{-1}	3.390×10^{-1}
b	-2.633×10^{-3}	1.566×10^{-2}	1.964×10^{-2}	-4.931×10^{-2}
c	9.718×10^{-3}	7.599×10^{-4}	5.935×10^{-4}	2.336×10^{-3}
d	4.144×10^{-5}	1.792×10^{-5}	1.843×10^{-5}	1.836×10^{-6}

[a] Data from Poehn et al. (1985).

II.1.5   CRITICAL EXCITATION (ABSORPTION EDGE) WAVELENGTHS

Users of wavelength dispersive spectrometers find it more practical to visualize the data in the previous table in terms of wavelengths; i.e. what is the maximum wavelength that will excite atoms in any given shell and lead to the emission of characteristic x-radiations from that atom? Energies can be expressed in terms of wavelengths, in Å units, from the relationship

$$\lambda = \frac{12.3981}{\text{keV}}$$

Table II.1.3   Absorption edge wavelengths (Å)

Element, (Z)	K	L_1	L_2	L_3
Si (14)	6.738			
Cr (24)	2.070	17.8		
Fe (26)	1.743	14.65	17.20	17.53
Ni (28)	1.488	12.30	14.22	14.50
Mo (42)	0.620	4.326	4.723	4.920

Thus Table II.1.3 provides the same information as that in Table II.1.1 except that it is expressed in terms of wavelength.

II.1.6   ENERGIES AND WAVELENGTHS OF CHARACTERISTIC X-RADIATIONS

Given the energy binding electrons in the various levels, it is possible to calculate the energy of the characteristic X-rays emitted in the transition of outer electrons to inner levels. For example, the photon energies of the characteristic E_{K-L_3} line are equal to the difference in energy between the E_K and E_{L_3} levels, i.e.

$$E_{K-L_3} = E_K - E_{L_3} \qquad (\text{II.1.3})$$

Thus, for the element iron, substitution of E values from Table II.1.1 gives

$$F_{\text{Fe } K-L_3} = 7.112 - 0.708$$
$$= 6.404 \text{ keV}$$

Tables II.1.4 and II.1.5 list the energies and wavelengths of the characteristic K lines and some of the stronger L lines for the elements Si, Cr, Fe, Ni and Mo.

Table II.1.4   Characteristic line energies (keV)[a]

	E_{K-L_3}	E_{K-L_2}	$E_{K-L_{2,3}}$	E_{K-M_3}	$E_{L_3-M_5}$	$E_{L_3-M_4}$	$E_{L_2-M_4}$	$E_{L_3-N_5}$
Si	1.740	1.739	1.740	1.836				
Cr	5.415	5.406	5.412	5.947	0.573	0.573		
Fe	6.404	6.391	6.400	7.058	0.705	0.705		
Ni	7.478	7.461	7.473	8.265	0.851	0.851		
Mo	17.479	17.374		19.608	2.293	2.290	2.395	2.518

[a] Calculated from data in Table II.1.1.

Table II.1.5   Characteristic line wavelengths (Å)[a]

	λ_{K-L_3}	λ_{K-L_2}	$\lambda_{K-L_{2,3}}$	λ_{K-M_3}	$\lambda_{L_3-M_5}$	$\lambda_{L_3-M_4}$	$\lambda_{L_2-M_4}$	$\lambda_{L_3-N_5}$
Si	7.125	7.128	7.126	6.753				
Cr	2.290	2.294	2.291	2.085				
Fe	1.936	1.940	1.937	1.757	17.59			
Ni	1.658	1.662	1.659	1.500	14.56			
Mo	0.709	0.714	0.711	0.632	5.407	5.414	5.177	4.923

[a] Calculated from data in Table II.1.4.

Poehn *et al.* (1985) also proposed an algorithm to facilitate computation of the energies of characteristic X-radiations, namely

$$E_{\text{char.}} \text{ (keV unit)} = a + bZ + cZ^2 + dZ^3 + eZ^4 \tag{II.1.4}$$

i.e. in the case of the unresolved $K-L_3$ and $K-L_2$ lines for element i

$$E_{K-L_{2,3}} = a + bZ_i + cZ_i^2 + dZ_i^3 + eZ_i^4 \tag{II.1.5}$$

A list of a, b, c, d and e values to generate $E_{K-L_{2,3}}$ and $E_{L_{3}M_{3,4}}$ data is given in Table II.1.6. In the case of i = Fe, where $Z = 26$, substitution in Eqn (II.1.5) gives for $E_{K-L_{2,3}}$

$$
\begin{aligned}
a &= -6.654 \times 10^{-2} & &= -0.06654 \\
+\ bZ &= -8.609 \times 10^{-3} \times 26 & &= -0.2238 \\
+\ cZ^3 &= 9.630 \times 10^{-3} \times 26^2 & &= +6.510 \\
+\ dZ^3 &= 7.268 \times 10^{-6} \times 26^3 & &= +0.1277 \\
+\ eZ^4 &= 1.131 \times 10^{-7} \times 26^4 & &= \underline{+0.0517} \\
& & &= \quad 6.399 \quad \text{keV}
\end{aligned}
$$

Compared with the generally accepted value of 6.400 keV.

II.1.7.1 Experimentally measured spectral distributions

Although the basic equation of Kramers (1923) has been modified by Tertian and Broll (1984) and by Pella *et al.* (1985, 1991) to generate I_λ emanating from X-ray tubes as a function of target material, kV etc., the spectral distributions used in subsequent numerical examples of fundamental parameters calculations will be data measured experimentally for W, Cr and Rh target tubes, all operated at 45 kV (constant potential), and intensities measured at 0.02 Å intervals. The integrated intensities (in arbitrary units) in the spectrum of a W target OEG-50 tube (Table II.1.7) are from Gilfrich and Birks (1968). The integrated intensities for a Cr target OEG-50 tube (Table II.1.8) are from Birks

Table II.1.6 Fit parameters for characteristic line energies (keV) Eqn (II.1.4)[a]

Z range	$E_{K-L_{2,9}}$ 11–50	$E_{L_3-M_4}$ 30–83
a	-6.654×10^{-2}	4.776×10^{-2}
b	-8.609×10^{-3}	-1.706×10^{-2}
c	9.630×10^{-3}	1.523×10^{-3}
d	7.268×10^{-6}	4.414×10^{-6}
e	1.131×10^{-7}	-1.739×10^{-8}

[a] Data from Poehn *et al.* (1985).

Table II.1.7 Integrated intensity (arbitrary units) in the spectrum of a tungsten target tube at 45 kV (CP). $\Delta\lambda = 0.02$ Å

λ (Å)	$I_\lambda \Delta\lambda$	λ (Å)	$I_\lambda \Delta\lambda$	λ (Å)	$I_\lambda \Delta\lambda$	λ (Å)	$I_\lambda \Delta\lambda$	λ (Å)	$I_\lambda \Delta\lambda$
0.29	15.5	0.75	53.9	1.21	33.4	1.67	35.9	2.13	8.6
0.31	36.6	0.77	51.8	1.23	36.2	1.69	21.1	2.15	8.2
0.33	56.8	0.79	49.9	1.25	215.8	1.71	20.1	2.17	7.8
0.35	76.6	0.81	48.2	1.27	35.4	1.73	19.2	2.19	7.5
0.37	96.2	0.83	46.6	1.29	442.0	1.75	18.3	2.21	7.3
0.39	111.1	0.85	45.2	1.31	34.6	1.77	17.5	2.23	7.0
0.41	116.4	0.87	44.0	1.33	34.1	1.79	16.8	2.25	6.7
0.43	114.6	0.89	42.9	1.35	33.5	1.81	16.1	2.27	6.4
0.45	109.9	0.91	42.0	1.37	33.0	1.83	15.5	2.29	6.1
0.47	104.5	0.93	41.2	1.39	32.5	1.85	14.9	2.31	5.9
0.49	99.1	0.95	40.4	1.41	32.0	1.87	14.3	2.33	5.7
0.51	93.9	0.97	39.6	1.43	31.5	1.89	13.7	2.35	5.4
0.53	89.0	0.99	38.8	1.45	30.9	1.91	13.1	2.37	5.1
0.55	84.4	1.01	38.1	1.47	622.3	1.93	12.6	2.39	4.9
0.57	80.3	1.03	37.4	1.49	29.7	1.95	12.2	2.41	4.7
0.59	76.6	1.05	36.7	1.51	29.0	1.97	11.7	2.43	4.5
0.61	73.3	1.07	49.1	1.53	28.3	1.99	11.2	2.45	4.2
0.63	70.1	1.09	63.4	1.55	27.6	2.01	10.8	2.47	3.9
0.65	67.0	1.11	35.1	1.57	26.8	2.03	10.4	2.49	3.7
0.67	64.1	1.13	34.5	1.59	26.0	2.05	10.0	2.51	3.5
0.69	61.2	1.15	33.9	1.61	25.2	2.07	9.7	2.53	3.3
0.71	58.6	1.17	33.4	1.63	24.3	2.09	9.4	2.55	3.1
0.73	56.2	1.19	33.0	1.65	23.2	2.11	9.0	2.57	2.9

(1969), and those for a Rh target SEG-50H tube (Table II.1.9) are from Gilfrich *et al.* (1971). In all three tables, the intensities of the characteristic lines of the target element, which were measured and reported separately, have been added to the continuum data at the appropriate Δλ interval.

II.1.8.1 Experimental determination of μ

Table II.1.10 lists additional data from Dalton (1969) relating to the experimental determination of μ_{Er} for incident X-radiations Zn $K-L_{2,3}$, i.e. 1.436 Å. It comprises measurements made on three different foils with identical incident intensities and measurements made on the same foil at three different incident intensities. The values in the column headed μ_{Er} are obtained by substituting the appropriate data in

$$\mu_{Er} = \frac{\text{area (in cm}^2)}{\text{weight (in g)}} \ln \frac{I_0}{I_x} \qquad\qquad \text{(II.1.6)}$$

Table II.1.8 Integrated intensity (arbitrary units) in the spectrum of a chromium target tube at 45 kV (CP). $\Delta\lambda = 0.02$ Å

λ (Å)	$I_\lambda \, \Delta\lambda$	λ (Å)	$I_\lambda \, \Delta\lambda$	λ (Å)	$I_\lambda \, \Delta\lambda$	λ (Å)	$I_\lambda \, \Delta\lambda$	λ (Å)	$I_\lambda \, \Delta\lambda$
0.29	3.0	0.75	14.0	1.21	9.0	1.67	5.0	2.13	8.3
0.31	6.6	0.77	14.0	1.23	8.8	1.69	4.9	2.15	8.1
0.33	8.8	0.79	14.0	1.25	8.5	1.71	4.8	2.17	7.8
0.35	9.9	0.81	13.9	1.27	8.3	1.73	4.7	2.19	7.6
0.37	11.0	0.83	13.8	1.29	8.1	1.75	4.6	2.21	7.4
0.39	12.0	0.85	13.7	1.31	7.8	1.77	4.4	2.23	7.2
0.41	12.9	0.87	13.6	1.33	7.6	1.79	4.3	2.25	7.0
0.43	13.5	0.89	13.4	1.35	7.4	1.81	4.2	2.27	6.7
0.45	13.9	0.91	13.2	1.37	7.2	1.83	4.1	2.29	2517.5
0.47	13.8	0.93	13.0	1.39	7.0	1.85	4.0	2.31	6.3
0.49	13.8	0.95	12.7	1.41	6.8	1.87	3.9	2.33	6.1
0.51	13.8	0.97	12.4	1.43	6.7	1.89	3.8	2.35	5.9
0.53	13.8	0.99	12.1	1.45	6.5	1.91	3.7	2.37	5.7
0.55	13.8	1.01	11.8	1.47	6.3	1.93	3.6	2.39	5.6
0.57	13.8	1.03	11.4	1.49	6.2	1.95	3.5	2.41	5.4
0.59	13.8	1.05	11.1	1.51	6.0	1.97	3.4	2.43	5.2
0.61	13.8	1.07	10.9	1.53	5.9	1.99	3.3	2.45	5.0
0.63	13.8	1.09	10.6	1.55	5.8	2.01	3.2	2.47	4.9
0.65	13.8	1.11	10.3	1.57	5.6	2.03	3.1	2.49	4.7
0.67	13.9	1.13	10.0	1.59	5.5	2.05	3.0	2.51	4.6
0.69	13.9	1.15	9.8	1.61	5.4	2.07	5.9	2.53	4.4
0.71	14.0	1.17	9.5	1.63	5.3	2.09	423.8	2.55	4.3
0.73	14.0	1.19	9.3	1.65	5.1	2.11	8.5	2.57	4.2

II.1.8.2 Algorithms for generating μ values

The experimental measurement of mass attenuation coefficients for pure elements at specific monochromatic incident wavelengths provides valuable data *per se*, but it is the extension of these data to provide coefficients in algorithms that can generate μ values at any intermediate wavelength that is the more valuable. The process is described using the μ_{Fe} data in Table 1.9. The first step (see Table II.1.11) consists in expressing λ and μ_{Fe} in terms of their natural logarithms. These were the values plotted in Figure 1.5a, showing a linear relation between discontinuities. In the expression $\mu_\lambda = C\lambda^n$, the coefficient C represents the intercept, and the exponent n represents the slope when $\ln \mu$ is plotted on the y axis and $\ln \lambda$ is plotted on the x axis. From analytical geometry, the slope of a line is given by

$$\text{slope} = \frac{y_2 - y_1}{x_2 - x_1} = n \tag{II.1.7}$$

Table II.1.9   Integrated intensity (arbitrary units) in the spectrum of a rhodium target tube at 45 kV (CP). $\Delta\lambda = 0.02$ Å

λ (Å)	$I_\lambda \Delta\lambda$	λ (Å)	$I_\lambda \Delta\lambda$	λ (Å)	$I_\lambda \Delta\lambda$	λ (Å)	$I_\lambda \Delta\lambda$	λ (Å)	$I_\lambda \Delta\lambda$
0.29	8.0	0.75	49.8	1.21	38.4	1.67	24.0	2.13	13.0
0.31	25.0	0.77	49.6	1.23	37.6	1.69	23.6	2.15	12.6
0.33	33.6	0.79	49.2	1.25	36.8	1.71	23.0	2.17	12.2
0.35	38.2	0.81	48.8	1.27	36.0	1.73	22.4	2.19	11.8
0.37	41.2	0.83	48.6	1.29	35.4	1.75	22.0	2.21	11.4
0.39	43.4	0.85	48.2	1.31	34.6	1.77	21.6	2.23	11.1
0.41	44.4	0.87	47.8	1.33	34.0	1.79	21.0	2.25	10.8
0.43	44.8	0.89	47.6	1.35	33.2	1.81	20.4	2.27	10.4
0.45	45.2	0.91	47.2	1.37	32.8	1.83	20.0	2.29	10.1
0.47	44.8	0.93	46.8	1.39	31.8	1.85	19.4	2.31	9.8
0.49	44.4	0.95	46.4	1.41	31.2	1.87	18.8	2.33	9.5
0.51	43.6	0.97	46.0	1.43	30.4	1.89	18.4	2.35	9.2
0.53	46.0	0.99	45.6	1.45	29.8	1.91	18.0	2.37	8.9
0.55	157.2	1.01	45.2	1.47	29.4	1.93	17.4	2.39	8.6
0.57	52.0	1.03	44.8	1.49	28.8	1.95	17.0	2.41	8.3
0.59	51.8	1.05	44.4	1.51	28.4	1.97	16.4	2.43	8.1
0.61	496.6	1.07	43.6	1.53	27.8	1.99	16.0	2.45	7.8
0.63	51.4	1.09	43.2	1.55	27.2	2.01	15.7	2.47	7.6
0.65	51.0	1.11	42.4	1.57	26.6	2.03	15.5	2.49	7.3
0.67	50.8	1.13	41.6	1.59	26.0	2.05	14.8	2.51	7.1
0.69	50.4	1.15	40.8	1.61	25.4	2.07	14.4	2.53	6.8
0.71	50.2	1.17	40.0	1.63	25.0	2.09	13.9	2.55	6.7
0.73	50.0	1.19	39.2	1.65	24.6	2.11	13.4	2.57	6.5

Substitution of the ln values listed in Table II.1.11 for $\lambda = 0.8766$ Å and 1.659 Å yields

$$n = \frac{5.961 - 4.202}{0.506 - (-0.132)} = 2.757$$

for wavelengths shorter than the K absorption edge of iron.

The intercept C may then be obtained from the following relationship applied to any of the wavelengths

$$C = \frac{\mu_\lambda}{\lambda^n} \tag{II.1.8}$$

$$= \frac{66.83}{0.8766^{2.757}} = 96.1$$

For greater accuracy, one would use a least squares fit for many data points. In the present case, a least squares fit of the six data points at wavelengths

Table II.1.10 Data relating to the experimental determination of μ_{Er} for $\lambda = 1.436$ Å

Foil	Area, cm^2	Weight, g	I_0	I_x	μ_{Er}
1	5.5355	0.064512	12173	393.53	293.94
2	5.5333	0.060122	12173	522.06	289.83
3	5.5333	0.054293	12173	698.23	291.33
4	5.5357	0.070627	10788	261.40	291.58
4			11353	277.69	290.84
4			11485	277.49	291.81

Table II.1.11 Data used to illustrate the experimental determination of coefficients C and n for iron in Eqn (1.32)[a]

λ Å	μ_{Fe}	ln λ	ln μ_{Fe}
0.7476	43.10	-0.291	3.764
0.8766	66.83	-0.132	4.202
1.177	149.7	0.163	5.009
1.295	193.5	0.259	5.265
1.542	309.9	0.443	5.736
1.659	388.0	0.506	5.961
1.757	53.04	0.564	3.971
2.085	86.52	0.735	4.460
2.748	189.0	1.011	5.242
3.359	334.1	1.212	5.811
3.742	438.0	1.320	6.08

[a] Experimental data from Dalton and Goldak (1969).

shorter than the K absorption edge gave $C = 95.6$ and $n = 2.743$; thus in this context

$$\mu_{FE\lambda} = 95.6 \; \lambda^{2.743}$$

may be used to generate mass absorption coefficients of iron for wavelengths shorter than the iron K absorption edge, namely 1.743 Å. It may be noted that the C_K and n_K values in Table II.1.12, which were obtained by regression of similar data from a number of sources, are given as 95.8 and 2.72, respectively.

Dalton and Goldak (1969) reported C and n values of 11.14 and 2.794 for the μ_{Fe} data for λ between 1.743 and 14.65 Å, the K and L_1 absorption edges of iron. A similar treatment of the data for nickel in Table 1.9 gives the following:

for $\mu < K$ absorption edge of Ni (1.488 Å), $C = 118.2$, $n = 2.718$;
for K absorption edge $< \lambda < L_1$ edge, $C = 14.13$, $n = 2.804$.

Table II.1.12   Coefficients and exponents used in the general expression $\mu = C\lambda^n$, Eqn (1.32)[a]

| Element (and Z) | $\lambda < \lambda_{abs\ K}$ | | | $\lambda_{abs,K} < \lambda < \lambda_{abs,L_1}$ | | |
	C_K	n_K	$\lambda_{abs\ K}$	C_{KL}	n_{KL}	$\lambda_{abs\ L_1}$
(a)						
(14) Si	18.5	2.77	6.745	1.54	2.73	105.0
(22) Ti	62.1	2.73	2.497	7.00	2.73	23.4
(23) V	69.8	2.73	2.269	8.02	2.73	19.8
(24) Cr	78.0	2.73	2.070	9.18	2.73	17.84
(25) Mn	86.7	2.72	1.896	10.45	2.73	16.14
(26) Fe	95.8	2.72	1.743	11.75	2.73	14.65
(27) Co	105.5	2.71	1.608	13.25	2.73	13.33
(28) Ni	115.9	2.71	1.488	14.80	2.73	12.20
(29) Cu	126.8	2.71	1.380	16.45	2.73	11.27
(30) Zn	138.0	2.70	1.283	18.25	2.73	10.33
(34) Se	189.4	2.69	0.980	26.40	2.73	7.506
(37) Rb	235.5	2.68	0.816	33.90	2.73	5.998
(38) Sr	251.3	2.68	0.770	36.50	2.73	5.583
(39) Y	268.1	2.67	0.728	39.30	2.73	5.232
(b)						
(34) Se	192.4	2.85	0.980	25.15	2.73	7.503
(37) Rb	238.8	2.85	0.816	32.45	2.73	6.008
(38) Sr	255.7	2.85	0.770	35.19	2.73	5.592
(39) Y	273.2	2.85	0.728	38.04	2.73	5.217
(40) Zr	291.5	2.85	0.689	41.00	2.73	4.879
(41) Nb	310.4	2.85	0.653	44.11	2.73	4.575
(42) Ag	330.1	2.85	0.620	47.30	2.73	4.304
(47) Ag	440.0	2.85	0.486	65.63	2.71	3.256
(50) Sn	515.3	2.85	0.425	78.79	2.69	2.777

[a] (a) C and n data from Heinrich (1966); (b) C and n data computed from Leroux and Thinh (1977).

The fact that mass attenuation coefficients are used extensively in X-ray fluorescence spectrometry, electron probe microanalysis, gamma-ray spectrometry, etc., has led a number of workers to examine the topic in depth and to propose algorithms that can be used with confidence over ever widening energy ranges. Leroux (1961) proposed a method based on empirical equations and graphs to fill the enormous gaps between available experimental values. Following a comprehensive examination of published mass attenuation coefficient data, Heinrich (1966) noted the wide discrepancies in the values reported and proposed the simple expression $\mu = C\lambda^n$ along with a set of C and n values. In this instance $\mu_{Fe\ \lambda\ Ni\ K-L_{2,3}}$ is given by

$$\mu_{Fe\ \lambda\ Ni\ K-L_{2,3}} = 95.8 \times 1.659^{2.72}$$

Leroux and Thinh (1977) proposed revised tables of X-ray mass attenuation coefficients based on the expression

$$\mu = CE_{ab}(12.3981/E)^n \qquad\qquad (II.1.9)$$

where C is a universal constant which is distinct for each element, E_{ab} is the energy (keV) of the lower absorption edge of the two absorption edges between which is located λ or E, and n is an exponent having a distinct value between these two absorption edges.

Table II.1.12 lists typical values from Heinrich (1966) and Leroux and Thinh (1977). The coefficients C_K and n_K are used for wavelengths shorter than the K absorption edge of the element, whereas the coefficients C_{KL} and n_{KL} are used for wavelengths between the K and L_1 absorption edges. The C_K and C_{KL} coefficients listed under (b) are computed from Leroux and Thinh's data in order to provide a partial comparison with some Heinrich values.

II.1.9 GENERATING μ AS A FUNCTION OF Z AND λ

McMaster *et al.* (1969) published a comprehensive tabulation of photoelectric, coherent, incoherent and total mass attenuation coefficients for the range 1–1000 keV. Included are plots of cross-sections in barns/atom versus E; tables of τ, σ_{coh}, σ_{inc} and μ in both barns/atom and $cm^2\ g^{-1}$ units; absorption jump ratios and fit coefficients for τ, σ_{coh} and σ_{inc} in expressions of the type

$$\ln \tau = a + b \ln(12.3981/\lambda) + c \ln(12.3981/\lambda)^2 + \cdots \qquad (II.1.10)$$

The fit parameters listed in Table II.1.13 are those relating to iron and are used as follows to calculate the mass attenuation coefficient of iron for Ni $K-L_{2,3}$, $\lambda = 1.659$ Å:

(a) $\ln(12.3981/1.659) = 2.0113$

(b) $\ln \tau =$
$$
\begin{aligned}
& 14.3456 \\
-\,&1.23491 \times 2.0113 &&= -2.4838 \\
-\,&0.41878 \times 2.0113^2 &&= -1.6941 \\
&0.03217 \times 2.0113^3 &&= \underline{0.2617} \\
& &&= 10.4294
\end{aligned}
$$

(c) $\tau = 33\ 840\ cm^2\ g^{-1}$

(d) conversion factor $= \dfrac{\text{atomic wt.(g/mol)}}{\text{Avogadro number (a/mol)}} \times 10^{24} \qquad (II.1.11)$

$$= 92.74$$

(e) $\tau_{Fe\ \lambda\ Ni\ K-L_{2,3}} = 33\ 840/92.74 = 364.9\ cm^2\ g^{-1}$

Table II.1.13 Fit parameters for the calculation of τ_{Fe}, $\sigma_{coh\,Fe}$ and $\sigma_{inc\,Fe}$ as a function of E (keV)[a]

	τ_{Fe}	
Range	$E > E_K$	$E_{L_1} \leqslant E \leqslant E_K$
a	1.43456×10^1	1.36696×10^1
b	-1.23491×10^0	-2.39195×10^0
c	-4.18785×10^{-1}	-1.37648×10^1
d	3.21662×10^{-2}	0

	$\sigma_{Fe,coh}$	$\sigma_{Fe,inc}$
a	5.93292×10^0	-3.42379×10^{-1}
b	2.25038×10^{-1}	1.57245×10^0
c	-3.61748×10^{-1}	-2.53198×10^{-1}
d	1.93024×10^{-2}	9.85822×10^{-3}

[a] Data from McMaster et al. (1969).

(f) $\ln \sigma_{coh} =$

$$
\begin{array}{rcr}
& & 5.93292 \\
+0.22504 \times 2.0113 &=& 0.4526 \\
-0.36175 \times 2.0113^2 &=& -1.4634 \\
0.01930 \times 2.0113^3 &=& \underline{0.1571} \\
&=& 5.0792
\end{array}
$$

(g) $\sigma_{coh} = 160.65$

(h) $\sigma_{coh,Fe}\,\lambda\,Ni\,K-L_{2,3} = 160.65/92.74 = 1.73 \text{ cm}^2 \text{ g}^{-1}$

(i) $\ln \sigma_{inc} =$

$$
\begin{array}{rcr}
& & -0.34238 \\
+1.57245 \times 2.0113 &=& 3.1627 \\
-0.25320 \times 2.0113^2 &=& -1.0243 \\
0.00986 \times 2.0113^3 &=& \underline{0.0802} \\
& & 1.8761
\end{array}
$$

(j) $\sigma_{inc} = 6.528$

(k) $\sigma_{inc,Fe}\,\lambda\,Ni\,K-L_{2,3} = 6.528/92.74 = 0.070 \text{ cm}^2 \text{ g}^{-1}$

(l) $\mu_{Fe}\,\lambda\,Ni\,K-L_{2,3} = \tau_{Fe} + \sigma_{coh,Fe} + \sigma_{inc,Fe}$ (II.1.12)

$$
\begin{array}{l}
= 364.9 + 1.73 + 0.070 \\
+ 366.7 \text{ cm}^2 \text{ g}^{-1}
\end{array}
$$

Poehn et al. (1985) proposed the following algorithm to generate mass attenuation coefficients:

$$\mu_{Z\lambda} = \exp[k \ln(12.3981/\lambda) + d_Z] \quad\quad (II.1.13)$$

where

$$d_Z = a + bZ + cZ^2 + dZ^3 + eZ^4 \qquad\qquad (\text{II}.1.14)$$

The fit parameters k, a, b, c, d and e are listed in Table II.1.14.

In this instance, the mass attenuation coefficient of iron for the $K-L_{2,3}$ line of nickel is given as follows.

(a) $\ln(12.3981/1.659) = 2.0113$

(b) $d_Z =$ 5.955
    $+ 3.917 \times 10^{-1} \times 26 \ = \ 10.184$
    $- 1.054 \times 10^{-2} \times 26^2 = \ -7.125$
    $1.520 \times 10^{-4} \times 26^3 = \ 2.672$
    $- 8.508 \times 10^{-7} \times 26^4 = \ \underline{-0.389}$
    $= \ 11.297$

(c) $\mu_{\text{Fe } \lambda \text{ Ni } K-L_{2,3}} = \exp\left[(-2.685 \times 2.0113) + 11.297\right]$
    $= \exp[5.8966]$
    $= 363.8 \text{ cm}^2 \text{ g}^{-1}$

The mass attenuation coefficient of iron for the $K-L_{2,3}$ line of iron, $\lambda = 1.937$ Å, is computed in the same way, using the data in row L_1 (i.e., 1.937 Å > 1.743 Å $< \lambda_{L_1}$), and is given by

$$\mu_{\text{Fe } \lambda \text{ Fe } K-L_{2,3}} = \exp\left[(-2.669 \times 1.8563) + 9.236\right]$$
$$= 72.4 \text{ cm}^2 \text{ g}^{-1}$$

The fact that traditional exponential models do not adequately cover the energy region below 1 keV, and that even in the wavelength region generally used in X-ray fluorescence analysis μ values may differ by 10% or more, led Heinrich (1986) to propose a new model which aims to overcome these limitations. Mass attenuation coefficients of all elements were fitted for photon energies from 200 eV to 20 keV. In the region above 1 keV, the model was adjusted to available experimental results. No data generated prior to 1950

Table II.1.14   Fit parameters used for the calculation of mass attenuation coefficients, Eqns (II.1.13) and (II.1.14)[a]

	K	L_1	M_1	N
k	-2.685	-2.669	-2.514	-2.451
a	5.955	3.257	2.382	4.838
b	3.917×10^{-1}	3.936×10^{-1}	2.212×10^{-1}	4.911×10^{-2}
c	-1.054×10^{-2}	-8.483×10^{-3}	-2.028×10^{-3}	
d	1.520×10^{-4}	9.491×10^{-5}	6.891×10^{-6}	
e	-8.508×10^{-7}	-4.058×10^{-7}		

[a] Data from Poehn et al. (1985).

were employed and measurements lacking internal consistency were disregarded. Below the highest N level absorption edge, a slightly different model is proposed. The new model has the form

$$\mu_{Z\lambda} = C(Z^4/A)\lambda^n(1 - \exp\{[(-12398.1/\lambda) + b]/a\})   \tag{II.1.15}$$

where A is the atomic weight and C, n, a and b are parameters fitted specifically to cover eleven energy regions from $E > E_K$ to $E_N > E$. In this instance, the computation of $\mu_{Fe\ \lambda\ Ni\ K-L_{2,3}}$ is carried out as follows.

$$C = a + bZ + cZ^2 + dZ^3 + eZ^4   \tag{II.1.16}$$

$$
\begin{array}{rll}
= & & 5.253 \times 10^{-3} \\
+1.33257 \times 10^{-3} \times 26 & = & 3.4647 \times 10^{-2} \\
-7.5937 \times 10^{-5} \times 26^2 & = & -5.1333 \times 10^{-2} \\
1.69357 \times 10^{-6} \times 26^3 & = & 2.9766 \times 10^{-2} \\
-1.3975 \times 10^{-8} \times 26^4 & = & \underline{-6.386 \times 10^{-3}} \\
& & 1.1947 \times 10^{-2}
\end{array}
$$

$$
\begin{array}{rl}
n = 3.112 & 3.112 \\
-0.0121 \times 26 & \underline{-0.3146} \\
& 2.7974
\end{array}
$$

$$
\begin{array}{rll}
a = 0.0 & & 0.0 \\
+47.0 \times 26 & = & 1222.0 \\
+\ 6.52 \times 26^2 & = & 4407.5 \\
-\ 0.152624 \times 26^3 & = & \underline{-2682.5} \\
& & 2947.0
\end{array}
$$

$$b = 0$$

Thus the term with the exponential function is equal to

$$
\begin{aligned}
& 1 - \exp\{[(12398.1/\lambda_{Ni\ K-L_{2,3}}) + b]/a\} \\
& = 1 - \exp\{[(12398.1/1.659) + 0]/2947] \\
& = 0.9208
\end{aligned}
$$

(f) Substitution in Eqn (II.1.15) yields

$$
\begin{aligned}
\mu_{Fe} & = (1.1947 \times 10^{-2} \times 26^4/55.85) \times 1.659^{2.7974} \times 0.9208 \\
& = 370.9 \text{ cm}^2 \text{ g}^{-1}
\end{aligned}
$$

Thus, in summary, we have the following calculated values for $\mu_{Fe\ \lambda\ Ni\ K-L_{2,3}}$:

379.6 cm^2 g^{-1} from the Heinrich (1966) tabulation;
366.7 cm^2 g^{-1} calculated from McMaster et al. (1969);
387.1 cm^2 g^{-1} from the Leroux and Thinh (1977) tabulation;
363.8 cm^2 g^{-1} calculated from Poehn et al. (1985);

370.9 cm^2 g^{-1} calculated from Heinrich (1986);
388.0 cm^2 g^{-1} experimental value from Dalton and Goldak (1969).

This very limited examination of algorithms for generating $\mu_{Z\lambda}$ provides only a bird's eye view of the variations that may be encountered in the selection of algorithms for generating mass attenuation coefficients. For a more comprehensive study of uncertainties in mass attenuation coefficients, the reader is referred to Heinrich (1966, 1986) and to Vrebos and Pella (1988).

II.1.11.1 Absorption jump ratios; absorption jump factors

In subsequent developments, absorption jump ratios, symbol r, will be used to calculate absorption jump factors; i.e. the probability p_{λ_i} that an incident photon will eject electrons from a $K, L, M, \ldots$ energy level. For example, the probability that a K level electron of element i will be ejected rather than one from an L or M level is given by (see Section 2.3.2)

$$p_{K,i} = \frac{r_{K,i} - 1}{r_{K,i}} \tag{II.1.17}$$

Poehn et al. (1985) proposed the following algorithm for generating absorption jump ratio values: for $Z = 11-50$

$$r_{K,i} = a + bZ_i + cZ_i^2 + dZ_i^3 \tag{II.1.18}$$

and for $Z = 30-83$

$$r_{L_3,i} = a + bZ_i + cZ_i^2 + dZ_i^3 \tag{II.1.19}$$

along with the necessary fit parameters; e.g. in the case of iron $r_{K,\mathrm{Fe}}$ is given by

$$
\begin{aligned}
r_{K,\mathrm{Fe}} = & & 17.54 \\
& -0.6608 \times 26 = & -17.181 \\
& +0.01427 \times 26^2 = & 9.646 \\
& -0.00011 \times 26^3 = & \underline{-1.933} \\
& & 8.07
\end{aligned}
$$

Broll (1986) noted that the absorption jump factor $P_{K,i}$, being less sensitive than the K absorption jump ratio, could be represented in the range $Z = 10-70$ by the simple linear relationship

$$p_{K,i} = 0.915 - 0.0014 Z_i \tag{II.1.20}$$

In the case of iron the above expression yields

$$p_{K,\text{Fe}} = 0.915 - 0.0014 \times 26 = 0.879$$

and substitution in Eqn (II.1.17) in

$$p_{K,\text{Fe}} = \frac{8.07 - 1}{8.07} = 0.876$$

Table II.1.15 lists typical values for absorption jump ratios and probabilities for the ejection of K level electrons for the elements Si, Cr, Fe, Ni and Mo.

II.1.11.2 Relative line intensities within series

Once an electron has been ejected from an atomic level, the transition of another electron from an outer level takes place, leading to the emission of a characteristic line. The probability that a certain transition takes place is one of the three factors that enter into computing the probability that a given line is emitted. Based on experimental data from Heinrich et $al.$ (1979) and from Salem et $al.$ (1972), and on theoretical data from Scofield (1974), Schreiber and Wims (1982) developed algorithms for calculating transition probabilities for K, L and M levels. For example, in the atomic number range 11–60, $f_{K-L_{2,3}}$ is given by

$$Z = 11\text{–}19: \quad f_{K-L_{2,3}} = 1.052 - 4.39 \times 10^{-4} \times Z^2$$
$$Z = 20\text{–}29: \quad f_{K-L_{2,3}} = 0.896 - 6.575 \times 10^{-4} \times Z$$
$$Z = 30\text{–}60: \quad f_{K-L_{2,3}} = 1.0366 - 6.82 \times 10^{-3} \times Z + 4.815 \times 10^{-5} \times Z^2$$

Thus the probability that a $K-L_{2,3}$ line transition will take place in the case of iron atoms is given by

$$\begin{aligned} f_{K-L_{2,3},\text{Fe}} &= 0.896 - (6.575 \times 10^{-4} \times 26) \\ &= 0.879 \end{aligned}$$

Table II.1.15 Absorption jump ratios r_K and absorption jump factors p_K (K absorption edge)

Element (and Z)	$r_K{}^a$	$r_K{}^b$	$p_K{}^c$	$p_K{}^d$	$p_K{}^e$
(14) Si	10.442	10.78	0.904	0.907	0.895
(24) Cr	8.779	8.38	0.886	0.881	0.881
(26) Fe	8.221	8.07	0.878	0.876	0.879
(28) Ni	7.855	7.81	0.873	0.872	0.876
(42) Mo	6.968	6.81	0.856	0.853	0.856

[a] Data from McMaster et $al.$ (1969).
[b] Calculated using fit parameters from Poehn et $al.$ (1985).
[c] Calculated from data in column 2.
[d] Calculated from data in column 3.
[e] Calculated using $p_K = 0.915 - 0.0014\,Z$ (Broll, 1986).

Algorithms for $f_{K-L_{2,3}}$ and $f_{L_3-M_{4,5}}$ transition probabilities and fit parameters are given in Poehn et al. (1985) and Schreiber and Wims (1982).

II.1.11.3 Fluorescence yields

The probability that an electronic transition that has taken place will lead to a fluorescence emission rather than to the emission of an Auger electron is given by the fluorescence yield ω. More specifically, values are required for ω_K, ω_{L_1}, ω_{L_2}, ω_{L_3} (five ω_M values if M characteristic lines are selected for intensity measurement), it being recognized (Broll, 1986) that fortunately the $L_3-M_{4,5}$ lines relate to ω_{L_3} data, which are many and comparatively precise. Bambynek et al. (1972) proposed the following algorithm for generating ω values:

$$\omega = F/(1 + F) \qquad \text{(II.1.21)}$$

where

$$F = (a + bZ + cZ^3)^4 \qquad \text{(II.1.22)}$$

For ω_K, $a = 0.015$, $b = 0.0327$ and $c = 6.4 \times 10^{-7}$. Thus, in the case of iron, F is given by

$$
\begin{aligned}
F = (\qquad\qquad & 0.015 \\
+ 0.0327 \times 26 \quad &= +0.850 \\
- 6.4 \times 10^{-7} \times 26^3 &= \underline{-0.011})^4 \\
&= \quad 0.854^4 \\
&= \quad 0.532
\end{aligned}
$$

and

$$\omega_{K,Fe} = 0.532/(1 + 0.532) = 0.347$$

Broll (1986) noted that ω_K can be represented remarkably well by an earlier equation due to Wentzel, namely

$$\omega_K = \left(1 + \frac{a}{Z^4}\right)^{-1} \qquad \text{(II.1.23)}$$

where

for $Z = 10-30$: $a = 0.8 \times 10^6$
for $Z = 30-70$: $a = [0.8 + 0.01 \times (Z - 30)] \times 10^6$

In the case of iron, $\omega_{K,Fe}$ is given by

$$\omega_{K,Fe} = \left(1 + \frac{0.8 \times 10^6}{26^4}\right)^{-1}$$

$$= 0.364$$

Table II.1.16 Fluorescence yields ω_K

Element (and z)	$\omega_K{}^a$	$\omega_k{}^b$	$\omega_K{}^c$	$\omega_K{}^d$	$\omega_K{}^e$
(12) Mg	0.0215 ± 0.36	0.022	0.027	0.025	0.033
(14) Si		0.042	0.047	0.046	0.054
(22) Ti	0.209 ± 0.008	0.215	0.219	0.226	0.223
(24) Cr	0.279 ± 0.012	0.280	0.281	0.293	0.283
(26) Fc	0.350 ± 0.008	0.350	0.347	0.364	0.348
(28) Ni	0.432 ± 0.008	0.420	0.414	0.434	0.415
(30) Zn	0.499 ± 0.009	0.490	0.479	0.503	0.484
(38) Sr	0.699 ± 0.007	0.700	0.691	0.703	0.725
(42) Mo		0.765	0.764	0.772	0.784

[a] Data (experimental) from Freund (1975).
[b] Data from Broll (1986).
[c] Calculated using fit parameters from Bambynek et al. (1972).
[d] Calculated using Wentzel algorithm given in Broll (1986).
[e] Calculated using fit parameters from Poehn et al. (1985).

Poehen et al. (1985) proposed the following algorithm and fit parameters for generating ω_K and ω_{L_3} fluorescence yields:

$$\omega = a + bZ + cZ^2 + dZ^3 + eZ^4 \qquad (II.1.24)$$

Substitution of the fit parameters listed for ω_K for the range $Z = 11-42$ in the case of iron gives

$$
\begin{aligned}
\omega_{K,\text{Fe}} = \quad & 0.08502 \\
+(-1.458 \times 10^{-2} \times 26) = & -0.37908 \\
+(5.677 \times 10^{-2} \times 26^2) = & 0.38377 \\
+(3.147 \times 10^{-5} \times 26^3) = & 0.55786 \\
+(-6.559 \times 10^{-7} \times 26^4) = & \underline{-0.29975} \\
& 0.3478
\end{aligned}
$$

Table II.1.16 lists fluorescence yield data from five sources.

II.1.12 SUMMARY AND CONCLUSIONS

Applications relating to Chapter 1 consist mainly in having access to fundamental parameters data. Preferably these are obtained from more recent tabulations in the X-ray fluorescence literature, where authors have reviewed a host of previously published experimental and theoretical data. In order to reduce the very large set of fundamental parameters, which may overtax some computers, authors have also published least squares fits of fundamental parameters. These higher order models are simply functions of the element's atomic number and should be used with some caution.

II.2 X-RAY FLUORESCENCE EMISSIONS

II.2.1 TWO-THETA VALUES FOR CHARACTERISTIC LINES

As shown schematically in Figure 2.1, and implied in Eqn (2.2), the experimental measurement of the intensity of characteristic lines in wavelength dispersive spectrometry requires knowledge of the wavelength of the line and the d spacing of the crystal used for dispersion. Let us consider the case in which a LiF (200) crystal, $d = 2.0135$ Å, is used (the numbers 2, 0 and 0 are the Miller indices of the diffracting planes parallel to the surface) and the Fe $K-L_{2,3}$ line is to be measured. From Eqn (2.2),

$$\sin \theta = \frac{n\lambda}{2d}$$

$$= \frac{1 \times 1.937}{2 \times 2.0135} = 0.4810 \tag{II.2.1}$$

$$\theta = 28.75°$$

$$2\theta = 57.50°$$

Table II.2.1 lists 2θ values for the $K-L_{2,3}$ and $K-M_3$ characteristic lines of Cr, Fe, Ni and Mo and a few lines of Pb.

II.2.2 EXCITATION PROBABILITY FACTOR

The excitation probability factor $p_{i\lambda_i}$ (Eqns (2.14) and (2.15)) is constant for

Table II.2.1 Typical 2θ listings (in degrees) for LiF (200), $2d = 4.0285$ Å

	Cr	Fe	Ni	Mo
$K-L_{2,3}$	69.35	57.50	48.66	20.34
$K-M_3$	62.36	51.74	43.74	18.06

	n	$K-L_3$	$K-L_2$	$K-M_3$	L_3-M_5	L_2-M_4	L_3-N_5	L_2-N_4	L_1-M_3
Cr	1	69.29	62.43	62.36					
Fe	1	57.45	57.58	51.74					
Ni	1	48.61	48.71	43.74					
	2	110.82	111.12	96.28					
Mo	1	20.28	20.39	18.06					
	2	41.22	41.47	36.58					
	3	63.75	64.15	56.16					
	4	89.51	90.15	77.75					
Pb	1	4.70	4.84	4.15	33.92	28.22	28.25	24.07	27.84
	2	9.40	9.68	8.31	71.38	58.36	58.43	49.30	57.52
	3	14.12	14.55	12.49	122.12	94.00	94.13	77.46	92.39
	4	18.86	19.44	16.67				113.06	

any given line. For example, the probability that Fe $K-L_{2,3}$ photons are emitted is given by

$$p_{\text{Fe } K-L_{2,3}} = \left(\frac{r_{K,\text{Fe}} - 1}{r_{K,\text{Fe}}}\right)\left(\frac{I_{\text{Fe } K-L_{2,3}}}{I_{\text{Fe } K-L_{2,3}} + I_{\text{Fe } K-M_3}}\right)(\omega_{K \text{ Fe}}) \tag{II.2.2}$$

$$= p_{K \text{ Fe}} \times f_{\text{Fe } K-L_{2,3}} \times \omega_{K \text{ Fe}} \tag{II.2.3}$$

or, in general

$$p_{K-L_{2,3}} = (p_K f_{K-L_{2,3}} \omega_K)_i \tag{II.2.4}$$

Substitution of the parameters retained by Broll (1986) in the case of iron gives

$$p_{\text{Fe } K-L_{2,3}} = 0.878 \times 0.892 \times 0.350$$
$$= 0.2743$$

The probability $p_{\text{Fe } K-M_3}$ that the Fe $K-M_3$ photons are emitted is given by (only the two strongest K lines are taken into consideration):

$$p_{\text{Fe } K-M_3} = 0.878 \times (1 - 0.892) \times 0.350$$
$$= 0.0332$$

Should the assumption be made that only the $K-L_{2,3}$ line is used to represent the excitation factor, then $f_{\text{Fe } K-L_{2,3}}$ is equated to unity and the excitation probability factor is equal to 0.3074. This assumption is certainly not valid in many cases and is therefore not recommended as a general practice.

II.2.3 EXPERIMENTAL VERIFICATION OF PRIMARY EMISSION

Data listed in Table II.2.2 provide the material for a comprehensive experimental confirmation of Eqn (2.28), in that intensities were measured for five specimens (two pure metals and three binary alloys) at four different monochromatic incident wavelengths, namely 0.492, 0.711, 1.437 and 1.659 Å. The five specimens are from the larger suite of alloys used in the context of a polychromatic incident source by Rasberry and Heinrich (1974). Intensity data listed in Table II.2.2 were measured on an energy dispersive spectrometer, whereas the 1974 data were obtained on a wavelength dispersive spectrometer.

The following detailed calculations relate to analyte Ni, specimen 126B, incident wavelength $\lambda = 0.492$ Å. From Eqn (2.28), the emitted intensity of the $K-L_{2,3}$ line of nickel is given by

$$I_{Ni\lambda} = I_{(Ni)\lambda}C_{Ni} \frac{\mu'_{Ni} + \mu''_{Ni}}{\mu'_s + \mu''_s} \tag{II.2.5}$$

Table II.2.2 Data used to illustrate fundamental parameters computations; λ in Å, I in c/s

System: Fe–Ni; instrument geometry: $\psi' = \psi'' = 45°$

	$\lambda_{K-L_{2,3}}$	λ_{K-M_3}	C_K	n_K	λ_K	C_{KL}	n_{KL}	λ_{L_1}	$p_{\lambda_{K-L_{2,3}}}$
Fe	1.937	1.757	95.8	2.72	1.743	11.75	2.73	14.7	0.2742
Ni	1.659	1.500	115.9	2.71	1.488	14.80	2.73	12.2	0.3269

Specimen	Ni	986	1159	126B	Fe
C_{Fe}	0.0	0.3067	0.5100	0.6314	1.0
C_{Ni}	1.0	0.6931	0.4820	0.3599	0.0

(a) excitation source $\lambda = 0.492$

I_{Fe}	0.0	231.7	237.3	382.3	550.4
I_{Ni}	904.7	278.8	139.8	90.1	0.0

(b) excitation source $\lambda = 0.711$

I_{Fe}	0.0	259.2	368.6	431.2	626.2
I_{Ni}	952.9	345.4	108.9	116.9	0.0

(c) excitation source $\lambda = 1.437$

I_{Fe}	0.0	314.1	467.4	555.1	811.7
I_{Ni}	1017.8	581.7	360.0	252.7	0.0

(d) excitation source $\lambda = 1.659$

I_{Fe}	0.0	735.2	980.5	1086.8	1316.2

(a) Calculation of effective mass attenuation coefficients:

$$\mu'_{Ni} = 115.9 \times 0.492^{2.71} \times 1.4142 = 23.98 \text{ cm}^2 \text{g}^{-1}$$

$$\mu''_{Ni} = 14.8 \times 1.659^{2.73} \times 1.4142 = 83.36 \text{ cm}^2 \text{g}^{-1}$$

$$\mu'_{Fe} = 95.8 \times 0.492^{2.72} \times 1.4142 = 19.68 \text{ cm}^2 \text{g}^{-1}$$

$$\mu''_{Fe} = 95.8 \times 1.659^{2.72} \times 1.4142 = 536.86 \text{ cm}^2 \text{g}^{-1}$$

$$\mu'_s = (0.3599 \times 23.98) + (0.6315 \times 19.68) = 21.06 \text{ cm}^2 \text{g}^{-1}$$

$$\mu''_s = (0.3599 \times 83.36) + (0.6315 \times 536.86) = 369.03 \text{ cm}^2 \text{g}^{-1}$$

(b) Substitution in Eqn (II.2.5) yields

$$I_{Ni\ K-L_{2,3}} = 904.7 \times 0.3599 \times \frac{23.98 + 83.36}{21.06 + 369.03}$$

$$= 89.6 \text{ c/s}$$

which compares favourably with the experimentally measured intensity of 90.1 c/s. Similar calculations for incident $\lambda = 0.711$ Å give 119.2 c/s, versus 116.9 c/s measured experimentally. The remaining seven Ni intensities may be verified by carrying out similar computations.

II.2.4 EXPERIMENTAL VERIFICATION OF SECONDARY EMISSION

Data from Table II.2.2 may also be used for the confirmation of experimental data when enhancement is present, namely, in this instance, Fe in the system Fe–Ni. The following detailed calculations relate to analyte Fe, specimen 986, incident wavelength $\lambda = 0.711$ Å. From Eqn (2.28), the primary fluorescence emitted intensity is given by

$$I_{Fe,p} = \frac{I_{(Fe)}C_{Fe}(\mu'_{Fe} + \mu''_{Fe})}{\mu'_s + \mu''_s} \tag{II.2.6}$$

(a) Calculation of effective mass attenuation coefficients:

$$\mu_{Fe,\lambda} = 95.8 \times 0.711^{2.72} = 37.88$$

$$\mu'_{Fe} = 37.88 \times 1.4142 = 53.57 \text{ cm}^2 \text{g}^{-1}$$

$$\mu_{Ni,\lambda} = 115.9 \times 0.711^{2.71} = 45.99$$

$$\mu'_{Ni} = 45.99 \times 1.4142 = 65.04 \text{ cm}^2 \text{g}^{-1}$$

$$\mu''_{Fe} = 11.75 \times 1.937^{2.73} \times 1.4142 = 101.02 \text{ cm}^2 \text{g}^{-1}$$

$$\mu''_{Ni} = 14.80 \times 1.937^{2.73} \times 1.4142 = 127.24 \text{ cm}^2 \text{g}^{-1}$$

$$\mu_s' = (53.57 \times 0.3067) + (65.04 \times 0.6931) = 61.51 \text{ cm}^2\text{g}^{-1}$$

$$\mu_s'' = (101.02 \times 0.3067) + (127.24 \times 0.6931) = 119.17 \text{ cm}^2\text{g}^{-1}$$

(b) Substitution in Eqn (II.2.6) yields

$$I_{\text{Fe,p}} = \frac{(626.2 \times 0.3067) \times (53.57 + 101.02)}{61.51 + 119.17}$$

$$= 164.3 \text{ c/s}$$

From Eqn (2.47), the total emitted intensity (primary + secondary) is given by

$$I_{\text{Fe}} = I_{\text{Fe,p}} + I_{\text{Fe,p}} e_{\text{FeNi } K-L_{2,3}} C_{\text{Ni}} + I_{\text{Fe,p}} e_{\text{FeNi } K-M_3} C_{\text{Ni}} \qquad \text{(II.2.7)}$$

(c) Calculation of effective mass attenuation coefficients:

$$\mu_{\text{Fe}\lambda_{\text{Ni } K-L_{2,3}}} = 95.8 \times 1.659^{2.72} = 379.62 \text{ cm}^2\text{g}^{-1}$$

$$\mu_{\text{Ni}\lambda_{\text{Ni } K-L_{2,3}}} = 14.8 \times 1.659^{2.73} = 58.94 \text{ cm}^2\text{g}^{-1}$$

$$\mu_{\text{Fe}\lambda_{\text{Ni } K-M_3}} = 95.8 \times 1.500^{2.72} = 288.62 \text{ cm}^2\text{g}^{-1}$$

$$\mu_{\text{Ni}\lambda_{\text{Ni } K-M_3}} = 14.8 \times 1.500^{2.71} = 44.41 \text{ cm}^2\text{g}^{-1}$$

$$\mu_{\text{s}\lambda_{\text{Ni } K-L_{2,3}}} = (379.62 \times 0.3067) + (58.94 \times 0.6931)$$

$$= 157.28 \text{ cm}^2\text{g}^{-1}$$

$$\mu_{\text{s}\lambda_{\text{Ni } K-M_3}} = (288.62 \times 0.3067) + (44.41 \times 0.6931)$$

$$= 119.30 \text{ cm}^2\text{g}^{-1}$$

(d) The following relates to enhancement of Fe by the $K-L_{2,3}$ line of nickel. From Table II.2.2, the excitation probability factor for Ni is equal to 0.3269.

(e) the coefficients e_{Ni}, e_{Ni}' and e_{Ni}'' are given by substitution in Eqns (2.37)–(2.39), respectively:

$$e_{\text{Ni } K-L_{2,3}} = 0.5 \times 0.3269 \times 379.62 \times (45.99/37.88) = 75.33$$

$$e_{\text{Ni } K-L_{2,3}}' = \frac{1}{61.51} \ln\left(1 + \frac{61.51}{157.28}\right) = 0.005367$$

$$e_{\text{Ni } K-L_{2,3}}'' = \frac{1}{119.17} \ln\left(1 + \frac{119.17}{157.28}\right) = 0.004732$$

(f) the coefficient $e_{\text{FeNi } K-L_{2,3}}$ is given by Eqn (2.36):

$$e_{\text{FeNi } K-L_{2,3}} = 75.36 \times (0.005367 + 0.004732) = 0.761$$

(g) the following calculations relate to enhancement of Fe by the $K-M_3$ line of nickel:

$$p_{Ni\ K-M_3} = 0.873 \times 0.108 \times 0.420 = 0.0396$$

$$e_{Ni\ K-M_3} = 0.5 \times 0.0396 \times 288.62 \times (45.99/37.88) = 6.938$$

$$e'_{Ni\ K-M_3} = \frac{1}{61.51} \ln\left(1 + \frac{61.51}{119.30}\right) = 0.006761$$

$$e''_{Ni\ K-M_3} = \frac{1}{119.17} \ln\left(1 + \frac{119.17}{119.30}\right) = 0.005811$$

$$e_{FeNi\ K-M_3} = 6.938 \times (0.006761 + 0.005811) = 0.0872$$

(h) substitution in Eqn (II.2.7) yields

$$I_{Fe} = 164.3 + (164.3 \times 0.761 \times 0.6931) + (164.3 \times 0.0872 \times 0.6931)$$

$$= 260.9 \text{ c/s, versus } 259.2 \text{ c/s measured experimentally.}$$

The other iron intensities may be verified by carrying out similar calculations, bearing in mind that only the weight fractions will change from specimen to specimen. In the case of incident wavelength $\lambda = 1.659$ Å, as the nickel characteristic lines are not emitted, only the absorption of iron by nickel need be taken into consideration.

In summary, the emitted Ni intensity for specimen 126B for the hypothetical case $\mu_{Fe}^* = \mu_{Ni}^*$ would have been equal to $I_{Ni}C_{Ni} = 904.7 \times 0.3599 = 325.6$ c/s (Table II.2.2). The experimentally measured intensity is 90.1 c/s, and the calculated intensity from fundamental parameters expressions is 89.6 c/s, a difference of 0.5 c/s. Thus the value for nickel characteristic radiations absorbed by iron, -235.5 c/s, is *predicted* within 0.5 c/s in this specific context. As mentioned previously, theoretical emitted intensity calculations do not yield absolute values, due in part to the uncertainties in the values of the fundamental parameters, but also because of uncertainties in the experimental data and in the compositional data. There are also trace element effects which have not been included, as indicated by the fact that the sum of the weight fractions of iron and nickel is not quite unity.

At this stage it is also important for analysts to be aware that the tendency to reduce the very large set of fundamental parameter data by having recourse to fitted algorithms must be exercised with care. A problem may arise if the analyst is unaware that, in a given context, this practice may lead to a transposition of a line across an absorption edge and thus cause serious errors in any subsequent computations.

II.2.5 SUMMARY

Applications relating to Chapter 2 consist mainly in combining fundamental parameters data with instrumental parameters so as to generate effective parameters, i.e. adapt fundamental data to any given spectrometer configuration. Effective parameters were applied for the calculation of primary and secondary fluorescence emissions in the context of monochromatic sources and the values were compared to experimental data. These processes are extended to polychromatic excitation sources in Chapter II.3.

II.3 THE FUNDAMENTAL PARAMETERS APPROACH

II.3.2 CONCEPT OF THE FUNDAMENTAL PARAMETERS APPROACH

The major steps introduced in the following computations are: (i) the process deals with an incident polychromatic beam rather than being limited to monochromatic sources; (ii) applications will relate to three-component systems rather than being limited to binaries. The Cr–Fe–Ni system is retained and, for a number of numerical examples, the analytical context relating to experimental data from Rasberry and Heinrich (1974) will be the norm. Unless otherwise specified, the fundamental parameters listed in Table II.3.1 are retained in the software used to generate the theoretical data; the notation 'Analytical context: AC 1' is used to indicate this.

In order to provide a comprehensive treatment of the fundamental parameters approach for quantitative analysis, relative intensity data from Rasberry and Heinrich (1974) are used to generate the nominal intensities for seven Cr–Fe–Ni alloys as shown in Table II.3.2, based on assumed intensities for the pure elements. Data relating to four hypothetical specimens are also listed; specimen 5202 normalized to 1.0; specimen SS, an idealized 18–8 stainless steel; Fe14, so as to provide a link to Tables II.3.7–II.3.9; and specimen AVG, which is representative of the average of the seven real reference materials.

The nominal intensities listed in Table II.3.2 are used to generate the experimental relative intensities (Analytical context AC 1) given in Rasberry and Heinrich (1974) for the seven alloys selected. The same analytical context is retained to generate theoretical relative intensities for the seven real and four hypothetical alloys in Table II.3.3.

The close agreement between the experimental and the theoretically calculated relative intensities in Table II.3.4 is a clear demonstration of the power

Table II.3.1 Analytical context AC 1. This context is retained in many subsequent applications

Analytical context: AC 1
Excitation source: W target tube; 45 kV (CP); see Table II.1.7
Instrument geometry: $\psi' = 63°$, $\psi'' = 33°$
System: Cr–Fe–Ni

	$\lambda_{K-L_{2,3}}$	C_K	n_K	λ_K	C_{KL}	n_{KL}	λ_{L_1}	$p^a_{\lambda_i}$
Cr	2.291	78.0	2.73	2.070	9.18	2.73	17.8	0.2481
Fe	1.937	95.8	2.72	1.743	11.75	2.73	14.7	0.3073
Ni	1.659	115.9	2.71	1.488	14.80	2.73	12.2	0.3667

$p^a_{\lambda_i}$, $f_{K-L_{2,3}}$ equated to 1.0 in the computation of $p_{\lambda_i \, K-L_{2,3}}$

Table II.3.2 Nominal itensities generated from fundamental parameters expressions using the fundamental parameters listed in Table II.3.1. First seven samples relate to real reference materials, last seven relate to hypothetical specimens

Specimen	C_{Cr}	C_{Fe}	C_{Ni}	I_{Cr} (c/s)	I_{Fe} (c/s)	I_{Ni} (c/s)
5054	0.2577	0.7250	0.0015	4017.6	6564.5	10.8
5074	0.2425	0.6838	0.0498	3909.6	6315.4	365.4
5181	0.1988	0.6945	0.0996	3181.2	6959.4	748.8
5202	0.2130	0.6303	0.1480	3340.8	6272.0	1155.6
5321	0.1988	0.5919	0.2002	3098.4	6080.2	1616.4
5324	0.2696	0.5280	0.1927	3973.2	4940.6	1477.8
5364	0.2784	0.4721	0.2357	4033.2	4450.6	2007.0
5202N	0.215	0.636	0.149	3322.4	6284.5	1118.7
SS	0.18	0.74	0.08	2946.2	7551.1	568.8
Fe14	0.20	0.70	0.10	3188.4	6970.5	723.6
AVG	0.24	0.61	0.15	3636.0	5821.9	1132.9
Cr	1.0			12000		
Fe		1.0			14000	
Ni			1.0			18000

of the fundamental parameters approach proposed by Criss and Birks (1968).

Let us now consider an example of the first step in applying the fundamental parameters method: the calculation of the relative intensity R_i without measuring $I_{(i)}$ for the pure analyte i. The relative intensity R_i is given by Eqn (3.5)

$$R_i = \frac{I_{iu}}{I_{ir}} \times \left(\frac{P_i + S_i}{P_{(i)}} \right)_r$$

(II.3.1)

Specimen 5181 is considered to be the unknown and specimen 5202 is the

Table II.3.3 Theoretical emission entities for the Cr–Fe–Ni system. Analytical context AC 1

Specimen	$P_{Cr} + S_{Cr}$	$P_{Fe} + S_{Fe}$	P_{Ni}	P_{Fe}	P_{Cr}
5054	544.3	816.2	1.04	815.8	385.3
5074	524.6	783.9	35.5	771.8	374.1
5181	421.4	866.7	72.2	839.5	290.7
5202	438.5	782.2	111.5	745.2	309.7
5321	406.6	76.4	156.2	712.7	287.3
5324	524.9	615.0	152.0	577.0	389.9
5364	534.2	553.1	193.4	510.6	402.4
5202N	438.8	782.2	111.2	745.3	309.9
SS	389.1	939.9	56.6	916.4	262.3
Fe14	421.1	867.6	71.9	840.4	290.4
AVG	480.2	724.6	112.6	690.3	345.7
Cr 1	1584.7				1584.7
Fe 1		1742.5		1742.5	
Ni 1			1789.7		

Table II.3.4 Comparison of experimental and theoretical relative intensities[a]

Analytical context AC 1

Specimen	C_{Cr}	C_{Fe}	C_{Ni}		R_{Cr}	R_{Fe}	R_{Ni}
5054	0.2577	0.7250	0.0015	(a)	0.3348	0.4689	0.0006
				(b)	0.3434	0.4684	0.0006
5074	0.2525	0.6838	0.0498	(a)	0.3258	0.4511	0.0203
				(b)	0.3310	0.4499	0.0199
5181	0.1988	0.6945	0.0996	(a)	0.2651	0.4971	0.0416
				(b)	0.2659	0.4974	0.0403
5202	0.2130	0.6303	0.1480	(a)	0.2784	0.4480	0.0642
				(b)	0.2767	0.4489	0.0623
5321	0.1988	0.5919	0.2002	(a)	0.2582	0.4343	0.0898
				(b)	0.2566	0.4375	0.0873
5324	0.2696	0.5280	0.1927	(a)	0.3311	0.3529	0.0821
				(b)	0.3312	0.3530	0.0849
5364	0.2784	0.4721	0.2357	(a)	0.3361	0.3179	0.1115
				(b)	0.3371	0.3174	0.1081
5202N	0.215	0.636	0.149	(b)	0.2769	0.4489	0.0622
SS	0.18	0.74	0.08	(b)	0.2455	0.594	0.0316
Fe14	0.20	0.70	0.10	(b)	0.2657	0.4979	0.0402
AVG	0.24	0.61	0.15	(b)	0.3030	0.4159	0.0629

[a] (a) Experimental data from Rasberry and Heinrich (1974); (b) calculated from fundamental parameters expression.

reference material. The theoretical data relating to $P_i + S_i$ and $P_{(i)}$ are given in Table II.3.3. Substitution in Eqn (II.3.1) yields

$$R_{Cr5181} = \frac{3181.2}{3340.8} \times \left(\frac{438.5}{1584.7}\right) = 0.2635 \text{ vs } 0.2651(\text{expr})$$

$$R_{Fe5181} = \frac{6959.4}{6272.0} \times \left(\frac{782.2}{1742.5}\right) = 0.4981 \text{ vs } 0.4971(\text{expr})$$

$$R_{Ni5181} = \frac{748.8}{1155.6} \times \left(\frac{111.5}{1789.7}\right) = 0.0405 \text{ vs } 0.0416(\text{expr})$$

Alternatively, in the context of $C = KIM$, where R is equal to the product KI and where K is defined as in Eqn (3.50),

$$K_i = \frac{1}{I_{(i)}} = \left(\frac{P_i + S_i}{I_i P_{(i)}}\right)_r \tag{II.3.2}$$

For iron, substitution of data relating to specimen 5202 yields

$$K_{Fe} = \frac{782.2}{6272.0 \times 1742.5} = 7.157 \times 10^{-5}$$

and for specimen 5181

$$R_{Fe5181} = 7.157 \times 10^{-5} \times 6959.4 = 0.4981$$

Table II.3.5 Determination of Cr, Fe and Ni by iteration using a fundamental parameters expression

Analytical context AC 1

Specimen: 5181 (see Table II.3.3)
Given: $R_{Cr} = 0.2651$; $R_{Fe} = 0.4971$; $R_{Ni} = 0.0416$

Iteration	C_{Cr}	C_{Fe}	C_{Ni}	R_{Cr}	R_{Fe}	R_{Ni}
(a)						
1	0.3333	0.3333	0.3333	0.3796	0.2148	0.1646
2	0.2139	0.7087	0.0774	0.2832	0.4910	0.0308
3	0.1958	0.7018	0.1024	0.2608	0.5026	0.0412
4	0.1997	0.6965	0.1038	0.2650	0.4961	0.0418
5	0.1996	0.6972	0.1032	0.2649	0.4966	0.0416
6	0.1996	0.6973	0.1031	0.2649	0.4967	0.0415
(b)						
1	0.3298	0.6184	0.0518	0.4045	0.3653	0.0207
2	0.1966	0.7010	0.1024	0.2617	0.5014	0.0412
3	0.1994	0.6974	0.1033	0.2648	0.4966	0.0416

Alternatively, from Eqn (3.48),

$$K_{Fe} = \left(\frac{C_{Fe}}{I_{Fe}M_{Fe}}\right)_r \qquad (II.3.3)$$

where, from Eqn (3.49),

$$M_{Fe} = \left(\frac{C_{Fe}P_{(Fe)}}{P_{Fe} + S_{Fe}}\right)_r \qquad (II.3.4)$$

Substitution of 5202 data leads to

$$M_{Fe} = \frac{0.6303 \times 1742.5}{782.2} = 1.4041$$

$$K_{Fe} = \frac{0.6303}{6272.0 \times 1.4041} = 7.157 \times 10^{-5}$$

The value for K_{Fe} computed directly from the nominal value for $I_{(Fe)}$ is

$$K_{Fe} = \frac{1}{14000} = 7.143 \times 10^{-5}$$

As suggested by Criss and Birks (1968), bias that may be introduced by using a single reference may be minimized by using more than one standard reference. R_{Fe5181} computed using 5074, 5321 and 5364 yielded 0.4958, 0.5008 and 0.4963, respectively. Hence the average of the four R_{Fe5181} values, 0.4979, would be considered a better value for R_{Fe}. The computation of concentrations given R_{Cr}, R_{Fe} and R_{Ni} values proceeds by iteration.

II.3.2.3 The iterative process

Briefly stated, the iterative process involves:

- making an assumption of specimen composition;
- calculating relative intensities from theoretical principles;
- comparing with relative intensities measured experimentally;
- making a better assumption of specimen composition;
- repeating this process to convergence.

Three examples are provided. They relate to specimen 5181 in Table II.3.2, and it is assumed that only the three experimentally measured relative intensities are known, as would be the case if this specimen were submitted for analysis. In the first example, the process followed is knowingly not very effective in the sense that more iteration steps are required. This is exemplified by the first assumption of composition made; there being three elements, it is assumed initially that all three are present in equal concentrations. Taking 0.3333 as a first estimate of concentrations, three relative intensities are calculated. In the case of iron, $R_{Fe} = 0.2148$, clearly indicating that the estimated concentration of iron is too low. The ratio $R_{Fe,measured}/R_{Fe,calculated}$ provides an adjustment factor for generating a better estimate. Thus,

$$C_{Fe,second\ estimate} = 0.3333 \times (0.4971/0.2148) = 0.7711$$

Similar treatment of the Cr and Ni data yields adjusted concentrations of 0.2328 and 0.0843, respectively. Since the three adjusted weight fractions total 1.0884, they are normalized to 1.0, yielding 0.2139, 0.7087 and 0.0774, which then become the next estimated concentrations as shown in Table II.3.5(a). The process is continued and, in this instance, the R values converge to 0.2649, 0.4967 and 0.0415; i.e. relative intensities from the sixth iteration are essentially identical to those from the fifth, which in turn are quite similar to the experimental values. Thus, the estimated concentrations used for the sixth iteration would be considered as the composition of specimen 5181. As expected, these are in good agreement with the known values, the measured and calculated relative intensities being very similar, but the fact that the concentrations were known *a priori* did not enter into the iteration process.

The second example of the iteration process is that proposed by Criss and Birks (1968), namely

$$C_{i,2} = \frac{R_{i,m}C_{i,1}(1 - R_{i,c})}{R_{i,m}(C_{i,1} - R_{i,c}) + R_{i,c}(1 - C_{i,1})} \quad\quad (II.3.5)$$

where $C_{i,2}$ is the new estimate of concentration, $R_{i,m}$ is the measured relative intensity, $C_{i,1}$ is the present estimate of concentration, and $R_{i,c}$ is the calculated relative intensity.

As in the previous example, the concentrations are normalized to unity prior to each iteration. The initial estimate of concentrations is the measured relative intensities normalized to 1.0. As shown in Table II.3.5(b), fewer iterations are required to reach convergence.

Another option proposed by Criss *et al.* (1978) uses α coefficients as an intermediate step to shorten the computation process. For example, a glance at Table II.4.2 indicates that the the following values can be considered good approximations:

$$\alpha_{CrFe} = -0.34 \quad\quad \alpha_{CrNi} = -0.03$$
$$\alpha_{FeCr} = 2.11 \quad\quad \alpha_{FeNi} = -0.22$$
$$\alpha_{NiCr} = 1.26 \quad\quad \alpha_{NiFe} = 1.77$$

This third example involves substitution in the Lachance–Traill model, keeping the experimental relative intensities constant and using them as the first estimate of composition:

first iteration: $C_{Cr} = 0.222$, $C_{Fe} = 0.726$, $C_{Ni} = 0.107$

second iteration: $C_{Cr} = 0.199$, $C_{Fe} = 0.694$, $C_{Ni} = 0.103$

Only one pass is then required if the second iteration values are used as the initial estimate of composition and substituted in the fundamental parameters algorithm.

II.3.4 COMPREHENSIVE TABULATION $P_i + S_i$, A_{ij} AND E_{ij} DATA

The entities $P_i + S_i$, A_{ij} and E_{ij} provide a simple means for presenting a comprehensive examination of absorption and enhancement effects in X-ray fluorescence analysis. Table II.3.3 lists the theoretical emission entities $P_i + S_i$ and P_i for the specimens in Table II.3.2, and Table II.3.6 lists the values of A_{ij} and E_{ij} for the same specimens.

The data presented in Tables II.3.7–II.3.9 are directly linked to the experimental data of Tables II.3.3 and II.3.6, having been generated for the same analytical context AC 1. Thus in each case, the data in Tables II.3.7–II.3.9 may be regarded as equivalent to having had to prepare and experimentally measure intensities in three suites of 64 specimens of known composition. It will be shown subsequently that theoretically calculated intensities can be linked directly to experimental intensities via a proportionality constant.

Tables II.3.7–II.3.9 relate to analytes Cr, Fe and Ni in the system Cr–Fe–Ni for analytical context AC 1. The analyte concentrations range from 100% to 0.1%, the concentrations 70, 50, 30 and 10% having been selected for more detailed coverage. Data from Tables II.3.7–II.3.9 will be used extensively in a number of developments in this and subsequent chapters, e.g. to compute:

- R_i, M_i and M_{ij};
- a_{ij}, e_{ij} and m_{ij} influence coefficients;

Table II.3.6 A_{ij} and E_{ij} data relating to specimens listed in Table II.3.2

Specimen	A_{CrFe}	A_{CrNi}	E_{CrFe}	E_{CrNi}	A_{FeCr}	A_{FeNi}	E_{FeNi}	A_{NiCr}	A_{NiFe}
5054	40.0	103.0	219.0	104.1	1785.8	89.4	249.3	1.3	1.8
5074	38.6	98.5	212.6	103.6	1688.3	82.5	242.6	44.7	62.6
5181	27.6	73.0	176.7	80.5	1849.8	89.5	273.4	91.1	127.4
5202	30.2	77.7	183.4	89.3	1635.0	76.5	249.9	141.0	197.3
5321	27.5	70.5	172.7	85.3	1564.2	71.0	248.4	198.3	277.2
5324	41.5	100.1	212.5	118.9	1253.1	56.1	197.3	192.9	269.7
5364	43.5	102.6	216.2	127.6	1105.6	47.9	180.5	246.3	344.2
5202N	30.3	77.7	181.9	88.6	1635.1	76.5	247.7	140.7	196.8
SS	24.0	65.3	163.7	70.6	2028.3	99.9	294.0	71.3	99.7
Fe14	27.6	73.0	175.3	79.9	1851.9	89.6	271.7	90.8	127.0
AVG	35.1	88.1	195.8	99.9	1508.6	70.0	228.2	142.5	199.3

Table II.3.7 Theoretical emission, absorption and enhancement data for system Cr–Fe–Ni. Analytical context AC 1, analyte = Cr

	C_{Cr}	C_{Fe}	C_{Ni}	$P_{Cr} + S_{Cr}$	A_{CrFe}	E_{CrFe}	A_{CrNi}	E_{CrNi}
Cr 1	1.000	0.000	0.000	1584.77	274.32	461.64	543.21	542.68
2	0.999	0.001	0.000	1583.37	273.90	461.38	542.47	541.88
3	0.999	0.000	0.001	1583.19	273.84	461.27	542.24	542.15
4	0.99	0.01	0.00	1570.81	270.17	459.08	535.82	534.67
5	0.99	0.00	0.01	1568.96	269.62	457.99	533.55	537.39
6	0.90	0.10	0.00	1446.47	234.03	435.81	471.20	465.66
7	0.90	0.05	0.05	1437.06	231.80	430.77	461.44	477.75
8	0.90	0.00	0.10	1430.23	229.75	426.04	451.82	491.14
Cr 9	0.80	0.20	0.00	1310.35	196.40	409.05	403.27	335.08
10	0.80	0.10	0.10	1291.43	192.93	399.82	386.98	416.21
11	0.80	0.00	0.20	1282.03	190.04	391.77	371.07	442.12
12	0.70	0.30	0.00	1175.18	161.37	380.84	339.25	330.31
13	0.70	0.25	0.05	1163.75	159.94	376.36	332.47	338.53
14	0.70	0.20	0.10	1154.22	158.63	372.15	325.78	347.59
15	0.70	0.15	0.15	1146.72	157.44	368.21	319.17	357.59
16	0.70	0.10	0.20	1141.37	156.36	364.54	312.62	368.67
Cr17	0.70	0.05	0.25	1138.34	155.41	361.15	306.13	380.99
18	0.70	0.00	0.30	1137.86	154.59	358.04	299.67	384.74
19	0.50	0.50	0.00	901.40	99.17	317.20	222.48	216.05
20	0.50	0.45	0.05	899.25	98.40	313.31	218.18	221.22
21	0.50	0.40	0.10	878.46	97.69	309.65	213.94	226.88
22	0.50	0.30	0.20	861.21	96.49	302.98	205.59	239.97
23	0.50	0.25	0.25	854.91	96.00	299.98	201.46	247.57
24	0.50	0.20	0.30	850.31	95.58	297.22	197.35	256.00
Cr25	0.50	0.10	0.40	846.84	94.99	292.41	189.17	275.95
26	0.50	0.05	0.45	848.41	94.82	290.38	185.08	287.85
27	0.50	0.00	0.50	852.58	94.73	288.62	180.97	301.36
28	0.30	0.70	0.00	606.19	48.26	235.06	120.66	119.15
29	0.30	0.67	0.03	599.48	48.09	233.17	119.31	120.76
30	0.30	0.64	0.06	593.08	47.93	231.34	117.97	122.47
31	0.30	0.60	0.10	585.05	47.73	228.99	116.19	124.88
32	0.30	0.30	0.20	567.55	47.23	223.58	111.81	131.75
Cr33	0.30	0.40	0.30	553.93	47.07	218.86	197.48	140.09
34	0.30	0.35	0.35	548.71	47.00	216.76	105.33	144.94
35	0.30	0.30	0.40	544.67	46.97	214.85	103.18	150.36
36	0.30	0.20	0.50	540.54	47.03	211.60	98.91	163.25
37	0.30	0.10	0.60	542.88	47.28	209.18	94.49	179.97
38	0.30	0.06	0.64	546.09	47.43	208.47	92.71	188.02
39	0.30	0.03	0.67	549.55	47.58	208.05	91.36	194.80
40	0.30	0.00	0.70	554.03	47.74	207.71	90.00	202.29

Table II.3.7 (*continued*)

	C_{Cr}	C_{Fe}	C_{Ni}	$P_{Cr} + S_{Cr}$	A_{CrFe}	E_{CrFe}	A_{CrNi}	E_{CrNi}
Cr41	0.20	0.80	0.00	438.98	27.75	180.29	75.70	76.25
42	0.20	0.40	0.40	383.72	27.43	163.61	64.94	95.69
43	0.20	0.00	0.80	390.82	28.73	156.93	53.91	146.24
44	0.10	0.90	0.00	245.26	11.32	107.75	35.22	36.63
45	0.10	0.85	0.05	239.15	11.31	106.06	34.60	37.44
46	0.10	0.80	0.10	233.43	11.32	104.06	33.98	38.32
47	0.10	0.70	0.20	223.12	11.34	101.52	32.76	40.34
48	0.10	0.60	0.30	214.36	11.40	98.93	31.55	42.76
Cr49	0.10	0.50	0.40	207.23	11.50	96.70	30.34	45.72
50	0.10	0.45	0.45	204.32	11.57	95.72	29.73	47.45
51	0.10	0.40	0.50	201.89	11.64	94.84	29.12	49.39
52	0.10	0.30	0.60	198.65	11.82	93.39	27.88	54.06
53	0.10	0.20	0.70	198.06	12.06	92.40	26.60	60.20
54	0.10	0.10	0.80	201.16	12.37	91.95	25.28	68.68
55	0.10	0.05	0.85	204.70	12.55	91.96	24.59	74.30
56	0.10	0.00	0.90	210.18	12.75	92.16	23.88	81.33
Cr57	0.05	0.95	0.00	131.75	4.93	60.20	16.88	17.96
58	0.05	0.475	0.475	106.32	5.20	52.72	14.15	23.64
59	0.05	0.00	0.95	109.94	5.95	50.93	11.16	43.47
60	0.01	0.99	0.00	28.23	0.86	13.37	3.25	3.54
61	0.01	0.495	0.495	22.09	0.94	11.53	2.71	4.72
62	0.01	0.00	0.99	22.93	1.12	11.19	2.10	9.25
63	0.001	0.999	0.000	2.87	0.08	1.37	0.32	0.35
64	0.001	0.000	0.999	2.32	0.11	1.14	0.21	0.94

Table II.3.8 Theoretical emission, absorption and enhancement data for system Cr–Fe–Ni. Analytical context AC 1, analyte = Fe

	C_{Fe}	C_{Cr}	C_{Ni}	$P_{Fe} + S_{Fe}$	A_{FeCr}	A_{FeNi}	E_{FeNi}
Fe 1	1.000	0.000	0.000	1742.48	4089.11	224.93	565.14
2	0.999	0.001	0.000	1736.67	4073.46	223.97	563.02
3	0.999	0.000	0.001	1741.08	4084.29	224.55	564.83
4	0.99	0.01	0.00	165.69	3936.97	215.65	544.43
5	0.99	0.00	0.01	1728.47	4040.99	221.16	562.06
6	0.90	0.10	0.00	1278.99	2892.45	152.88	400.03
7	0.90	0.05	0.05	1422.32	3208.21	168.38	458.37
8	0.90	0.00	0.10	1602.77	3616.90	188.44	533.80
Fe 9	0.80	0.20	0.00	966.54	2137.24	109.11	294.33
10	0.80	0.10	0.10	1164.80	2539.16	127.04	374.37
11	0.80	0.00	0.20	1463.29	3163.67	154.57	501.11
12	0.70	0.30	0.00	737.77	1606.55	79.130	220.16

(*continued*)

Table II.3.8 (*continued*)

	C_{Fe}	C_{Cr}	C_{Ni}	$P_{Fe} + S_{Fe}$	A_{FeCr}	A_{FeNi}	E_{FeNi}
13	0.70	0.25	0.05	77.73	1719.91	84.25	243.72
14	0.70	0.20	0.10	867.57	1851.90	89.57	271.72
15	0.70	0.15	0.15	950.00	2007.99	95.75	305.49
16	0.70	0.10	0.20	1048.89	2196.21	103.07	346.96
Fe17	0.70	0.05	0.25	1170.06	2429.11	112.00	399.04
18	0.70	0.00	0.30	1322.68	2727.71	123.34	466.49
19	0.50	0.50	0.00	421.42	899.65	42.48	122.43
20	0.50	0.45	0.05	449.30	947.55	44.15	133.23
21	0.50	0.40	0.10	480.77	1001.11	45.95	145.62
22	0.50	0.30	0.20	557.58	1129.98	50.05	176.71
23	0.50	0.25	0.25	605.10	1208.66	52.39	196.49
24	0.50	0.20	0.30	60.77	1300.11	54.97	220.14
Fe25	0.50	0.10	0.40	806.79	1537.65	61.08	284.36
26	0.50	0.05	0.45	905.40	1697.37	64.80	329.29
27	0.50	0.00	0.50	1030.59	1900.76	69.23	387.92
28	0.30	0.70	0.00	211.13	445.17	20.25	60.51
29	0.30	0.67	0.03	218.19	456.47	20.60	63.23
30	0.30	0.64	0.06	225.69	468.39	20.95	66.16
31	0.30	0.60	0.10	236.46	485.31	21.43	70.40
32	0.30	0.50	0.20	267.95	533.75	22.72	83.09
Fe33	0.30	0.40	0.30	308.05	593.50	24.10	99.80
34	0.30	0.35	0.35	32.50	629.01	24.82	110.26
35	0.30	0.30	0.40	360.76	669.34	25.54	122.58
36	0.30	0.20	0.50	432.93	769.39	26.92	155.04
37	0.30	0.10	0.60	537.64	908.92	27.94	204.25
38	0.30	0.06	0.64	594.09	982.15	28.10	231.65
39	0.30	0.03	0.87	644.42	1046.59	28.06	256.52
40	0.30	0.00	0.70	703.68	1121.65	27.81	286.29
Fe41	0.20	0.80	0.00	130.08	273.02	12.23	37.17
42	0.20	0.40	0.40	213.91	393.63	14.83	71.99
43	0.20	0.00	0.80	513.12	744.64	12.99	218.77
44	0.10	0.90	0.00	60.47	126.42	5.59	17.26
45	0.10	0.85	0.05	63.53	131.01	5.71	18.46
46	0.10	0.80	0.10	66.88	135.95	5.83	19.80
47	0.10	0.70	0.20	74.77	147.09	6.08	22.97
48	0.10	0.60	0.30	84.27	160.30	6.34	27.00
Fe49	0.10	0.50	0.40	96.41	176.24	6.57	32.28
50	0.10	0.45	0.45	103.75	185.54	6.67	35.55
51	0.10	0.40	0.50	112.19	195.93	6.76	39.38
52	0.10	0.30	0.60	133.47	220.96	6.83	49.35
53	0.10	0.20	0.70	163.66	254.08	6.62	64.10
54	0.10	0.10	0.80	209.69	300.53	5.78	87.64
55	0.10	0.05	0.85	243.09	331.93	4.84	105.36
56	0.10	0.00	0.90	288.34	372.14	3.23	130.00

Table II.3.8   (*continued*)

	C_{Fe}	C_{Cr}	C_{Ni}	$P_{Fe} + S_{Fe}$	A_{FeCr}	A_{FeNi}	E_{FeNi}
Fe57	0.05	0.95	0.00	29.21	60.96	2.68	8.34
58	0.05	0.475	0.475	51.19	90.20	3.17	17.72
59	0.05	0.00	0.95	155.20	186.35	0.68	72.34
60	0.01	0.495	0.00	5.69	11.86	0.52	1.63
61	0.01	0.495	0.495	10.14	17.65	0.61	3.54
62	0.01	0.00	0.99	33.29	37.35	−0.03	15.99
63	0.001	0.999	0.000	0.57	1.18	0.05	0.16
64	0.001	0.000	0.999	3.39	3.74	−0.01	1.64

Table II.3.9   Theoretical emission and absorption data for the Cr–Fe–Ni system. Analytical context AC 1; analyte = Ni

	C_{Ni}	C_{Cr}	C_{Fe}	P_{Ni}	A_{NiCr}	A_{NiFe}
Ni 1	1.000	0.000	0.000	1789.65	2679.70	3674.62
2	0.999	0.001	0.000	175.19	2671.55	3663.66
3	0.999	0.000	0.001	174.20	2669.74	3661.23
4	0.99	0.01	0.00	1745.75	2599.91	3567.30
5	0.99	0.00	0.01	1736.31	2582.87	3544.36
6	0.90	0.10	0.00	1408.98	2017.01	2779.69
7	0.90	0.05	0.05	1377.06	1964.28	2708.12
8	0.90	0.00	0.10	1346.65	1914.41	2640.38
Ni 9	0.80	0.20	0.00	1120.73	1554.97	2150.63
10	0.80	0.10	0.10	1077.09	1487.68	2058.65
11	0.80	0.00	0.20	1036.82	1426.17	1974.48
12	0.70	0.30	0.00	890.59	1207.22	1674.23
13	0.70	0.25	0.05	874.69	1183.83	1642.10
14	0.70	0.20	0.10	859.36	1161.35	1611.20
15	0.70	0.15	0.15	844.57	1139.73	1581.47
16	0.70	0.10	0.20	830.29	1118.91	1552.85
Ni17	0.70	0.05	0.25	8113.50	1098.86	1525.26
18	0.70	0.00	0.30	803.16	1079.53	1498.65
19	0.50	0.50	0.00	540.38	708.89	987.01
20	0.50	0.45	0.05	532.26	697.74	971.57
21	0.50	0.40	0.10	524.39	686.94	956.61
22	0.50	0.30	0.20	509.32	666.31	928.04
23	0.50	0.25	0.25	502.11	656.46	914.39
24	0.50	0.20	0.30	495.11	646.90	901.13
Ni25	0.50	0.10	0.40	481.66	628.59	875.76
26	0.50	0.05	0.45	475.21	619.82	863.60
27	0.50	0.00	0.50	468.93	611.30	851.78
28	0.30	0.70	0.00	282.85	363.92	506.68

(*continued*)

Table II.3.9 (*continued*)

	C_{Ni}	C_{Cr}	C_{Fe}	P_{Ni}	A_{NiCr}	A_{NiFe}
29	0.30	0.67	0.03	280.62	359.99	502.61
30	0.30	0.64	0.06	278.43	357.11	498.60
31	0.30	0.60	0.10	275.56	353.34	493.35
32	0.30	0.50	0.20	268.63	344.26	480.70
Ni33	0.30	0.40	0.30	262.04	335.63	468.69
34	0.30	0.35	0.35	258.86	331.48	462.90
35	0.30	0.30	0.40	255.76	327.43	457.26
36	0.30	0.20	0.50	249.78	319.61	446.38
37	0.30	0.10	0.60	244.08	312.17	436.00
38	0.30	0.06	0.64	241.87	309.28	431.98
39	0.30	0.03	0.67	240.24	307.16	429.02
40	0.30	0.00	0.70	238.63	305.06	426.10
Ni41	0.20	0.80	0.00	177.40	225.66	315.39
42	0.20	0.40	0.40	161.34	204.96	286.51
43	0.20	0.00	0.80	147.95	187.75	262.48
44	0.10	0.90	0.00	83.77	105.77	147.96
45	0.10	0.85	0.05	82.80	104.53	146.24
46	0.10	0.80	0.10	81.85	103.22	144.55
47	0.10	0.70	0.20	80.01	100.99	141.29
48	0.10	0.60	0.30	78.26	98.76	138.17
Ni49	0.10	0.50	0.40	76.58	96.63	135.19
50	0.10	0.45	0.45	75.76	95.59	133.74
51	0.10	0.40	0.50	74.97	94.58	132.33
52	0.10	0.30	0.60	73.42	92.62	129.50
53	0.10	0.20	0.70	71.94	90.75	126.96
54	0.10	0.10	0.80	70.52	88.94	124.44
55	0.10	0.05	0.85	69.83	88.06	123.22
56	0.10	0.00	0.90	69.15	87.21	122.02
Ni57	0.05	0.95	0.00	40.76	51.29	71.78
58	0.05	0.475	0.475	36.77	46.25	64.73
59	0.05	0.00	0.95	33.49	42.11	58.94
60	0.01	0.99	0.00	7.98	10.02	14.02
61	0.00	0.495	0.495	7.18	9.02	12.62
62	9.91	0.00	0.99	6.53	8.20	11.48
63	0.001	0.999	0.000	0.79	1.00	1.39
64	0.001	0.000	0.999	0.65	0.81	1.14

- m_{ijn} influence coefficients for the de Jongh algorithm;
- a_{ijp} and e_{ijp} influence coefficients for the Broll–Tertian and for the Rousseau algorithms;
- α_{ij} influence coefficients for the Lachance–Traill model;
- α_j, α_{jj} and α_{ijk} coefficients for the Claisse–Quintin model;

- α_1, α_2, α_3 and α_{ijk} coefficients for the COLA model;
- the value $1 + \varepsilon_i$ for Tertian's unified model.

The following experimental data from Rousseau and Bouchard (1986) are selected in order to examine the quantification of matrix effects in the Cr–Fe–Ni system, but in an analytical context different from that for the experimental data in Table II.3.2. Not only is the incident X-radiation source different, but the incidence and emergence angles are also different. Listed in sequential order in Table II.3.10 are:

- concentrations relating to three specimens and the corresponding A_{ij} and E_{ij} data for analyte Cr;
- net experimental and theoretical intensities for the three specimens and for specimens of pure Cr, Fe and Ni;
- experimental and theoretical relative intensities.

Table II.3.10 Comparison of experimental and theoretically calculated relative intensities[a]

Analytical context
Excitation source: Rh target tube; 45 kV (CP); see Table II.1.9
Instrument geometry: $\psi' = 57°$, $\psi'' = 40°$
System: Cr–Fe–Ni; Fundamental parameters as in AC 1

Specimen	C_{Cr}	C_{Fe}	C_{Ni}	A_{CrFe}	E_{CrFe}	A_{CrNi}	E_{CrNi}
1159	0.0006	0.5100	0.4820	-0.07	0.53	-0.09	0.723
D847	0.2372	0.6203	0.1326	-8.83	136.83	15.11	62.10
D850	0.0299	0.7132	0.2480	-3.37	25.34	-0.48	7.66

	Experimental (c/s)			Theoretical		
Specimen	I_{Cr}	I_{Fe}	I_{Ni}	I_{Cr}	I_{Fe}	I_{Ni}
Cr	488056			1133.2		
Fe		620644			1142.8	
Ni			651528			1114.3
1159	554	383549	150650	1.1	708.2	259.1
D847	158658	260296	33379	368.1	482.0	56.3
D850	24655	44755	64467	56.7	828.6	109.9

	Experimental			Theoretical		
Specimen	R_{Cr}	R_{Fe}	R_{Ni}	R_{Cr}	R_{Fe}	R_{Ni}
1159	0.0011	0.6180	0.2312	0.0010	0.6197	0.2325
D847	0.3251	0.4194	0.0512	0.3248	0.4218	0.0506
D850	0.0505	0.7118	0.0989	0.0500	0.7251	0.0986

[a] Experimental data from Rousseau and Bouchard (1986).

For the present, data from Table II.3.10 will be used to examine three concepts, as follows.

(a) Calculating the proportionality constant g_i

The constant g_i links measured intensities to theoretically calculated intensities in the expression (Eqn (2.1))

$$I_{i,\text{measured}} = g_i \times I_{i,\text{calculated}} \qquad\qquad\qquad (II.3.6)$$

$$= g_i(P_i + S_i) \qquad\qquad\qquad (II.3.7)$$

$$g_{Fe} = 620\ 644/1142.8 = 543.1$$
$$383\ 549/708.2 = 541.6$$
$$260\ 296/482.0 = 540.0$$
$$441\ 755/828.6 = 533.1$$
$$\text{Average} = 539.5$$

$$g_{Ni} = 651\ 528/1114.3 = 584.7$$
$$150\ 650/259.1 = 581.4$$
$$33\ 379/56.3 = 592.9$$
$$64\ 467/109.9 = 584.6$$
$$\text{Average} = 586.4$$

(b) Calculating R_i using a multielement reference

The calculation of R_i from theory is given by Eqn (II.3.1); thus, the calculation of R_{Cr} for D850 using D847 as the reference material takes the form

$$R_{Cr} = \frac{24655}{158658} \times \frac{368.1}{1133.2}$$

$$= 0.0505$$

versus

$$R_{Cr} = \frac{24655}{488056} = 0.0505$$

when calculated using $I_{(Cr)}$ measures experimentally.

Similarly, the calculation of R_{Fe} in D847 using 1159 as reference entails

$$R_{Fe} = \left(\frac{260296}{383549}\right)\left(\frac{708.2}{1142.8}\right)$$

$$= 0.4206$$

versus 0.4194 measured experimentally, and the calculation of R_{Ni} in D847 using D850 as reference entails

$$R_{Ni} = \left(\frac{33379}{64467}\right)\left(\frac{109.9}{1114.3}\right)$$

$$= 0.0511$$

versus 0.0512 measured experimentally.

(c) Calculating experimental intensities due to absorption and enhancement by matrix elements individually

Given g_i and A_{ij} data, it is possible to quantify the absorption and enhancement effect of each element in a given specimen. The following calculations dissect, as it were, the absorption and enhancement of Cr $K-L_{2,3}$ X-radiations by Fe and Ni in specimens D847 and D850. The intensity that would be emitted by direct proportionality for D847 and D850 is given by $I_{(Cr)}C_{Cr}$:

$$488056 \times 0.2372 = 115767 \text{ c/s}$$

and

$$488056 \times 0.0299 = 14593 \text{ c/s}$$

respectively. The constant g_{Cr} is given by

$$g_{Cr} = 488056/1133.2 = 430.7$$
$$158658/368.1 = 431.0$$
$$24655/56.7 = 434.8$$
$$\text{Average} = 432.2$$

In the case of D847, the difference between the measured intensity and the intensity predicted by direct proportionality is equal to $158\,658 - 115\,767$, a difference of 42 891 c/s. This can be construed as the result of the products $(A$ or $E) \times g \times C$:

absorption by Fe: $-(-8.83 \times 432.2 \times 0.6203) = 2367.3$ c/s
enhancement by Fe: $+(136.83 \times 432.2 \times 0.6203) = 36\,683.3$ c/s
absorption by Ni: $-(15.11 \times 432.2 \times 0.1326) = -865.9$ c/s
enhancement by Ni: $+(62.10 \times 432.2 \times 0.1326) = 3558.9$ c/s
$$\Sigma = 41\,744 \text{ c/s}$$

In the case of D850, the difference is 10 062 c/s, construed as the result of:

absorption by Fe: $-(-3.37 \times 432.2 \times 0.7132) = 1038.8$ c/s
enhancement by Fe: $+(25.34 \times 432.2 \times 0.7132) = 7810.9$ c/s

absorption by Ni: $-(-0.48 \times 432.2 \times 0.2480) = 51.4$ c/s
enhancement by Ni: $+(7.66 \times 432.2 \times 0.2480) = 821.0$ c/s

$$\Sigma = 9722.1 \text{ c/s}$$

The differences between the computed and measured values are quite acceptable, given that the presence of minor constituents is not taken into consideration in the above computations.

II.3.6   SUMMARY AND CONCLUSIONS

Applications relating to Chapter 3 consist in using the fundamental parameters approach proposed by Criss and Birks (1968), thus providing analysts with the means to:

(a) compute intensities relative to the pure analyte without having to physically measure $I_{(i)}$ (Eqn(II.3.1));

(b) convert R_i data to C_i for specimens of unknown composition by iteration, with the option of using influence coefficients in an intermediate step to shorten overall computation time;

(c) generate A_{ij} and E_{ij} data used to compute fundamental influence coefficients (the topic of the next chapter).

The fact that $I_{(i)}$ can be calculated for a polychromatic excitation source given $I_{i,\text{experimental}}$ for one (or preferably more) multielement specimens of known composition (if the pure elements are not available for direct measurement) underlies the fundamental parameters approach. It is important to recall, however, that a number of premises underlie the application of theoretical methods, namely that the specimens be *thick*, *flat* and *homogeneous* that essentially all of the elements be determined (by X-ray fluorescence or other methods); and that, for calibration purposes, the concentrations of *all* the elements in the reference materials be known (having been determined by alternative methods).

II.4 FUNDAMENTAL INFLUENCE COEFFICIENTS

As shown in Chapter 4, a number of algorithms have been proposed in the fundamental influence coefficient domain, using influence coefficients which are defined explicitly from fundamental X-ray fluorescence theory. Application will consist in using the entities $P_{(i)}$, P_i, S_i, A_{ij} and E_{ij} calculated for real and hypothetical specimens of known composition and tabulated in Chapter II.3, to quantify the fundamental influence coefficients m_{ij}, a_{ijp}, e_{ijp} and m_{ijn}. These will provide an alternative process to correct for matrix effects in the context of AC 1 (see Table II. 3.1).

II.4.6.1 Broll–Tertian formalism

Recalling Eqns (4.30), (4.35) and (4.37):

$$a_{ijp} = \frac{A_{ij}}{P_i} \tag{II.4.1}$$

$$e_{ijp} = \frac{E_{ij}}{P_i} \tag{II.4.2}$$

$$C_i = R_i\left[1 + \sum_j \left(a_{ijp} - e_{ijp}\frac{C_i}{R_i}\right)C_j\right] \tag{II.4.3}$$

the substitution of P_i values from Table II.3.3 and A_{ij} and E_{ij} values from Table II.3.6 leads to values for the *effective* coefficients a_{ijp} and e_{ijp} listed in Table II.4.1. For example, in the case of specimens 5202N and Fe14, respectively

$a_{CrFep} = 30.3/309.9 = 0.0978$

$e_{CrFep} = 181.9/309.9 = 0.5870$

Table II.4.1 Influence coefficient a_{ijp} and e_{ijp} values for specimens in Table II.3.2

Specimen	a_{CrFep}	a_{CrNip}	e_{CrFep}	e_{CrNip}	a_{FeCrp}	a_{FeNip}	e_{FeNip}
5054	0.1038	0.2672	0.5685	0.2701	2.1889	0.1096	0.3056
5074	0.1032	0.2634	0.5684	0.2770	2.1874	0.1069	0.3143
5181	0.0949	0.2513	0.6078	0.2770	2.2035	0.1066	0.3256
5202	0.0976	0.2508	0.5922	0.2883	2.1940	0.1026	0.3354
5321	0.0959	0.2453	0.6012	0.2969	2.1948	0.0996	0.3485
5324	0.1064	0.2568	0.5450	0.3051	2.1718	0.0972	0.3420
5364	0.1081	0.2552	0.5378	0.3175	2.1655	0.0939	0.3535
5202N	0.0976	0.2509	0.5869	0.2857	2.1940	0.1026	0.3324
SS	0.0915	0.2489	0.6243	0.2691	2.2134	0.1090	0.3208
Fe14	0.0949	0.2513	0.6037	0.2750	2.2036	0.1066	0.3233
AVG	0.1014	0.2548	0.5664	0.2891	2.1853	0.1014	0.3309

and

$$a_{CrFep} = 27.6/290.4 = 0.0950$$
$$e_{CrFep} = 175.3/290.4 = 0.6037$$

An example of the application of the effective coefficient approach is given by Broll (1986) for the determination of Si, Cr, Mn, Fe and Ni in austenitic stainless steels (absolute error less than 0.3%, relative error on all elements less than 0.9%). Data are also given illustrating the determination of Al, Si, Ti, Cr (0–25%), Mn (0–2%), Fe (0–75%), Co (0–17%), Ni (0–60%), Nb (0–5%) and Mo (0–10%) in stainless steels and high-temperature alloys with the stipulation that, for industrial utilization, it is necessary to make a careful choice of the standard used for calibration; i.e. one must restrict the composition range and centre it on the standard. The approach was also used to determine P, Ni, Cu, Zn and Sn in phosphor bronzes, and Al_2O_3, SiO_2, SO_3, K_2O, CaO, TiO_2 and Fe_2O_3 in cements in the form of fused discs (1 g of cement to 7.5 g of $Li_2B_4O_7$).

II.4.6.2 Rousseau formalism

It is recalled that the values of the absorption and enhancement coefficients listed in Table II.4.1 are also applicable in the Rousseau algorithm under the conditions of AC 1

$$C_i = R_i \left[\frac{1 + \sum_j a_{ijp} C_j}{1 + \sum_j e_{ijp} C_j} \right] \qquad (II.4.4)$$

Rousseau and Bouchard (1986) demonstrated the experimental validity of the above expression by analysing 28 standard reference materials for 15 elements. The concentration ranges varied very widely, e.g. Al $0.05-5\%$, Cr $O-24\%$, Fe $0.02-98\%$ Ni $0.01-80\%$ and Zn $5-95\%$. The experimental conditions: analyte line, crystal, filter collimation, detector, tube target, voltage and current, counting time and 2θ values for peaks and two background regions are listed, as well as all measured net intensities. Two standards (high–low), different for each analyte, were used to obtain the intercept and the calibration slopes, which are tabulated along with the mean absolute error. Johnson and Fleming (1987) applied Eqn (II.4.4) in the context of energy dispersive X-ray fluorescence spectrometry of massive sulphide specimens (pressed pellets) and essentially monochromatic excitation from secondary targets. The concentrations of the nine elements in the six standards varied widely; e.g., Si $0.2-19\%$, S $12-31\%$, Fe $2-24\%$, Cu $0.1-2.1\%$, Zn $2-57\%$, Pb $2-7\%$, Ag $90-1100$ ppm, Cd $600-7800$ ppm, and Sn $0.006-2.4\%$. The average difference ($\pm6\%$) between certified values and X-ray values was deemed satisfactory, considering that many of the certified values for trace elements had greater relative standard deviations and that mineralized powdered specimens cannot be considered as strictly homogeneous.

II.4.6.3 Lachance formalism

According to Tertian (1988), the simplest and most convenient influence coefficient algorithm is the conventional Lachance–Traill formulation, which in the fundamental influence coefficient domain takes the form

$$C_i = R_i\left[1 + \sum_j m_{ij}C_j\right] \tag{II.4.5}$$

in which the m_{ij} coefficients are defined as

$$m_{ij} = \frac{A_{ij} - E_{ij}}{P_i + S_i} = \frac{M_{ij}}{P_i + S_i} \tag{II.4.6}$$

where

$$M_{ij} = A_{ij} - E_{ij} \tag{II.4.7}$$

The m_{ij} and m_{ik} values for the analytes Cr, Fe and Ni for the eleven multi-element specimens of Table II.3.2 are listed in Table II.4.2. These are obtained by substituting the appropriate data in Tables II.3.3 and II.3.6 in Eqn (II.4.6). For example, in the case of specimens 5202N and Fe14, respectively:

$$m_{CrFe} = \frac{30.3 - 181.9}{438.8} = -0.3456$$

$$m_{CrNi} = \frac{77.7 - 88.6}{438.8} = -0.0246$$

Table II.4.2. Influence coefficient m_{ij} and m_{ik} values for specimens in Table II.3.2

Specimen	i = Cr		i = Fe		i = Ni	
	j = Fe	k = Ni	j = Cr	k = Ni	j = Cr	k = Fe
5054	−0.3290	−0.0021	2.1879	−0.1959	1.2544	1.7563
5074	−0.3317	−0.0097	2.1537	−0.2042	1.2578	1.7604
5181	−0.3538	−0.0178	2.1343	−0.2122	1.2614	1.7648
5202	−0.3493	−0.0265	2.0903	−0.2218	1.2653	1.7696
5321	−0.3571	−0.0365	2.0517	−0.2327	1.2698	1.7752
5324	−0.3257	−0.0358	2.0375	−0.2297	1.2694	1.7746
5364	−0.3234	−0.0468	1.9989	−0.2397	1.2737	1.7798
5202N	−0.3456	−0.0246	2.0905	−0.2189	1.2653	1.7696
SS	−0.3591	−0.0136	2.1580	−0.2066	1.2598	1.7629
Fe14	−0.3509	−0.0164	2.1348	−0.2099	1.2614	1.7648
AVG	−0.3348	−0.0247	2.0820	−0.2186	1.2654	1.7698

and

$$m_{CrFe} = \frac{27.6 - 175.3}{421.1} = -0.3509$$

$$m_{CrNi} = \frac{73.0 - 79.9}{421.1} = -0.0164$$

Similar calculations can thus be used to compute m_{ij} coefficients for all the hypothetical specimens in Tables II.3.7–II.3.9. A glance at the m_{ij} and m_{ik} coefficients in Table II.4.2 clearly indicates that the variations are such that a single value cannot be selected and used as a constant directly unless unknown and reference are quite similar; e.g. Fe14 coefficients used for specimen 5181. An estimate can be made of the error introduced if, for example, the coefficients m_{FeCr} and m_{FeNi} for specimen 5364 are used *directly* to correct for matrix effects in 5181:

$$C_{Fe5181} = R_{Fe5181}\left[1 + \sum m_{ij5364}C_{j5181}\right]$$
$$= 0.4971 \; [1 + (1.9989 \times 0.1988) + (-0.2397 \times 0.0996)]$$
$$= 0.6828$$

versus the known value of 0.6945.

The calculation of a_{ij} and e_{ij} values listed in Table II.4.3 is carried out only to illustrate that the total or net influence coefficient m_{ij} is equal to the absorption minus enhancement matrix effects, namely

$$m_{ij} = a_{ij} - e_{ij} \tag{II.4.8}$$

Table II.4.3. Influence coefficient a_{ij} and e_{ij} values for specimens in Table II.3.2

Specimen	a_{CrFe}	a_{CrNi}	e_{CrFe}	e_{CrNi}	a_{FeNi}	e_{FeNi}
5054	0.0735	0.1891	0.4024	0.1912	0.1095	0.3055
5074	0.0736	0.1878	0.4053	0.1975	0.1053	0.3095
5181	0.0655	0.1733	0.4193	0.1911	0.1032	0.3154
5202	0.0689	0.1771	0.4182	0.2036	0.0977	0.3195
5321	0.0677	0.1733	0.4248	0.2098	0.0931	0.3258
5324	0.0790	0.1907	0.4047	0.2265	0.0912	0.3209
5364	0.0814	0.1921	0.4047	0.2389	0.0866	0.3263
5202N	0.0690	0.1772	0.4145	0.2018	0.0978	0.3167
SS	0.0617	0.1678	0.4208	0.1814	0.1062	0.3128
Fe14	0.0654	0.1733	0.4163	0.1896	0.1032	0.3132
AVG	0.0730	0.1835	0.4078	0.2081	0.0966	0.3142

provided that a_{ij} and e_{ij} are defined

$$a_{ij} = \frac{A_{ij}}{P_i + S_i} \tag{II.4.9}$$

and

$$e_{ij} = \frac{E_{ij}}{P_i + S_i} \tag{II.4.10}$$

Substitution of the appropriate values from Tables II.3.3 and II.3.6 in the case of specimen 5202N gives

$$a_{CrFe} = 30.3/438.8 = 0.0690$$
$$e_{CrFe} = 181.9/438.8 = 0.4145$$
$$m_{CrFe} = 0.0690 - 0.4145 = -0.3455$$

and for specimen Fe14

$$a_{CrFe} = 27.6/421.1 = 0.0655$$
$$e_{CrFe} = 175.3/421.1 = 0.4163$$
$$m_{CrFe} = 0.0655 - 0.4163 = -0.3508$$

II.4.6.4 The de Jongh formalism

The concept of m_{ijn} influence coefficients was proposed by de Jongh (1973, 1979) to take advantage of the fact that any one element can be selected for elimination in the conventional model Eqn (II.4.5)), which can be expressed as

$$C_i = R_i[1 + m_{ii}C_i + m_{ij}C_j + m_{ik}C_k] \tag{II.4.11}$$

for a three element system where, by definition

$$m_{ii} = \frac{\mu_i^* - \mu_i^*}{\mu_i^*} = 0 \tag{II.4.12}$$

i.e. the analyte i is the eliminated element. Let us consider the system Cr–Fe–Ni in the context of ferrous alloys, in which, traditionally, it is customary not to determine the element iron. In the context of Eqn (II.4.5), the expressions for C_{Cr} and C_{Ni} take the form

$$C_{Cr} = \frac{I_{Cr}}{I_{(Cr)}} [1 + m_{CrFe}C_{Fe} + m_{CrNi}C_{Ni}] \tag{II.4.13}$$

$$C_{Ni} = \frac{I_{Ni}}{I_{(Ni)}} [1 + m_{NiFe}C_{Fe} + m_{NiCr}C_{Cr}] \tag{II.4.14}$$

Substitution of $(1 - C_{Cr} - C_{Ni})$ for C_{Fe} in the two above expressions and taking the terms $1 + m_{CrFe}$ and $1 + m_{NiFe}$ in factor yields

$$C_{Cr} = \frac{1 + m_{CrFe}}{I_{(Cr)}} I_{Cr}[1 + m_{CrCrFe}C_{Cr} + m_{CrNiFe}C_{Ni}] \tag{II.4.15}$$

$$C_{Ni} = \frac{1 + m_{NiFe}}{I_{(Ni)}} I_{Ni}[1 + m_{NiCrFe}C_{Cr} + m_{NiNiFe}C_{Ni}] \tag{II.4.16}$$

where

$$m_{CrCrFe} = \frac{m_{CrCr} - m_{CrFe}}{1 + m_{CrFe}} \tag{II.4.17}$$

$$m_{CrNiFe} = \frac{m_{CrNi} - m_{CrFe}}{1 + m_{CrFe}} \tag{II.4.18}$$

$$m_{NiNiFe} = \frac{m_{NiNi} - m_{NiFe}}{1 + m_{NiFe}} \tag{II.4.19}$$

$$m_{NiCrFe} = \frac{m_{NiCr} - n_{NiFe}}{1 + m_{NiFe}} \tag{II.4.20}$$

Substitution of the appropriate m_{ij} values from Table II.4.2 into the above expressions gives, in the case of specimen 5202N,

$$m_{CrCrFe} = \frac{0.0 - (-0.3456)}{1.0 + (-0.3456)} = 0.5281$$

$$m_{CrNiFe} = \frac{-0.0246 - (-0.3456)}{1.0 + (0.3456)} = 0.4905$$

$$m_{\text{NiNiFe}} = \frac{0.0 - 1.7696}{1.0 + 1.7696} = -0.6389$$

$$m_{\text{NiCrFe}} = \frac{1.2653 - 1.7696}{1.0 + 1.7696} = -0.1821$$

and in the case of specimen Fe14,

$$m_{\text{CrCrFe}} = \frac{0.0 - (-0.3509)}{1.0 + (-0.3509)} = 0.5406$$

$$m_{\text{CrNiFe}} = \frac{-0.0164 - (-0.3509)}{1.0 + (-0.3509)} = 0.5153$$

$$m_{\text{NiNiFe}} = \frac{0.0 - 1.7648}{1.0 + 1.7648} = -0.6383$$

$$m_{\text{NiCrFe}} = \frac{1.2614 - 1.7648}{1.0 + 1.7648} = -0.1821$$

The calibration constants are given by

$$K_{\text{Cr}} = \frac{1 + m_{\text{CrFe}}}{I_{(\text{Cr})}} = \frac{1 + (-0.3509)}{12000} = 5.409 \times 10^{-5}$$

$$K_{\text{Ni}} = \frac{1 + m_{\text{NiFe}}}{I_{(\text{Ni})}} = \frac{1 + 1.7648}{18000} = 1.536 \times 10^{-4}$$

The concentration of iron C_{Fe} which appeared in Eqns (II.4.13) and (II.4.14) does not appear in Eqns (II.4.15) and (II.4.16); hence, in the latter, the determination of Cr an Ni involves corrections for the matrix effect of iron without having to measure I_{Fe} or having knowledge of C_{Fe}.

As exemplified in Table II.4.4, de Jongh (1973) applied Eqns (II.4.15) and (II.4.16) for analysis of the elements Si, Cr, Mn, Ni, Co, Cu, Nb and Mo, Nb in stainless steel type alloys. Following de Jongh, the influence coefficients are calculated in the context of a reference material having 20% Cr, 70% Fe and 10% Ni for analyte Ni; e.g. $m_{\text{NiCr}} = 14.36/9.51 = 1.5100$, $m_{\text{NiFe}} = 2.0694$, $m_{\text{NiNi}} = 0.0$ etc., and are listed under the heading m_{ij}. The coefficients m_{ijn} are obtained from these values by equating the eliminated element n to Fe; e.g. substitution in Eqn (II.4.18) yields

$$m_{\text{NiCrFe}} = \frac{1.5100 - 2.0694}{1 + 2.0694} = -0.182$$

Table II.4.4. Theoretical data used to generate influence coefficients for the de Jongh
algorithm

Analytical context

Excitation source: Cr target tube; 45 kV (CP); see Table II.1.8
Instrument geometry: $\psi' = 64°$; $\psi'' = 40°$
System: Cr–Fe–Ni; Analyte = Ni
Reference specimen: $C_{Cr} = 0.20$; $C_{Fe} = 0.70$; $C_{Ni} = 0.10$
$P_{(Ni)} = 261.62$; $P_{Ni} = 9.51$; $M_{NiCr} = 14.36$;
$M_{NiFe} = 19.68$

m_{ij}	m_{ijFe}	$m_{ijFe}{}^{a}$
$m_{NiSi} = -0.3941$	$M_{NiSiFe} = -0.803$	-0.793
$m_{NiCr} = 1.5100$	$M_{NiCrFe} = -0.182$	-0.182
$m_{NiMn} = 1.7772$	$M_{NiMnFe} = -0.095$	-0.094
$m_{NiFe} = 2.0694$	$M_{NiFeFe} = 0.0$	0.0
$m_{NiNi} = 0.0$	$M_{NiNiFe} = -0.674$	-0.673
$m_{NiCo} = -0.0957$	$M_{NiCoFe} = -0.705$	-0.704
$m_{NiCu} = -0.2137$	$M_{NiCuFe} = -0.744$	-0.707
$m_{NiNb} = 0.3754$	$M_{NiNbFe} = -0.552$	-0.581
$m_{NiMo} = -0.4828$	$M_{NiMoFe} = -0.517$	-0.543

Analysed as unknown:

specimen 655-3; $C_{Ni6553,app} = 13.094\%$

Si (%)	Cr (%)	Mn (%)	Ni (%)	Cu (%)	Nb (%)	Mo (%)
0.60	18.54	1.59	11.50	0.088	0.59	0.052

^{a} de Jongh (1973) values.

Similarly

$$m_{NiFeFe} = \frac{2.0694 - 2.0694}{1 + 2.0694} = 0.0$$

$$m_{NiNiFe} = \frac{0.0 - 2.0694}{1 + 2.0694} = -0.674$$

The column headed $m_{ijFe}{}^{a}$ lists the equivalent values obtained by de Jongh.
The agreement is very good considering that the fundamental parameters used
were not identical, having been obtained from different sources.

II.4.6.5 Typical analytical procedure

Assuming that the data in Table II.4.5 are representative of a given analytical
context AC 1, i.e. measured net intensities for six reference materials, one
reference specimen (No. 5181) treated as an unknown, and the intensity for
a counting monitor to compensate for instrumental drift, the following is

Table II.4.5. Data used to illustrate fundamental influence coefficient methods: $I_{\text{Fe,monitor}} = 3500$ c/s

Specimen	C_{Fe}	I_{Fe}(c/s)	M_{ir}	$I_{(\text{Fe})}$(c/s)	$R_{(\text{Fe}),\text{mon}}$
(a) Used for calibration					
5054	0.7250	6564.5	1.5635	14157	4.0445
5074	0.6838	6315.4	1.5336	14164	4.0468
5202	0.6303	6272.0	1.4124	14055	4.0156
5321	0.5919	6080.2	1.3631	13984	3.9954
5324	0.5280	4940.6	1.5051	14084	4.0239
5364	0.4721	4450.6	1.4940	14084	4.0239
			Average	14088	4.0250
(b) Unknown					
5181	(0.6945)	6959.4			

illustrative of a typical procedure for the determination of iron in specimen 5181. Firstly, the calibration step is carried out, followed by analysis:

(a) The values of the correction term M_i for matrix effects for the six reference materials are computed from fundamental influence coefficient expressions, namely

$$M_{\text{ir}} = \left[1 + \sum_j \left\{ a_{\text{ijp}} - e_{\text{ijp}} \frac{C_i}{R_i} \right\} C_j \right]_r \tag{II.4.21}$$

$$M_{\text{ir}} = \left[\frac{1 + \sum_j a_{\text{ijp}} C_j}{1 + \sum_j e_{\text{ijp}} C_j} \right]_r \tag{II.4.22}$$

$$M_{\text{ir}} = \left[1 + \sum_j m_{\text{ij}} C_j \right]_r \tag{II.4.23}$$

Substitution of data for reference material 5202 (Tables II.4.1 and II.4.2) in the above equations yields (respectively)

$$M_{\text{Fe}} = 1 + (2.1940 \times 0.213) + \left(\left\{ 0.1026 - 0.3354 \times \frac{0.6303}{0.4489} \right\} \times 0.148 \right)$$

$$= 1.4124$$

$$M_{\text{Fe}} = \left[\frac{1 + (2.1940 \times 0.213) + (0.1026 \times 0.148)}{1 + (0.3354 \times 0.148)} \right]$$

$$= 1.4124$$

$$M_{\text{Fe}} = 1 + (2.0903 \times 0.213) + (-0.2218 \times 0.148)$$

$$= 1.4124$$

Similar calculations give the values for the remaining reference materials in the column headed M_{ir}.

(b) The intensities corrected for matrix effects for the six reference materials are then given by the product I_i times their respective matrix effect correction terms. The calculated intensity for the pure analyte is given by

$$I_{(i)} = \left(\frac{I_i M_i}{C_i}\right)_r \tag{II.4.24}$$

Thus, in the case of reference material 5202

$$I_{(Fe)} = \frac{6272.0 \times 1.4124}{0.6303}$$

$$= 14\ 055\ c/s$$

Any one standard can be use for calibration but, as is evident from the variations observed in the column headed $I_{(Fe)}$, it is generally preferable to use an average value. Alternatively, one may plot C_{ir} versus the product $(I_i M_i)_r$, in which case the slope through the origin yields $K_{(i)} = 1/I_{(i)}$.

(c) Intensities may also be expressed relative to a counting monitor $R_{i,mon}$ which, in fact, is simply 'intensity expressed on a different scale', i.e.

$$R_{i,mon} = \frac{I_i}{I_{i,mon}} \tag{II.4.25}$$

and

$$R_{(i),mon} = \frac{R_{i,mon} M_{ir}}{C_{ir}} \tag{II.4.26}$$

Thus, in the case of specimen 5202,

$$R_{Fe,mon} = \frac{6272.0}{3500} = 1.7920$$

and

$$R_{(Fe),mon} = \frac{1.7920 \times 1.4124}{0.6303} = 4.0156$$

$R_{(Fe),mon}$ values for all six calibration specimens are listed under that heading in Table II.4.5, along with the average value 4.0250. The latter is by definition equal to (14088/3500).

(d) R_{Fe5181} may now be calculated by either

$$R_{Fe5181} = 6959.4/14088 = 0.4940$$

or

$$R_{Fe5181} = \frac{6959.4/3500}{4.025} = 0.4940$$

which compares favourably with the value of 0.4971 measured experimentally, listed in Table II.3.4. A similar process leads to R_{Cr5181} and R_{Ni5181} values, which are then converted to concentrations using any fundamental parameters algorithm with its appropriate influence coefficients.

II.4.6.6 Interrelation between fundamental influence coefficients

The fact that fundamental influence coefficients can be defined as functions of the same fundamental entities implies that they can also be defined in terms of each other. Data listed in Table II.3.6 relating to analyte Cr in specimen 5202N will serve to illustrate this. From Tables II.3.3, II.3.4 and II.3.6

Analytical context: AC 1

Specimen: 5202N; $P_{Cr} + S_{Cr} = 438.77$

$C_{Cr} = 0.215$, $C_{Fe} = 0.636$, $C_{Ni} = 0.149$

$A_{CrFe} = 30.26$, $E_{CrFe} = 181.88$, $A_{CrNi} = 77.74$, $E_{CrNi} = 88.55$

(a) a_{ij} and e_{ij} from fundamental entities (Eqns (4.54) and (4.55) and Eqns (II.4.9) and (II.4.10)):

$$a_{CrFe} = 30.26/438.77 = 0.06897$$

$$e_{CrFe} = 181.88/438.77 = 0.41452$$

$$a_{CrNi} = 77.74/438.77 = 0.17718$$

$$e_{CrNi} = 88.55/438.77 = 0.20181$$

(b) m_{ij} from a_{ij} and e_{ij} (Eqn (4.56) or Eqn (II.4.8)):

$$m_{CrFe} = 0.06897 - 0.41452 = -0.34555$$

$$m_{CrNi} = 0.17718 - 0.20181 = -0.02463$$

(c) $(P_i + S_i)$ corrected for total matrix effects (Eqn (4.20)):

$$\begin{aligned}P_{(Cr)}C_{Cr} &= 438.77 + (30.26 \times 0.636) - (181.88 \times 0.636) + (77.74 \times 0.149) \\ &\quad - (88.55 \times 0.149) \\ &= 340.73\end{aligned}$$

(d) $P_{(i)}$ from $(P_i + S_i)$_{corrected for matrix effects} $\div C_i$

$P_{(Cr)} = 340.73/0.215 = 1584.79$

(e) Primary fluorescence emission P_i (Eqn (4.21)):

$P_{Cr} = 340.73 - (30.26 \times 0.636) - (77.74 \times 0.149)$

$\qquad = 309.90$

(f) Intensity relative to pure analyte (Eqn (4.1)):

$R_{Cr} = 438.77/1584.79 = 0.27689$

(g) a_{ijp} and e_{ijp} from fundamental entities (Eqns (4.30) and (4.35) or Eqns (II.4.1) and (II.4.2)):

$a_{CrFep} = 30.26/309.90 = 0.09764$

$e_{CrFep} = 181.88/309.90 = 0.58690$

$a_{CrNip} = 77.74/309.90 = 0.25086$

$e_{CrNip} = 88.55/309.90 = 0.28574$

(h) a_{ijp} and e_{ijp} from a_{ij} and e_{ij}:

$a_{CrFep} = (0.06897/309.90) \times 438.77 = 0.09765$

$e_{CrFep} = (0.41452/309.90) \times 438.77 = 0.58690$

$a_{CrNip} = (0.17718/309.90) \times 438.77 = 0.25086$

$e_{CrNip} = (0.20181/309.90) \times 438.77 = 0.28573$

(j) a_{ij} and e_{ij} from a_{ijp} and e_{ijp}:

$$a_{CrFe} = \frac{0.09765}{1 + (0.58690 \times 0.636) + (0.28573 \times 0.149)}$$

$\qquad = 0.09765/1.41585$

$\qquad = 0.06897$

$e_{CrFe} = 0.58690/1.41585$

$\qquad = 0.41452$

$a_{CrNi} = 0.25086/1.41585$
$\qquad = 0.17718$

$e_{CrNi} = 0.28573/1.41585$

$\qquad = 0.20181$

(k) m_{ijn} from m_{ij} and m_{in}; $i = Cr$, $j = Ni$, $n = Fe$:

$$m_{CrCrFe} = \frac{0.0 - (-0.34555)}{1 + (-0.34555)}$$

$$= 0.52800$$

$$m_{CrNiFe} = \frac{(-0.2463) - (0.34555)}{1 + (-0.34555)}$$

$$= 0.49037$$

(l) m_{ij} and m_{in} from m_{ijn} (this is a two step process):
Firstly, calculation of m_{in}

$$m_{CrFe} = \frac{-(0.52800)}{1 + 0.52800} = -0.34555$$

Secondly, calculation of m_{ij}:

$$m_{CrNi} = 0.49037 + (-0.34555 \times (1 + 0.49037))$$

$$= -0.02463$$

(m) The relative intensity or apparent concentration for the de Jongh algorithm is given by:

$$R_{Cr} = 0.27689 \times (1 + (-0.34555))$$

$$= 0.18121$$

Thus the calculation of C_{Cr}, i.e. substitution of the appropriate data in the Broll–Tertian algorithm (Eqn (II.4.3)), takes the form

$$C_{Cr} = 0.27689\left[1 + \left(0.09765 - 0.58690 \times \frac{0.215}{0.27689}\right)0.636\right.$$

$$\left. + \left(0.25086 - 0.28573 \times \frac{0.215}{0.27689}\right)0.149\right]$$

$$= 0.215$$

Substitution in the Rousseau algorithm (Eqn (II.4.4)) gives

$$C_{Cr} = 0.27689\left[\frac{1 + (0.09765 \times 0.636) + (0.25086 \times 0.149)}{1 + (0.58690 \times 0.636) + (0.28573 \times 0.149)}\right]$$

$$= 0.215$$

Substitution in the Lachance algorithm (Eqn (II.4.5)):

$$C_{Cr} = 0.27689 \; [1 + (-0.34555 \times 0.636) + (-0.02463 \times 0.149)]$$

$$= 0.215$$

Substitution in the de Jongh algorithm (Eqn II.4.15):

$$C_{Cr} = 0.18121 \ [1 + (0.52800 \times 0.215) + (0.49037 \times 0.149)]$$
$$= 0.215$$

II.4.6.7   Summary and conclusions

The current state of the art in quantitative X-ray fluorescence analysis is such that analysts have a wide choice of equivalent (from a theoretical point of view fundamental algorithms for converting intensities to concentrations; i.e. the algorithms are equivalent provided that the sum of the weight fractions is equal to unity, as required by theory. Thus, given identical fundamental data, the classical formalism proposed by Criss and Birks (1968) is equivalent to the fundamental influence coefficient formalisms examined in Chapter 4, as expressed by

$$K_{(i)} = \frac{1}{I_{(i)}} = \left(\frac{P_i + S_i}{I_i P_{(i)}} \right)_r = \left(\frac{C_i}{I_i M_i} \right)_r \qquad (II.4.27)$$

which implies that, within the reproducibility due to the uncertainties in the fundamental parameters, instrumental parameters, reference material composition and net experimental intensities, plot of $R_{ir,theoretical}$ versus $I_{ir,experimental}$ should be linear with slope $K_{(i)}$. Examples of the small anomalies that are introduced when the sum is not unity can be assessed by comparing data for specimen 5202 with hypothetical specimen 5202N and specimen 5181 with Fe14. In contrast to empirical influence coefficient methods, in which the experimental intensities from reference materials are used to *compute* the values of the coefficients, the fundamental influence coefficient approach calculates the values from *theory* and confirms their validity by using them to linearise a scatter of experimental intensity data.

Analysis based on fundamental parameters methods must incorporate an iterative process, which involves making initial estimates of specimen composition which are gradually refined during the process. Because iteration over the effective wavelength range may be somewhat time consuming when using less powerful computing facilities, Criss et al. (1978) proposed using semi-empirical influence coefficient models in an intermediate step to obtain a fairly close estimate of specimen composition, which then serves as input for the classical fundamental parameters algorithm. The preceding treatment should dispel the myths that the latter 'requires' measurement intensities for pure element specimens, or that it is a 'standardless' method.

II.5 ALPHA INFLUENCE COEFFICIENTS

In contrast to the fundamental influence coefficient domain, where the emphasis is on retaining *theoretical exactness* in defining and computing the values of the coefficients, the concept underlying the alpha influence coefficient domain is that of accepting some limitations to the applicable concentration ranges in return for greater versatility. *Exact* fundamental influence coefficients can be computed only *provided* that the composition of the specimen is known (reference materials) or, in the case of samples submitted for analysis, can be estimated very closely only as a result of a prior treatment of the intensity data. The computation of alpha coefficients, on the other hand, is in the main independent of *a priori* knowledge of specimen composition. Broadly speaking, the current concept in the alpha coefficient domain is to compute constants that are combined as part of the iteration process to generate α_{ij} values that *self-adjust* as a function of specimen composition. In essence, application is illustrated by:

- firstly, computing alpha influence coefficient values from the data generated from fundamental parameters expressions (Chapters 3 and 4), as listed in Tables II.3.7–II.3.9;
- secondly, using those coefficients for converting intensities to concentrations for the seven alloys in Table II.3.2 using the conventional formalisms

$$C_i = K_i I_i \left[1 + \sum_j \alpha_{ij} C_j \right]$$

$$= R_i \left[1 + \sum_j \alpha_{ij} C_j \right] \tag{II.5.1}$$

where the influence coefficients α_{ij} are defined in any one of

(a) the Lachance-Traill algorithm, in which

$$\alpha_{ij} = m_{ijr} \tag{II.5.2}$$

(b) the modified Claisse–Quintin algorithm, in which

$$\alpha_{ij} = \alpha_j + \alpha_{jj}C_M + \sum_k \alpha_{ijk}C_k \tag{II.5.3}$$

and

(c) the COLA algorithm, in which

$$\alpha_{ij} = \alpha_1 + \frac{\alpha_2 C_M}{1 + (1 - C_M)\alpha_3} + \sum_k \alpha_{ijk}C_k$$

$$= \alpha_{ij,hyp} + \sum_k \alpha_{ijk}C_k \tag{II.5.4}$$

where the term $\sum_k \alpha_{ijk}C_k$ compensates for third element effects.

(d) the Tertian algorithm

$$C_i = \frac{R_i}{1 + \varepsilon_i}\left[1 + \sum_j \alpha_{ij,hyp}C_j\right] \tag{II.5.5}$$

where

$$\alpha_{ij} + \gamma_1 \frac{\gamma_2 C_i}{1 + \gamma_3(1 - C_i)} \tag{II.5.6}$$

and the term $1 + \varepsilon_i$ compensates for third element effects.

II.5.4 THE LACHANCE–TRAILL ALGORITHM

Following Lachance (1970), the coefficients are treated as constants, but are equated to binary fundamental influence coefficients (absorption and enhancement) for a polychromatic excitation source *rather than* equated to $\alpha_{ij\lambda}$, which implies a monochromatic excitation source for binary systems in the absence of enhancement, as originally suggested by Lachance and Traill (1966); i.e.

$$\alpha_{ij} = m_{ij,bin} \tag{II.5.7}$$

This corresponds to the m_{ij} coefficients (polychromatic excitation) computed for $C_i = 0.50$ in Table II.5.1.

A better approach, however, is to equate α_{ij} to m_{ij} computed for a type material or average values expected in the unknowns. The fact that the influence coefficient is treated as a constant in both cases certainly limits the concentration ranges in which Eqn (II.5.1) can be applied with accuracy. This is clearly demonstrated in Table II.5.2, where the α values taken at $C_i = 0.5$ in Table II.5.1 are used as constants in the iteration process. In Table II.5.3,

Table II.5.1 Binary m_{ij} influence coefficients for systems Cr–Fe, Cr–Ni and Fe–Ni for analytical context AC 1; see Table II.3.1

C_i	m_{CrFe}	m_{CrNi}	m_{FeCr}	m_{FeNi}	m_{NiCr}	m_{NiFe}
0.999	−0.1180	0.0004	2.3456	−0.1954	1.4965	2.0520
0.99	−0.1202	−0.0024	2.3355	−0.1972	1.4893	2.0413
0.90	−0.1395	−0.0275	2.2615	0.2155	1.4315	1.9607
0.80	−0.1633	−0.0555	2.2112	−0.2368	1.3875	1.9044
0.70	−0.1868	−0.0836	2.1776	−0.2594	1.3555	1.8659
0.60	−0.2132	−0.1121	2.1532	−0.2835	1.3311	1.8379
0.50	−0.2419	−0.1412	2.1348	−0.3092	1.3118	1.8164
0.40	−0.2733	−0.1713	2.1203	−0.3370	1.2961	1.7994
0.30	−0.3082	−0.2027	2.1085	−0.3673	1.2831	1.7856
0.20	−0.3475	−0.2363	2.0988	−0.4010	1.2720	1.7741
0.10	−0.3932	−0.2734	2.0906	−0.4397	1.2625	1.7645
0.01	−0.4431	−0.3119	2.0843	−0.4813	1.2551	1.7570
0.001	−0.4488	−0.3162	2.0837	−0.4861	1.2544	1.7563
C–Q α_j	−0.1006	0.0048	2.2487	−0.1821	1.4260	1.9478
α_{jj}	−0.3086	−0.3014	−0.1874	−0.2737	−0.1924	−0.2170
COLA α_1	−0.1180	0.0004	2.3456	−0.1954	1.4965	2.0520
α_2	−0.3309	−0.3166	−0.2619	−0.2906	−0.2421	−0.2957
α_3	0.6699	0.2356	−0.7578	0.5542	−0.6889	−0.7446

Table II.5.2 Application of Lachance–Traill model using $\alpha_{ij} = m_{ij,binary}$ at $C_i = 0.5$, see Table II.5.1

Iteration		$C_{i,est}$	α_{ij}	α_{ik}	[L–T, 0.5]	$C_{i,calc}$	$C_{i,chem}$
Specimen 5054							
1	Cr	0.3348	−0.2419	−0.1412	0.8430	0.2822	
	Fe	0.4689	2.1348	−0.3092	1.7146	0.8040	
	Ni	0.0006	1.3118	1.8164	2.2908	0.0014	
2	Cr	0.2822			0.8052	0.2696	
	Fe	0.8040			1.6332	0.7658	
	Ni	0.0014			2.8494	0.0017	
7	Cr	0.2745			0.8198	0.2745	0.2577
	Fe	0.7436			1.5855	0.7434	0.7250
	Ni	0.0016			2.7105	0.0016	0.0015
Specimen 5181							
1	Cr	0.2651			0.8738	0.2317	
	Fe	0.4971			1.5531	0.7721	
	Ni	0.0416			2.2505	0.0936	
2	Cr	0.2317			0.8000	0.2121	
	Fe	0.7721			1.4657	0.7286	
	Ni	0.0936			2.7060	0.1126	
7	Cr	0.2156			0.8132	0.2156	0.1988
	Fe	0.7096			1.4272	0.7095	0.6945
	Ni	0.1070			2.5715	0.1070	0.0996

Table II.5.3 Application of Lachance–Traill model using $\alpha_{ij} = m_{ij}$ calculated for specimen AVG, see Table II.4.2

Iteration		$C_{i,est}$	α_{ij}	α_{ik}	[L–T, AVG]	$C_{i,calc}$	$C_{i,chem}$
Specimen 5054							
1	Cr	0.3348	− 0.3348	− 0.0247	0.8430	0.2822	
	Fe	0.4689	2.0820	− 0.2186	1.6969	0.7957	
	Ni	0.0006	1.2654	1.7698	2.2535	0.0014	
2	Cr	0.2822			0.7336	0.2456	
	Fe	0.7957			1.5873	0.7443	
	Ni	0.0014			2.7653	0.0017	
7	Cr	0.2543			0.7598	0.2544	0.2577
	Fe	0.7174			1.5291	0.7170	0.7250
	Ni	0.0016			2.5915	0.0016	0.0015
Specimen 5181							
1	Cr	0.2651			0.8325	0.2207	
	Fe	0.4971			1.5428	0.7669	
	Ni	0.0416			2.2152	0.0922	
2	Cr	0.2207			0.7409	0.1964	
	Fe	0.7669			1.4394	0.7155	
	Ni	0.0922			2.6366	0.1097	
7	Cr	0.2026			0.7645	0.2027	0.1988
	Fe	0.6958			1.3992	0.6956	0.6945
	Ni	0.1035			2.4879	0.1035	0.0996

the α values used are equated to the m_{ij} values computed for specimen AVG, a hypothetical material representative of the seven real samples 5054–5364; hence these values also do not change during the iteration process.

Clearly, the results in Table II.5.2 are semi-quantitative at best, and usually can be used to provide only a first estimate of concentration if required for subsequent treatment of data by a fundamental parameters approach, as suggested by Criss *et al.* (1978).

II.5.5   THE MODIFIED CLAISSE–QUINTIN ALGORITHM

Briefly stated, the innovation of Claisse–Quintin was proposing a model in which α_{ij} influence coefficients vary as a function of specimen composition, thereby extending the range of application. This is quite evident when α_{ij} is defined as in Eqn (II.5.3), wherein the coefficient α_{ij} varies as a linear function of C_M (C_j in the original Claisse–Quintin model), and also as a function of the sum of the products $\alpha_{ijk}C_k$. The latter term is introduced to compensate in part for the so-called 'third element effect'. The substitution of C_M for C_j results in negligible α_{ijk} values in the absence of enhancement, and much smaller values in the presence of enhancement than is the case if C_j is used; thus

$$\alpha_{ij} = \alpha_j + \alpha_{jj}C_M + \sum_k \alpha_{ijk}C_k \qquad (II.5.8)$$

is preferable to

$$\alpha_{ij} - \alpha_j + \alpha_{jj}C_j \mid \sum_k \alpha_{ijk}C_k \qquad (II.5.9)$$

Tables II.3.7–II.3.9 provide the source data for the computation of α influence coefficients for analytical context AC 1. Data in Table II.3.8 will be used for the computation of α influence coefficients for analyte Fe in the Cr–Fe–Ni system where j = Cr and k = Ni. For example, the following data are used to calculate α_{Fe} and α_{FeFe} for the Claisse–Quintin algorithm in Table II.5.1.

No.	C_{Fe}	$P_{Fe} + S_{Fe}$	A_{FeCr}
Fe1	1.0	1742.48	
Fe9	0.8	966.54	2137.4
Fe41	0.2	130.08	273.02

(a) calculation of $m_{ij,bin}$ coefficients from fundamental entities A_{FeCr} and $P_{Fe} + S_{Fe}$:

$m_{FeCr,bin(C_{Fe} = 0.8)} = 2137.4/966.54 = 2.2114$

$m_{FeCr,bin(C_{Fe} = 0.2)} = 273.02/130.08 = 2.0989$

(b) calculation of α_{Cr} and α_{CrCr}:

$2.2114 = \alpha_{Cr} + 0.2\alpha_{CrCr}$

$2.0989 = \alpha_{Cr} + 0.8\alpha_{CrCr}$

solving for α_{Cr} and α_{CrCr} yields

$\alpha_{Cr} = 2.2489 \ (2.2487)$

$\alpha_{CrCr} = -0.1875 \ (-0.1874)$

where the values in parentheses (from Table II.5.1) were obtained with no rounding off of the original data. Similar computation using Fe11 and Fe43 data yields

$\alpha_{Ni} = -0.1821$

$\alpha_{NiNi} = -0.2737$

Table II.5.1 lists the above α_j and α_{jj} values as part of a more comprehensive listing of binary m_{ij}, α_j and α_{jj} coefficients for the Cr–Fe–Ni system.

In addition to Fe1, the following data from Table II.3.8 are used for the computation of α_{FeCrNi} in Table II.5.4:

No.	C_{Fe}	C_{Cr}	C_{Ni}	$P_{Fe} + S_{Fe}$	A_{FeCr}	A_{FeNi}	E_{FeNi}
Fe28	0.30	0.70	0.00	211.13	445.17		
Fe34	0.30	0.35	0.35	332.50			
Fe40	0.30	0.00	0.70	703.68	1121.65	27.81	286.29

(c) calculation of $m_{ij,bin}$ from fundamental entities:

$$m_{FeCr,bin} = 445.17/211.13 = 2.1085$$
$$m_{FeNi,bin} = (27.81 - 286.29)/703.68 = -0.3673$$

(d) calculation of the correction term M_i using $m_{ij,bin}$ coefficients:

$$M = [1 + (2.1085 \times 0.35) + (-0.3673 \times 0.35)]$$
$$[1.6094]$$

Table II.5.4 Data relating to third element effects for system Cr–Fe–Ni; analytical context AC 1; analyte is Fe[a]

	C_{Fe}	C_{Cr}	C_{Ni}	[FP]	[hyp]	$1 + \varepsilon_{Fe}$	Δ	α_{FeCrNi}
Fe28	0.30	0.70	0.00	2.4760	2.4760			
29	0.30	0.67	0.03	2.3959	2.4017	1.0024	-0.0058	-0.290
Fe30	0.30	0.64	0.06	2.3162	2.3274	1.0048	-0.0112	-0.292
31	0.30	0.60	0.10	2.2107	2.2284	1.0080	-0.0176	-0.294
32	0.30	0.50	0.20	1.9509	1.9808	1.0153	-0.0299	-0.299
33	0.30	0.40	0.30	1.6969	1.7332	1.0214	-0.0363	-0.302
34	0.30	0.35	0.35	1.5722	1.6094	1.0237	-0.0372	-0.304
35	0.30	0.30	0.40	1.4490	1.4856	1.0253	-0.0366	-0.305
36	0.30	0.20	0.50	1.2075	1.2381	1.0253	-0.0306	-0.306
37	0.30	0.10	0.60	0.9723	0.9905	1.0187	-0.0182	-0.303
38	0.30	0.06	0.64	0.8799	0.8914	1.0131	-0.0115	-0.300
39	0.30	0.03	0.67	0.8112	0.8172	1.0074	-0.0060	-0.298
Fe40	0.30	0.00	0.70	0.7429	0.7429			

[a] [FP] = C_{Fe}/R_{Fe}; R_{Fe} is calculated from fundamental expression;

[hyp] = $1 + \sum \alpha_{ij,hyp}C_j$;

$1 + \varepsilon_{Fe}$ = [hyp]/[FP];
Δ = [FP] − [hyp];
α_{FeCrNi} = $\Delta/C_{Cr}C_{Ni}$.

whereas the theoretical value of the correction term [FP] for Fe34 from fundamental parameters is given by C_i/R_i; i.e.

$$[FP] = \frac{0.3}{(332.50/1742.48)}$$

$$= [1.5722]$$

(e) the coefficient α_{FeCrNi} for Fe34 is given by Eqns (5.55) and (5.56)

$$\alpha_{FeCrNi} = ([1.5722] - [1.6094])/(0.35 \times 0.35)$$

$$= -0.304$$

Similar calculations for the equivalent data for analyte Cr (Table II.3.7) yield $\alpha_{CrFeNi} = 0.371$, listed for Cr34 in Table II.5.5.

The fact that the values of α_{ij} and α_{ik} 'self-adjust' to specimen composition during the iteration process in the Claisse–Quintin model is shown in Table II.5.6.

Table II.5.5 Data relating to third element effects for system Cr–Fe–Ni; analytical context AC 1; analyte is Cr[a]

	C_{Cr}	C_{Fe}	C_{Ni}	[FP]	[hyp]	$1 + \varepsilon_{Cr}$	Δ	α_{CrFeNi}
Cr28	0.30	0.70	0.00	0.7843	0.7843			
29	0.30	0.67	0.03	0.7931	0.7874	0.993	0.0057	0.284
Cr30	0.30	0.64	0.06	0.8016	0.7906	0.986	0.0112	0.286
31	0.30	0.60	0.10	0.8126	0.7948	0.978	0.0176	0.297
32	0.30	0.50	0.20	0.8377	0.8054	0.961	0.0323	0.323
33	0.30	0.40	0.30	0.8583	0.8159	0.951	0.0424	0.353
34	0.30	0.35	0.35	0.8665	0.8211	0.948	0.0454	0.371
35	0.30	0.30	0.40	0.8729	0.8264	0.947	0.0465	0.387
36	0.30	0.20	0.50	0.8796	0.8370	0.952	0.0426	0.426
37	0.30	0.10	0.60	0.8758	0.8476	0.968	0.0282	0.470
38	0.30	0.06	0.64	0.8706	0.8518	0.978	0.0188	0.490
39	0.30	0.03	0.67	0.8651	0.8550	0.988	0.0101	0.502
Cr40	0.30	0.00	0.70	0.8581	0.8581			

[a] $[FP] = C_i/R_i$; R_i is calculated from fundamental expression;

$[hyp] = 1 + \sum \alpha_{ij,hyp} C_j$;

$1 + \varepsilon_i = [hyp]/[FP]$;
$\Delta = [FP] - [hyp]$;
$\alpha_{CrFeNi} = \Delta/C_{Fe}C_{Ni}$.

Table II.5.6 Application of Claisse–Quintin model. α_j and α_{jj} coefficients from Table II.5.1; α_{ijk} coefficients for Cr34 and Fe34 from Tables II.5.4 and II.5.5

Iteration		$C_{i,est}$	α_{ij}	α_{ik}	[C–Q]	$C_{i,calc}$	$C_{i,chem}$
Specimen 5054							
1	Cr	0.3348	−0.2455	−0.1463	0.8849	0.2963	
	Fe	0.4689	2.1858	−0.2739	1.7316	0.8119	
	Ni	0.0006	1.2714	1.7734	2.2572	0.0014	
2	Cr	0.2963	−0.3516	−0.2499	0.7146	0.2392	
	Fe	0.8119	2.1929	−0.2636	1.6492	0.7733	
	Ni	0.0014	1.2128	1.7073	2.7456	0.0016	
7	Cr	0.2542	−0.3273	−0.2262	0.7601	0.2545	0.2577
	Fe	0.7330	2.2008	−0.2521	1.5590	0.7310	0.7250
	Ni	0.0016	1.2360	1.7336	2.5850	0.0016	0.0015
Specimen 5181							
1	Cr	0.2651	−0.2668	−0.1672	0.8680	0.2301	
	Fe	0.4971	2.1912	−0.2662	1.5665	0.7787	
	Ni	0.0416	1.2794	1.7824	2.2252	0.0926	
2	Cr	0.2301	−0.3695	−0.2674	0.7142	0.1893	
	Fe	0.7787	2.1882	−0.2704	1.4720	0.7318	
	Ni	0.0926	1.2319	1.7289	2.6298	0.1094	
7	Cr	0.2006	−0.3486	−0.2470	0.7572	0.2007	0.1988
	Fe	0.7003	2.1918	−0.2653	1.4060	0.6989	0.6945
	Ni	0.1032	1.2527	1.7523	2.4784	0.1031	0.0996

II.5.6; II.5.7.1 THE COLA ALGORITHM

Basically, the COLA algorithm retains the concept of cross-coefficients proposed by Claisse and Quintin, but uses a hyperbolic function to approximate $m_{ij,bin}$ arrays instead of the linear function in the Claisse–Quintin algorithm; thus

$$\alpha_{ij} = \alpha_{ij,hyp} + \sum \alpha_{ijk}C_k \qquad (II.5.10)$$

wherein data for three binaries are required to calculate $(\alpha_1, \alpha_2 \text{ and } \alpha_3)_{ij}$ in order to compute $\alpha_{ij,hyp}$, for example in the case of analyte Fe in the analytical context AC 1, Table II.3.8:

No.	C_{Fe}	$P_{Fe} + S_{Fe}$	A_{FeCr}	I_{Fe} (nominal c/s)
Fe2	0.999	1736.67	4073.46	13 995.2
Fe19	0.5	421.42	899.65	3385.9
Fe63	0.001	0.5654	1.1781	4.5

A glance at the nominal iron intensities (calculated using 14 000 c/s for the pure analyte) clearly shows the difficulties that one encounters in calculating

$\alpha_{ij,hyp}$ using experimental data. Firstly, there is the problem of preparing the two end-member alloys, namely binaries containing exactly 0.1% of the matrix element and 0.1% of the analyte. Secondly, one is dealing with ratios of two similarly large numbers and with relatively low intensities in the case of $C_i = 0.001$; e.g. 320 c/s even if the emission intensity of the pure analyte emitted is 1.0×10^6 c/s. It is therefore much more reliable to deal with theoretical data, as follows (value in parentheses was obtained with no rounding off of the original data)

(a) $m_{ij,bin}$ coefficients are given by:

$$m_{FeCr,bin(C_{Fe} = 0.999)} = 4073.46/1736.67 = 2.3456$$

$$m_{FeCr,bin(C_{Fe} = 0.5)} = 899.65/421.42 = 2.1348$$

$$m_{FeCr,bin(C_{Fe} = 0.001)} = 1.1781/0.5654 = 2.0837$$

(b) α_1, α_2 and α_3 are obtained from:

$$\alpha_1 = m_{FeCr,bin(C_{Fe} = 0.999)} = 2.3456$$

$$\alpha_2 = m_{FeCr,bin(C_{Fe} = 0.001)} - \alpha_1$$
$$= 2.0837 - 2.3456$$
$$= -0.2619$$

$$\alpha_3 = \frac{\alpha_2}{m_{FeCr,bin(C_{Fe} = 0.5)} - \alpha_1} - 2$$

$$= \frac{-0.2619}{2.1348 - 2.3456} - 2$$

$$= -0.7576 \, (-0.7578)$$

Table II.5.7 Comparison of $m_{ij,binary}$ and $\alpha_{ij,hyperbolic}$ influence coefficient arrays; analytical context AC 1; binary system Cr–Fe

C_i	$m_{CrFe,bin}$	$\alpha_{CrFe,hyp}$	$m_{FeCr,bin}$	$\alpha_{FeCr,hyp}$
0.999	−0.1180	−0.1182	2.3456	2.3445
0.99	−0.1202	−0.1199	2.3355	2.3351
0.90	−0.1395	−0.1386	2.2615	2.2632
0.80	−0.1633	−0.1610	2.2112	2.2126
0.70	−0.1868	−0.1855	2.1776	2.1783
0.60	−0.2132	−0.2124	2.1532	2.1535
0.50	−0.2419	−0.2419	2.1348	2.1348
0.40	−0.2733	−0.2745	2.1203	2.1201
0.30	−0.3082	−0.3108	2.1085	2.1084
0.20	−0.3475	−0.3514	2.0988	2.0987
0.10	−0.3932	−0.3970	2.0906	2.0906
0.01	−0.4431	−0.4433	2.0843	2.0844
0.001	−0.4488	−0.4483	2.0837	2.0838

Examples of the close agreement between $m_{FeCr,bin}$ and $\alpha_{FeCr,hyp}$ and between $m_{CrFe,bin}$ and $\alpha_{CrFe,hyp}$ ($m_{NiCr,bin}$ and $\alpha_{NiCr,hyp}$ are negligible) for analytical context AC 1 are given in Table II.5.7.

The calculation of the α_{ijk} coefficients, Eqns (5.55) and (5.56), is identical to that used for the Claisse–Quintin algorithm (Eqn II.5.3) except that it is

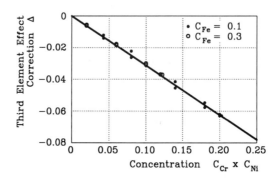

Figure II.5.1 Plot of $\Delta = [FP] - [hyp]$ versus the product $C_{Cr}C_{Ni}$ for the system Cr–Fe–Ni in the context of $C_{Fe} = 0.1$ and 0.3; the slope is equal to α_{FeCrNi}

Table II.5.8 Illustration of the iterative process for the COLA model; α_1, α_2 and α_3 from Table II.5.1; α_{ijk} for Cr34 and Fe34 from Tables II.5.4 and II.5.5

Iteration		$C_{i,est}$	α_{ij}	α_{ik}	[COLA]	$C_{i,cor}$	C_{chem}
Specimen 5054							
1	Cr	0.3348	−0.2326	−0.1317	0.8909	0.2983	
	Fe	0.4689	2.1686	−0.2666	1.7258	0.8092	
	Ni	0.0006	1.2715	1.7737	2.2574	0.0014	
2	Cr	0.2983	−0.3560	−0.2453	0.7121	0.2384	
	Fe	0.8092	2.1784	−0.2581	1.6493	0.7734	
	Ni	0.0014	1.2469	1.7488	2.7871	0.0017	
7	Cr	0.2549	−0.3244	−0.2185	0.7623	0.2552	0.2577
	Fe	0.7331	2.1917	−0.2482	1.5581	0.7306	0.7250
	Ni	0.0016	1.2553	1.7572	2.6081	0.0016	0.0015
Specimen 5181							
1	Cr	0.2651	−0.2542	−0.1534	0.8749	0.2319	
	Fe	0.4971	2.1764	−0.2598	1.5628	0.7769	
	Ni	0.0416	1.2758	1.7781	2.2221	0.0924	
2	Cr	0.2319	−0.3825	−0.2666	0.7048	0.1868	
	Fe	0.7769	2.1715	−0.2640	1.4727	0.7321	
	Ni	0.0924	1.2537	1.7556	2.6547	0.1104	
7	Cr	0.2000	−0.3526	−0.2425	0.7552	0.2002	0.1988
	Fe	0.8991	2.1772	−0.2591	1.4023	0.6971	0.6945
	Ni	0.1036	1.2626	1.7645	2.4861	0.1034	0.0996

Table II.5.9 Comparison of results obtained using various influence coefficient models

Specimen		Chem	[L–T, 0.5]	[L–T, AVG]	[C–Q]	[COLA]
				Concentrations		
5054	Cr	0.2577	0.2745	0.2544	0.2545	0.2552
	Fe	0.7250	0.7434	0.7170	0.7310	0.7306
	Ni	0.0015	0.0016	0.0016	0.0016	0.0016
5074	Cr	0.2525	0.2680	0.2510	0.2516	0.2523
	Fe	0.6838	0.7018	0.6817	0.6918	0.6908
	Ni	0.0498	0.0583	0.0513	0.0512	0.0515
5181	Cr	0.1988	0.2156	0.2027	0.2007	0.2002
	Fe	0.6945	0.7095	0.6956	0.6989	0.6971
	Ni	0.0996	0.1070	0.1035	0.1031	0.1034
5202	Cr	0.2130	0.2287	0.2180	0.2170	0.2168
	Fe	0.6303	0.6448	0.6362	0.6357	0.6342
	Ni	0.1480	0.1587	0.1542	0.1538	0.1540
5321	Cr	0.1988	0.2127	0.2050	0.2033	0.2030
	Fe	0.5919	0.6030	0.5998	0.5938	0.5926
	Ni	0.2002	0.2132	0.2085	0.2078	0.2076
5324	Cr	0.2696	0.2787	0.2700	0.2721	0.2734
	Fe	0.5280	0.5418	0.5368	0.5346	0.5341
	Ni	0.1927	0.1929	0.1882	0.1884	0.1885
5364	Cr	0.2784	0.2846	0.2794	0.2802	0.2816
	Fe	0.4721	0.4864	0.4856	0.4772	0.4771
	Ni	0.2357	0.2516	0.2468	0.2465	0.2462

recommended that a larger suite of data be taken into account, as shown in Tables II.5.4 and II.5.5. The slope of a linear least squares fit through the origin of Δ (Δ = [FP] – [hyp]) plotted versus the product $C_j C_k$ is equal to α_{ijk}. Figure II.5.1 illustrates how closely the single value α_{FeCrNi} calculated for $C_{Fe} = 0.3$, $C_{Cr} = C_{Ni} = 0.35$ approximates α_{FeCrNi} values for the contexts $C_{Fe} = 0.1$ and 0.3.

Table II.5.8 provides an example of the 'self-adjustment' of α_{ij} coefficients to specimen composition during the iteration process using the COLA algorithm.

Table II.5.9 provides a comparison of the results obtained for the first seven specimens in Table II.3.2. In all cases, the first estimates of concentrations are the experimental relative intensities from Rasberry and Heinrich (1974) processed as in Tables II.5.2, II.5.3, II.5.6 and II.5.8, respectively.

II.5.6, II.5.7.2   THE TERTIAN ALGORITHM

Tertian (1988) proposed the following algorithm:

$$C_i = \frac{R_i}{1 + \varepsilon_i}\left[1 + \sum_j \alpha_{ij,\text{hyp}}C_j\right]$$ (II.5.11)

where $\alpha_{ij,\text{hyp}}$ is defined by Eqn (II.5.6) and

$$1 + \varepsilon_i = \frac{R_i}{C_M}\sum_j \left(\frac{C_j}{R_{ij}}\right)_{\text{bin}}$$ (II.5.12)

or

$$1 + \varepsilon_i = \left[1 + \sum m_{ij,\text{bin}}C_j\right]/[\text{FP}]$$ (II.5.13)

The three $m_{ij,\text{bin}}$ values previously calculated in the context of the COLA algorithm are used to calculate γ_1, γ_2 and γ_3, namely

$$\gamma_1 = 2.0837$$
$$\gamma_2 = 2.3456 - 2.0837$$
$$= 0.2619$$
$$\gamma_3 = \frac{(2.0837 + 2.3456) - (2 \times 2.1348)}{2.1348 - 2.0837}$$
$$= 3.1254$$

and data for Fe28, Fe34 and Fe40 in Table II.3.8 are used to calculate $1 + \varepsilon_{\text{Fe}}$ for specimen Fe34 using Eqn (II.5.12):

$$R_{\text{Fe}} = 332.50/1742.48 = 0.1908$$
$$C_M = 0.35 + 0.35 = 0.7$$
$$1 + \varepsilon_{\text{Fe}} = \frac{0.1908}{0.7} \times \left(\frac{0.35}{(211.13/1742.48)} + \frac{0.35}{(703.68/1742.48)}\right)$$
$$= 0.27257 \times \left(\frac{0.35}{0.1212} + \frac{0.35}{0.4038}\right)$$
$$= 1.0237 \text{ (see Table II.5.4)}$$

where the relative intensity R_{FeCr} is obtained directly from fundamental data. This value may also be found from the knowledge of γ coefficients, namely

$$R_{\text{FeCr}} = \frac{C_{\text{Fe}}}{1 + \alpha_{\text{FeCr,hyp}}C_{\text{Cr}}}$$ (II.5.14)

where

$$\alpha_{\text{FeCr,hyp}} = 2.0837 + \frac{(0.2619 \times 0.3)}{1 + (3.1252 \times 0.7)}$$

$$= 2.1083$$

Thus

$$R_{\text{FeCr}} = \frac{0.3}{1 + (2.1083 \times 0.7)}$$

$$= 0.1212$$

The term $1 + \varepsilon_i$ can also be computed from Eqn (II.5.13):

$$1 + \varepsilon_{\text{Fe}} = 1.6094/1.5722$$

$$= 1.0237$$

Figure II.5.2 illustrates the variation of $1 + \varepsilon_{\text{Fe}}$ values as Cr replaces Ni in the system Fe–Cr–Ni.

The versatility of a practical procedure based on Eqn (II.5.11) is outlined by Tertian (1988), who distinguishes two cases:

(a) the unknown is categorized and can be related to some definite reference material, in which case the values of $\alpha_{ij,\text{hyp}}$ and $1 + \varepsilon_i$ are little different for unknown and reference, thus yielding excellent compositional data for unknowns;

(b) the unknown is not categorized, in which case a reference specimen is selected with the only condition that it contain all elements of interest in sufficient proportion. The first step is identical to (a), except that the concentration

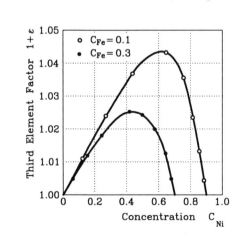

Figure II.5.2 Plot of $1 + \varepsilon_{\text{Fe}}$ versus C_{Ni} as Cr replaces Ni for the system Cr–Fe–Ni in the context of $C_{\text{Fe}} = 0.1$ and 0.3

values obtained are considered as first estimates and are used in a second step to calculate $\alpha_{ij,hyp}$ and $1 + \varepsilon_i$ corresponding to the specimen at hand. The process is simple, given that for a spectrometer operating under fixed conditions the values γ_1, γ_2 and γ_3 are constants and can be used to calculate R_{ij} values at any concentration C_i, and hence to compute $1 + \varepsilon_i$ values for any given specimen.

In summary, the application of Eqn (II.5.11) in the first case is most useful for the analysis of type materials and offers a rapid method of calculation. In the second case, it is theoretically equivalent to fundamental parameters calculations, thus offering a more general or 'universal' scope of applications.

II.6 EMPIRICAL INFLUENCE COEFFICIENTS

The basis of empirical influence coefficient methods rests on the statistical treatment of concentration and intensity data. It therefore follows that the choice of any one model is somewhat subjective in practice. It cannot be otherwise, given that the choice depends on how well a purely statistical treatment can be representative of a given analytical context; i.e. establishing a definite correlation between concentrations and characteristic line intensities. More specifically, empirical influence coefficient methods are the next simplest step when the scatter observed in a plot of concentration versus intensity is such that the plot cannot be represented by a linear or second order equation: hence the attempt to relate the magnitude of the deviations, either above or below a linear fit, to the abundance of the other elements present in the specimen. It is therefore understandable that many variants have been proposed in order to find the 'best' solution.

II.6.3 THE LACHANCE–TRAILL MODEL: Eqn (6.7)

$$C_i = R_i \left[1 + \sum_j r_{ij} C_j \right]$$
(II.6.1)

Budesinsky (1979) used a multivariate least squares method to compute empirical influence coefficients for the correction of matrix effects in the context of alloys (Cr–Fe–Ni) and pressed powder specimens. Copper mattes were analysed for Fe, Cu and S; copper concentrates for Al, Si, S, Ca, Fe and Cu; and slags for Al, Si, Ca, Fe and Cu. The multivariate least squares method was also used (Budesinsky, 1980) in the context of systems Cu–Fe–Zn and Cu–Fe–Zn–Pb for specimens in 3.5% aqueous nitric acid. Klimasara (1992) proposed a method that automatically selects the closest standards to compute empirical alpha coefficients from experimental data in the context of stainless

Table II.6.1 Experimental data used to demonstrate application of empirical regression methods[a]

Specimen	Concentration (%)				Intensity (c/s)		
	Cr	Fe	Ni	balance	Cr	Fe	Ni
x101	24.78	61.77	10.60	2.85	3391	4699	549
x102	25.13	54.86	17.57	2.44	3371	4288	933
x103	25.39	58.75	13.07	2.79	3388	4561	691
x104	24.46	52.51	20.64	2.39	3174	4218	1138
x107	18.60	53.08	25.10	3.22	2593	4633	1438
x108	16.40	44.01	36.63	4.96	2488	4144	2168
x109	15.75	41.17	39.60	3.48	2129	4203	2362
x112	15.25	43.96	37.66	3.13	2306	4130	2276
x115	13.50	24.30	58.48	3.72	2084	2705	4122
x119	12.95	9.10	76.70	1.25	2253	914	6188
x120	14.45	7.89	77.11	0.55	2022	869	6422
x121	19.75	1.57	75.77	2.91	2609	88	6443
InvI		64.13	35.87	0.0		7576	2037
InvII		63.83	35.51	0.66		7651	2010
Cr	100.0				10000		
Fe		100.0				10000	
Ni			100.0				10000

[a] Data from Noakes (1954).

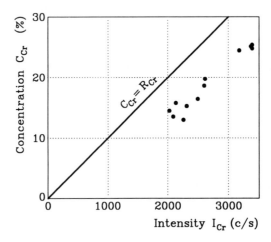

Figure II.6.1 Plot of C_{Cr}(%) versus I_{Cr} for the system Cr–Fe–Ni illustrating net enhancement effects; data from Table II.6.1

steels and ceramic tiles. The software is adaptable to commercially available software packages; it automatically recognizes the resolution of the CRT display and selects the best possible display for the available equipment. A variant of the Lachance–Traill model proposed by Zimmerman and Ingles (1972) consisted of adding a term for 'self-absorption', namely

$$C_i = K_i I_i [1 + r_{ii} C_i + R_{ij} C_j + \cdots]$$ (II.6.2)

in the context of the determination of Fe, Co, Ni and Cu in aqueous solutions.

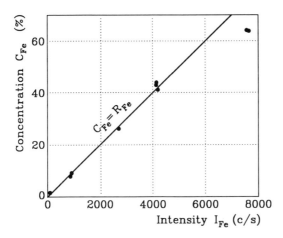

Figure II.6.2 Plot of $C_{Fe}(\%)$ versus I_{Fe} for the system Cr–Fe–Ni illustrating net absorption and net enhancement effects; data from Table II.6.1

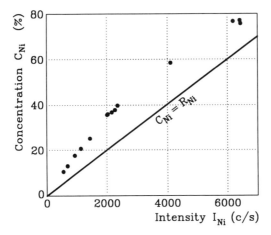

Figure II.6.3 Plot of $C_{Ni}(\%)$ versus I_{Ni} for the system Cr–Fe–Ni illustrating absorption effects; data from Table II.6.1

II.6.3.1 Numerical example

The intensity data in Table II.6.1, taken from Noakes (1954), will be used to illustrate an application of Eqn (6.7) in the context of Cr–Fe–Ni alloys. A plot of C versus I clearly shows that neither a linear nor a second order regression is possible in the case of Cr and Fe (Figures II.6.1 and II.6.2), respectively. In the case of Ni (Figure II.6.3) a second order regression could be considered, as shown in Table II.6.2. Also shown in Table II.6.2 are results obtained using empirical coefficients obtained assuming an equivalent wavelength (c) and by regression (d). The results show that this is a case of overextending the range of concentrations if constant values are used for the influence coefficients.

Table II.6.2 Comparison of results obtained by regression methods[a]

Specimen		C_{Cr}	Δ	C_{Fe}	Δ	C_{Ni}	Δ
x101	(a)	24.78		61.77		10.60	
	(b)					10.83	0.23
	(c)	25.06	0.28	62.07	0.30	10.77	0.17
	(d)	25.16	0.38	61.99	0.22	10.72	0.12
x102	(a)	25.13		54.86		17.57	
	(b)					17.47	−0.10
	(c)	24.83	−0.30	54.75	−0.11	18.14	0.57
	(d)	24.99	−0.14	55.14	0.39	17.44	−0.13
x103	(a)	25.39		58.75		13.07	
	(b)					13.32	0.25
	(c)	25.06	−0.33	59.11	0.36	13.35	0.28
	(d)	25.10	−0.29	59.59	0.84	13.29	0.22
x104		24.46		52.51		20.64	(a)
	(b)					20.88	0.24
	(c)	23.55	−0.91	52.32	−0.19	20.89	0.25
	(d)	23.42	−1.04	52.62	0.11	20.78	0.14
x107	(a)	18.60		53.08		25.10	
	(b)					25.70	0.60
	(c)	18.73	0.13	53.25	0.17	25.95	0.85
	(d)	18.57	−0.03	53.43	0.35	25.84	0.74
x108	(a)	16.40		44.01		36.63	
	(b)					36.60	−0.03
	(c)	17.88	1.48	45.09	1.08	36.81	0.18
	(d)	17.56	1.16	45.12	1.11	36.60	−0.03
x109	(a)	15.75		41.17		39.60	
	(b)					39.30	−0.30
	(c)	15.25	−0.50	43.72	2.55	39.16	−0.44
	(d)	14.94	−0.81	43.70	2.53	38.97	−0.63

Table II.6.2 (*continued*)

Specimen		C_{Cr}	Δ	C_{Fe}	Δ	C_{Ni}	Δ
x112	(a)	15.25		43.96		37.66	
	(b)					38.11	0.45
	(c)	16.57	1.32	43.94	−0.02	38.07	0.41
	(d)	16.26	1.01	43.95	−0.01	37.86	0.20
x115	(a)	13.50		24.30		58.48	
	(b)					59.99	1.51
	(c)	14.92	1.42	25.41	1.11	59.22	0.74
	(d)	14.44	0.94	25.47	1.17	58.73	0.25
x119	(a)	12.95		9.10		76.70	
	(b)					75.54	1.16
	(c)	16.49	3.54	8.09	−1.01	76.60	−0.10
	(d)	15.59	2.64	8.05	−1.05	75.65	−1.05
x120	(a)	14.45		7.89		77.11	
	(b)					76.70	−0.41
	(c)	14.76	0.31	7.43	−0.46	77.74	0.63
	(d)	13.92	−0.53	7.38	−0.51	77.14	0.03
x121	(a)	19.75		1.57		75.77	
	(b)					76.80	1.03
	(c)	19.86	0.11	0.85	−0.72	75.33	−0.44
	(d)	18.80	−0.95	0.82	−0.75	74.79	−0.98
InvI	(a)			64.13		35.87	
	(b)					34.73	−1.14
	(c)			63.60	−0.53	35.90	0.03
	(d)			63.24	−0.89	35.97	0.10
InvII	(a)			63.93		35.51	
	(b)					34.34	−1.17
	(c)			64.20	0.27	35.60	0.09
	(d)			63.97	0.04	35.68	0.17

[a] (a) Chemical values;
(b) $C_{Ni} = 0.7728 + (1.8923 \times 10^{-2} \times I_{Ni}) + (1.1054 \times 10^{-6} \times I_{Ni}^2)$;
(c) data from Traill and Lachance (1965);
(d) data from Lachance and Traill (1966).

II.6.4 THE RASBERRY–HEINRICH MODEL: Eqn (6.11)

$$C_i = R_i \left[1 + \sum_j r_{ij} C_j + \sum_k \frac{r_{ik}}{1 + C_i} C_k \right] \qquad \text{(II.6.3)}$$

The results obtained by Rasberry and Heinrich (1974) in the context of the Cr–Fe–Ni system are listed in Table II.6.3. Intensity measurements from 31 reference materials were used to generate the r_{ij} (absorption) and r_{ik}

Table II.6.3 Application of the Rasberry–Heinrich model. Data from Rasberry and Heinrich (1974)[a]

Specimen	C_{Cr}	C_{Fe}	C_{Ni}
5040 (a)	0.2577	0.7250	0.0015
(b)	0.2522	0.7229	0.0015
5202 (a)	0.2130	0.6303	0.1480
(b)	0.2090	0.6277	0.1507
5364 (a)	0.2784	0.4721	0.2357
(b)	0.2675	0.4750	0.2409
3987 (a)	0.0000	0.3431	0.6552
(b)	0.0000	0.3378	0.6512
1188 (a)	0.1540	0.0660	0.7265
(b)	0.1421	0.0655	0.7236

[a] (a) Chemical value;
(b) by X-ray using empirical influence coefficients.

(enhancement) influence coefficients, which then served to analyse five reference materials treated as unknowns. The results showed better agreement with the chemical values than did the application of the Lachance–Traill model to the same data. Budesinsky (1979, 1980) treated the same experimental data (see Section II.6.3) using the Rasberry–Heinrich model and concluded that both models yielded similar acceptable results when dealing with three element systems. However when the number of elements was four or more, in the case of the Rasberry–Heinrich model, the whole system of equations became gradually distuned and led to unacceptable results. Ji *et al.* (1980) applied Eqn (6.11) for the correction of matrix effects for the determination of Sr, Nb, Ba and Ce in cerium-doped strontium–barium niobate single crystals. The specimens were prepared as fused discs with borax as the fluxing agent, with the addition of SiO_2 to help the glass formation. Excellent agreement was obtained on comparing X-ray results to chemical values for eight samples. Schroeder *et al.* (1980) applied Eqn (II.6.3) in the context of geological materials; in the form of fused discs for the determination of 10 major constituents, and in the form of pressed pellets in the case of 12 trace elements. Enhancement effects were assumed to be negligible; hence only absorption influence coefficients were computed by regression of experimental data from 23 igneous rocks.

Two variants for the Rasberry–Heinrich model that have been proposed in order to make it more versatile are:

$$C_i = R_i \left[1 + \sum_j \left(r_{ija} + \frac{r_{ije}}{1 + C_i} \right) C_j \right]$$ (II.6.4)

envisaged by Rasberry and Heinrich (1974), and

$$C_i = R_i \left[1 + \sum_j \left(r_{ij_1} + \frac{r_{ij_2}}{r_{ij_3} + C_i} \right) C_j \right] \qquad (II.6.5)$$

proposed by Mainardi *et al.* (1981). The relationship was derived for a monochromatic source and the influence coefficients were linked with fundamental parameters. The analyst is left with the alternative of evaluating the coefficients either numerically or empirically for polychromatic excitation.

II.6.5 THE LUCAS-TOOTH AND PRICE MODEL: Eqn (6.14)

Gillieson *et al.* (1965) applied Eqn (6.14) for the determination of Ti, Cr, Mn, Fe, Co, Ni, Zr, Nb, Mo and W in high-temperature alloys. Intensity measurements from 25 alloys were used to compute the empirical influence coefficients, which were then used to:

● illustrate the 'goodness of fit' by treating these intensity data as unknown;
● illustrate the application using 13 different reference materials as unknowns.

The model was modified by adding an exponent to the intensity in the denominator of the least squares difference expression to achieve better precision in the large percentages by reducing unnecessary precision in the very small percentages. Two sets of calculations were made, one with background correction and the second without. The results of the latter proved better, and indicated that the time consuming measurement of backgrounds was of doubtful practical value. It was observed that one condition stood out: *results cannot be extrapolated by this method.* The wide range of concentrations in the 13 unknowns; e.g. Fe 1–66%; Co 0.4–52%; Ni 9–64%; Mo 1–24%; W 0.0–16%, resulted in errors of the order of 10% or more. The total of X-ray results was 92.6% in one sample and 108.6% in another; hence the authors' conclusion that it is preferable to limit the method to more restricted percentage ranges. Eich *et al.* (1967) applied Eqn (6.14) for the determination of Au, Ag, Cu, Pd, Pd and Zn in gold and and platinum group alloys, with excellent results when the specimens were prepared by polishing with 600 grit silicon carbide.

One variant of the Lucas-Tooth and Price model is expressed as

$$C_i = a + bI_i \left[1 + \sum_j r_{ij} I_j \right] \qquad (II.6.6)$$

II.6.6   THE LUCAS-TOOTH AND PYNE MODEL: Eqn (6.17)

Basically, the Lucas-Tooth and Pyne model is equivalent to Eqn (6.14), the only difference being that apparent concentrations are used instead of intensities, given that

$$C_{i,app} = a + bI_i \tag{II.6.7}$$

which is a normalizing step to avoid having to recalculate the influence coefficients in the case of instrumental drift. This modification was introduced to avoid having to deal with coefficients having unwieldy powers of ten in manual calculations, which is no problem in the case of computer data reduction.

II.6.8   DISCUSSION

In contrast to theoretical approaches, which can be regarded as 'general' in the sense that calculations made for a given analytical context are 'transferable', i.e. directly applicable to any other identical instrumentation, empirical approaches by their very nature are 'customized' to each specific application. This is all the more so in the case of intensity models in that the quantification of influence coefficients is *directly* linked to:

- the nature of the elements in the reference materials;
- their composition;
- the specific reference materials used;
- the instrumental operating conditions.

All of these factors have an important bearing on a successful application. Conversely, the application of intensity models does not involve an iterative process and can be limited to the determination of one or two elements, and the determination of each analyte is independent of any other. This is very much in contrast to concentration models.

Another aspect that can come into play in the application of empirical models is that of taking into consideration some basic principles of X-ray fluorescence theory. Thus, one can try regression on major constituents only, as they are likely to have the larger matrix effects. Alternatively, one can try regression on the strongest absorber(s) and/or enhancer(s) only in order to minimize the possibility of an 'accidental' correlation between the magnitude of the correction term and a trace constituent. Some of these criteria apply also in the case of concentration models.

Consider for example the semi-empirical method proposed by Mosichev *et al.* (1984), who set as their only final goal a mathematically adequate description of the analytical process and obtaining correct analytical results. This path was chosen because it was felt that semi-empirical methods are sufficiently simple for practical use while permitting a substantial reduction of the number of reference materials required for calibration. Based on the premise that it is natural to assume the magnitude of the enhancement effect to be directly proportional to the enhancing line intensity, the theoretical expression

$$C_i = K_i I_i \left[\frac{1 + \sum_j a_{ij\lambda} C_j}{1 + \sum_j e_{ij\lambda} C_j} \right] \qquad (II.6.8)$$

is transformed to

$$C_i = K_i I_i \left[1 + \sum_j r_{ija} C_j + \sum_j r_{ije} I_j \right] \qquad (II.6.9)$$

which, incidentally, is similar to Eqn (4.57), and where r_{ija} is an influence coefficient that accounts for absorption whereas r_{ije} accounts for enhancement. Conscious that influence coefficients are dependent on composition, the authors recommended combining theoretical data based on hypothetical reference materials with experimental intensity data from available reference materials in order to cover the compositional region where the variance of the quantities of interest is greatest. The theoretical verification of Eqn (II.6.9) was performed on the Cr–Fe–Ni, Cr–Fe–Co–Ni and Al–Ti–Cr–Mn–Fe –Co–Ni–Mo systems by generating theoretical intensities from fundamental parameters in the context of a polychromatic source. A comparison of the distributions of systematic deviations in Cr for the Cr–Fe–Ni system showed a marked improvement over the Lachance–Traill and Lucas-Tooth and Price models. Molchanova *et al.* (1992) examined the dependence of the influence coefficients on specimen composition in the context of steel samples, and proposed a model that is described as combining the ideology of the COLA algorithms, Eqns (5.48) and (5.49), and those of JIS Eqns (6.19) and (6.20). In contrast to the COLA approach, the variations of the binary influence coefficients are approximated by a quadratic function rather than a hyperbolic function.

II.7 GLOBAL MATRIX EFFECT CORRECTION METHODS

II.7.2 DOUBLE DILUTION METHOD: Eqn (7.20)

The attraction of the double dilution method is that, if a flux is chosen as the diluent and fused disc specimens are prepared, the method can be considered as 'universal' for the determination of major and minor constituents. Solid samples must be finely ground or easily converted to that form. The following outline of the special care that should be taken in practical applications, along with advantages of the method, are taken from Sections 14.4–14.7 in Tertian and Claisse (1982).

(a) Given that the intensities I_{i_1} and I_{i_2} are roughly in the same range and thus have a similar relative precision, $\phi_{i_1} = \phi_{i_2}$, the precision of the corrected intensity D_i is given by

$$D_i = \phi_{i_1} \frac{I_1 \times I_2}{I_1 - I_2} \tag{II.7.1}$$

and the relative error of D_i as a function of the intensity ratio (I_{i_1}/I_{i_2}) is illustrated in Figure II.7.1.

(b) The optimal concentration ratio (C_d/C_s) depends on the particular diluent; a value of four or five is convenient in the case of lithium tetraborate and a value of three with sodium tetraborate, the reason being that a_{sd} is higher for the lithium salt than for the sodium salt.

(c) Concentrations and dilutions must be expressed in terms of weight.

(d) The diluted specimens must be infinitely thick with respect to the fluorescent element.

(e) The loss on fusion or the gain on fusion, if present, need not be known; the result is the concentration C_i in the original sample.

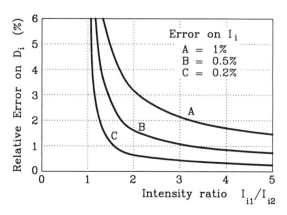

Figure II.7.1   Relative error of the corrected intensity D_i as a function of the intensity ratio (I_{i_1}/I_{i_2})

(f) Dead time is automatically compensated for, so that it is possible to work at high counting rates with no loss of accuracy.

(g) Greater care must be taken in weighing and intensity measurement.

In summary, the double dilution method offers a practical solution for those 'one of a kind' materials that tend to be submitted on a more or less regular basis in most laboratories.

II.7.3   COMPTON (INCOHERENT) SCATTER METHODS: Eqn (7.30)

The attraction of methods based on incoherent scatter is that they may be used on samples 'as submitted', or pelletized in the case of finely ground powdered samples. The measured intensity of incoherent scatter may be used directly to compensate for matrix effects or, alternatively, to quantify $\mu_s^\star$ (Willis, J. P., 1989) or background intensity at any appropriate background (Feather and Willis, 1976), which may then also be used to correct for matrix effects.

Harvey (1992) proposed an iterative approach to Compton scatter corrections for matrix absorption in the context of the determination of trace elements in geological materials. He started from the well known fact that the ratio of the mass absorption coefficient of a sample to that of a reference monitor remains essentially constant provided that the difference in composition between sample and monitor does not change significantly owing to crossing absorption edges. Harvey was concerned that, in most geological samples, absorption edge corrections for abnormally high trace element concentrations are not normally performed. The algorithm proposed is theoretically based and allows absorption edge corrections to be made for any element occurring in significant concentrations. In practice, processing is performed in wavelength order away from the measured Compton wavelength, with iteration over each absorption edge encountered. Briefly stated, the critical step in the correction process is the calculation of zonal ratios, illustrated in Figure II.7.2.

In outline, the method is based on the following relationships:

(a) the relation between $I_{\lambda_{inc}}$ and $\mu_{\lambda_{inc}}$ is modelled by the function

$$\mu_{\lambda_{inc}} = \exp(\gamma_0 - \gamma_1 \ln I_{inc}) \qquad (II.7.2)$$

(b) the Compton ratio for unknown to reference is given by

$$r_{inc} = (I_{inc,r}/I_{inc,u}) \qquad (II.7.3)$$

(c) the initial mass absorption ratio used and modified as necessary during processing is

$$r_0 = r_{inc}^{-\gamma_1} \qquad (II.7.4)$$

(d) the ratio r_i is defined by

$$r_i = \frac{\overset{*}{\mu}_{u\lambda_i}}{\overset{*}{\mu}_{r\lambda_i}} \qquad (II.7.5)$$

(e) r_j is defined in terms of r_i

$$r_j = r_i + C_{iu}k_{ij} \qquad (II.7.6)$$

where k_{ij} is a constant for a given reference and pair of wavelengths i and j, and defined by

$$k_{ij} = \left(\frac{\overset{*}{\mu}_{\lambda_j}}{\overset{*}{\mu}_{r\lambda_j}} - \frac{\overset{*}{\mu}_{\lambda_i}}{\overset{*}{\mu}_{r\lambda_i}}\right) \qquad (II.7.7)$$

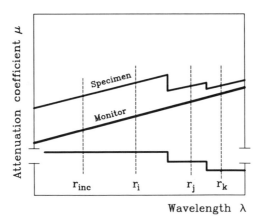

Figure II.7.2 Schematic representation of mass attenuation coefficient and zonal ratios plotted versus wavelength

Table II.7.1 Results obtained using an iterative approach to Compton scatter corrections for the determination of Nb, Zr and Sr present in higher than average concentrations[a]

| Element | Concentration (ppm) | | | r_j | |
	Initial	Final	Recommended	Initial	Final
Nb	1111.3	953.7	960	1.4885	1.4683
Zr	13333.7	10950.7	11000	1.4683	1.2746
Sr	5580.6	4586.7	4600	1.2746	1.2234

[a] Data from Harvey (1992).

(f) concentration is given by

$$C_{iu} = C_{iu,est}(r_j/r_i) \qquad (II.7.8)$$

In practice, processing is performed in wavelength order away from the Compton wavelength towards longer wavelengths. Initially, r_i is equated to the intensity ratio measured at the Compton wavelength r_0. This value is retained until an absorption edge is encountered, in which case r_i is replaced by r_j using Eqn (II.7.6). An initial concentration estimate $C_{iu,est}$ is obtained, assuming that r_i is equal to the intensity ratio measured at the Compton wavelength, and is used to generate a first estimate of r_j. This value is used in Eqn (II.7.8) to calculate C_{iu}, which provides a better estimate of C_{iu}. These steps are repeated, comparing C_{iu} with $C_{iu,est}$ and using successive r_j values, until convergence is reached. A detailed numerical example is given for a highly mineralized specimen, which led to the results shown in Table II.7.1.

Analysts anticipating using Compton scatter methods will find the detailed treatment of the subject in this reference invaluable.

II.7.5 STANDARD ADDITION METHOD

The standard addition method and the double dilution method can be visualized as complementary in the sense that the former provides for analyses at low concentrations whereas the latter is generally restricted to medium and high concentrations. In common with the double dilution method, standard addition combined with the fusion technique is ideal in that: (a) it overcomes the problems associated with particle size and mineralogical effects, and (b) it most likely guarantees that dead time is minimal and that one is operating in the linear C versus I region. With the advent of automated spectrometers and the resulting high efficiency when dealing with type materials, the standard addition method, like the double dilution method, has been relegated to handling 'one of a kind' submissions. It therefore follows that *reported* applications are comparatively few.

For powdered specimens, the particle size effect can be minimized by fine grinding. In the case of mineralogical effects, e.g. the added analyte is in the form of an oxide whereas it is present in the sample as a sulphide, sulphate, etc., the problem can be minimized if the mineralization of the material is known. Given a relatively pure source of a similar mineral, a finely ground portion can be diluted with an inert material to serve as the added standard. Thus, one adds a sulphide to a sulphide-bearing material, sulphate to a sulphate-bearing material, etc. Liquid samples understandably are ideal for standard addition; the standard solutions are easily prepared and mixing is no problem. This advantage may be counterbalanced by the problems introduced in handling solutions that are likely to be corrosive.

II.7.6 HIGH DILUTION–HEAVY ABSORBER METHOD

In the main, the same criteria apply to high dilution methods as to the double dilution method. Combined with the fusion technique, it offers a quick solution for the determination of major and minor constituents in variable matrices. However, the method lends itself to automation and hence can be contemplated for routine analysis.

II.7.7 MATRIX MATCHING METHODS

In the context of analysis of manufactured products, matrix matching offers the simplest process for converting intensities to concentrations. This will often be the case when the determination of elements occurring in low concentration in an otherwise relatively constant matrix is called for. This may require a preliminary sorting of the samples in two or three categories. Assuming that type reference materials are available for each category, a least squares fit of C versus I relative to a counting monitor will compensate for instrumental drift, thus providing a reliable calibration factor over an extended time period. Specific type alloys, cements, pigments, etc. are likely candidates for matrix matching. Samples may be in the form of solids, powders or liquids and, whenever possible, analysed 'as received' or, in the case of powders, prepared as pressed pellets.

Yager and Quick (1993) described a personal-computer-based program for energy dispersive X-ray spectra that is available in the public domain. The program has been applied primarily to determine the concentration of minor and trace elements heavier than Fe in rocks. First, the spectra are processed using least squares spectral deconvolution. The program then corrects for instrumental drift, background and peak overlap. For most elements, correction for matrix effects takes the form

$$C_{iu} = K_i C_{iu,app} \tag{II.7.9}$$

with

$$C_{iu,app} = C_{ir} \times b_i \times \left(\frac{I_{inc,r}}{I_{inc,u}}\right) \tag{II.7.10}$$

where b_i is the fraction of the ith reference spectrum in the sample spectrum. A second order regression is another option offered for converting $C_{iu,app}$ to C_{iu}.

II.8 SPECIAL ANALYTICAL CONTEXTS

Even a cursory examination of the X-ray literature reveals examples of how X-ray fluorescence spectrometry is being used in contexts that can be classed as 'special' with little fear of contradiction. Consider the following.

Lankosz *et al.* (1993) evaluated Monte Carlo simulations in the context of quantitative analysis of individual particles ($\approx 150\ \mu$m) using a finely collimated incident beam (nominal $\approx 70\ \mu$m). Calculated and experimental relative intensities for Ca, Zn, Sr $K-L_{2,3}$ and Mo Compton scatter versus particle size (50–170 μm) were in good agreement.

McKee and Renault (1993) developed a method for the determination of Pb in paint chips in order to establish Pb levels in old painted surfaces, forming part of a legal requirement for urban redevelopment projects. The chip is weighed to the nearest 0.1 mg and cemented to a KCl disc. The Cl intensity is used to compute the area of the chip and the results are reported in mg cm^{-2} of Pb.

Ahlgren *et al.* (1980) developed a method for the *in vivo* determination of Cd in kidneys and Pb in fingers in the medical context of estimating the body burden of these toxic substances in patients. Collimated beams from radioisotopes were used as the energy source and scattered intensities were measured using solid state detectors. Pella and Soares (1991) explored secondary target X-ray excitation for *in vivo* measurement of Pb in bone (finger or tibia). The $K-L_{2,3}$ lines of Y, Zr, Nb and Mo were used as energy sources to excite the L_3-M_5 line of lead (10.55 keV). Minimum detection limits in simulated bone based on a 30 min measurement time ranged from 1–11 μg g^{-1}.

II.8.2 LOW Z ANALYTES AND MATRICES

In addition to the low fluorescence yields and the low energy limit of the available tube photons, Mantler (1993) examined additional unusual excitation possibilities in the case of analyte carbon in the Fe–Fe$_3$C system. These included excitation by Auger electrons and by $L_{2,3}$ photons after K shell ionization. Results showed that the $C_{K-L_{2,3}}$ intensity of pure carbon could be increased by a factor of 2.5–3 and the intensity ratios in Fe–C binary systems increased by a factor of 5–10, depending upon the experimental conditions. Weber *et al.* (1993) concluded that the emitted intensities of the $K-L_{2,3}$ carbon line in steels are a function not only of the total carbon content but also of graphite segregation and to some extent of the specimen preparation procedure (polishing down to 0.3 μm alumina powder and cleaning with soap and denatured alcohol).

Sieber (1993) developed a general procedure for the determination of additive elements in various greases. Stressing the advantage of having a single method for all types of greases encountered in the oil industry, the method has the following advantages:

(a) the specimen preparation method (absorption in a binder powder, 9 parts binder and 1 part grease) eliminates outgassing and overcomes differences among greases, e.g. varying oil content;

(b) without compressing to form a pellet (pressing squeezes oil out of the pellet), a fixed weight is transferred into a liquid sample cup and tamped to a fixed depth;

(c) the specimen preparation method allows the use of liquid oil standards for calibration (for P, S, Ca, Zn, Mo, Ba and Pb);

(d) separate correction schemes are used for energy and wavelength dispersive spectrometers;

(e) the complicated spectrum manipulations using energy dispersion required maintaining strict control over the energy calibration for the determination of P and S (less than a fraction of a channel, 40 eV per channel).

II.8.4 THIN SPECIMENS

The converse problem so to speak, is encountered in the analysis of 'thin films', i.e. specimens considered as falling in the linear region in Figure 8.4, if the films become 'too thick'. This problem was examined by Holmes (1981) in the context of the analysis of natural particulate matter. The author concluded that:

(a) the degree to which emitted intensities are absorbed far exceeds that which is generally assumed;

(b) matrix effects are of sufficient magnitude to render the term thin film inappropriate to the analysis of natural particulates.

A theoretical model was derived (and confirmed experimentally) which accounts for the signal loss.

Table II.8.1 is a partial listing of a database search compilation (23 elements) of critical specimen thickness in the context of precious metals

Table II.8.1 Compilation of a database search for critical specimen thickness for catalyst material[a]

Element		Line	λ (Å)	Critical thickness (μ)
12	P	$K-L_{2,3}$	6.158	12 ± 2
22	TI	$K-L_{2,3}$	2.750	74 ± 15
56	Ba	L_3-M_5	2.776	91 ± 14
26	Fe	$K-L_{2,3}$	1.937	197 ± 43
58	Ce	L_3-N_5	2.209	139 ± 22
30	Zn	$K-L_{2,3}$	1.436	404 ± 111
78	Pt	L_3-M_5	1.313	561 ± 148
35	Br	$K-L_{2,3}$	1.041	1099 ± 298
82	Pb	L_2-M_4	0.983	1153 ± 243
40	Zr	$K-L_{2,3}$	0.787	2212 ± 640
46	Pd	$K-L_{2,3}$	0.587	4550 ± 1298
50	Sn	$K-L_{2,3}$	0.492	6692 ± 2956

[a] Thickness data from Kunz and Belitz (1993).

automotive catalysts carried out by Kunz and Belitz (1993). A procedure is described for calculating the critical thickness and the volume analysed for Pt/Rh, Pd/Rh and Pd-only catalysts. Experimental measurements led the authors to conclude that quantitative results obtained from Pd/Rh and Pd-only catalyst samples weighing less than 6.0 g may be suspect.

II.8.2 THIN FILMS, FILM THICKNESS

Feng (1993) proposed a simple approach for multilayer thin film analysis based on the fundamental parameters expressions in the following contexts (energy dispersion):

(a) the determination of P in borophosphosilicate (BPSG) films on a Si substrate (verified versus wet chemical results);

(b) the determination of Ti and TiN layer thickness and of Al layer thickness on Si substrates (verified versus two independent sets of results).

Working with similar materials but using wavelength dispersion, Levine and Higgins (1991) developed a method that simultaneously measures both the phosphorus and the surface density in BPSG and phosphosilicate glass films on Si substrates. Given that the excitation source must be described in terms of monochromatic wavelength, it was deemed that the Cr $K-L_{2,3}$ line of a Cr-target X-ray tube was a good approximation, with the stipulation that this may not be the case for other X-ray tubes. Results correlated well with independent ICP-AES analyses for P and with ellipsometer measurements of thickness.

White and Huang (1989) described a technique for high-precision measurement of ultra-thin carbon coating thickness on Si and CoCrX alloy substrates. A synthetic crystal composed of alternating layers of tungsten and carbon with a $2d$ spacing of 16.0 nm was used.

II.8.8   FUSED DISC SPECIMENS

The inclusion of fused disc specimens in a chapter dealing with special situations certainly does not stem from the rarity of the use of the fusion technique in X-ray fluorescence analysis. The widespread use of the fusion technique in the cement industry and in the geological sciences for rock and mineral samples, to name but two, attests to that. What may be envisaged as special in these contexts is the many 'manipulations' introduced by analysts, i.e. modifications to influence coefficients in order to retain the conventional model

$$C_i = R_i \left[1 + \sum_j \alpha_{ij} C_j \right]$$
(II.8.1)

Consider the following 'adaptations' that have become more or less common practice:

(a) although X-ray fluorescence is in the domain of atomic spectrometry, i.e. electronic transitions at the elemental level, analysts generally express concentrations in terms of oxides;

(b) if influence coefficients are used for the matrix effect correction, analysts generally resort to modified coefficients, i.e. the correction term for the matrix effect of the flux is taken out of the summation by modifying all the remaining influence coefficients;

(c) although absorption and enhancement effects are a function of the concentrations of elements in the fused disc, i.e. equal to $C_j(1 - C_f)$, analysts generally prefer to work with concentrations in the original samples, which involves an additional modification of the influence coefficients;

(d) loss of volatile constituents or gain in weight as a result of oxidation during fusion may be compensated for by determining the weight loss or gain and applying a correction using appropriate semi-empirical influence coefficients;

(e) similar corrections for LOF and GOF may be made using theoretically calculated modified influence coefficients if the stoichiometry of the constituents (before and after fusion) is known.

II.9 ANALYTICAL STRATEGIES

Analysts cannot ignore the impact of reporting one (or a few) erroneous result(s) if this leads the user to reach an erroneous conclusion. This certainly is a major consideration in the detailed elaboration of a procedure. The question is: how much should one be concerned with the likelihood of an event that may occur 'once in a hundred, once in a thousand, . . .'? Thus, rather than discuss the factors K, I, M and S separately, it is preferable to examine in some detail how a number of experienced analysts have applied X-ray fluorescence spectrometry in various contexts, namely the compositional analysis of alloys, cements and geological materials.

No matter what the analytical context, the analyst and subsequent users should have some measure of the validity (precision and accuracy) of the analytical data reported. In many instances, Eqn (9.50)

$$C_i = C_i \pm (a + bC_i) \tag{9.50}$$

may be used to express levels of uncertainty associated with a given procedure (Figure II.9.1), where 'difference' may relate to (a) the difference between duplicate C_i values, where both are obtained as if they were independent samples; or (b) the difference between the known concentration (reference material not used for calibration) and the concentration found.

Thus the values of the intercept a and slope b may be so chosen as to represent the estimate of the number of data that will fall within a given percentile or standard deviation. For a more comprehensive treatment, the reader is referred to Thompson and Howarth (1976, 1978) and Howarth and Thompson (1976).

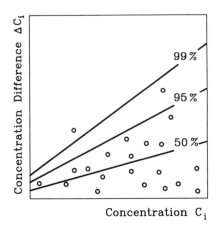

Figure II.9.1 Schematic representation of percentiles of absolute difference versus concentration curves

II.9.1 STEELS, NICKEL-BASE, COPPER-BASE, BROAD MIXTURES

Having already developed a quantitative method for the analysis of silicates using influence coefficients for the correction of matrix effects, Gunicheva *et al.* (1990) proposed extending the method to the analysis of steels and alloys, for which the matrix effects (especially enhancement) are larger than for silicates. Specimens are prepared from cylindrical samples by an abrasive-cutting machine, then treated with an emery cutter of 50 units grain size and emery paper. The excitation source was a Pd target tube operated at 30 kV and 40 mA. Counting time was 40 s. The authors considered a slightly modified Claisse–Quintin model as an optimum form for the matrix effect correction equation, and used polynomial expressions to calculate theoretical influence coefficients for that algorithm. The experimental validity of the algorithm was demonstrated and compared using previously published experimental data.

Two independent studies were carried out by Griffiths and Webster (1986) with the aim of developing wide-range calibrations for 16 elements in nickel-base alloys using theoretically calculated influence coefficients in the de Jongh algorithm. The average concentration values used to calculate the influence coefficients are listed in Table II.9.1. Calibrations in one study were based on a selection of reference materials of different alloys (21 types, 68 reference materials) covering the concentration range over which determinations were to be made, whereas in the other study, several reference materials of each type of alloy were included in the calibrations. Chromium target tubes were used in both studies. Lines selected were $K-L_{2,3}$, L_3-M_5, L_2-M_4 and L_3-N_5; operating conditions were 40, 50 and 60 kV; 10, 20, 30, 45 and 50 mA;

Table II.9.1 Average values used to calculate alpha coefficients and ranges of nickel-base alloys, from Griffiths and Webster (1986)

Element	Avg. (%)	Range (%)	Element	Avg. (%)	Range (%)
Al	4.0	0–7	Ni	44.0	47–75
Si	1.0	0–1	Cu	1.0	0–0.5
Ti	2.0	0–6	Zr	0.5	0–0.2
V	1.0	0–1	Nb	1.0	0–3
Cr	14.5	5–25	Mo	5.0	0–11
Mn	1.0	0–1	Hf	1.0	0–2
Fe	1.0	0–4	Ta	2.0	0–9
Co	15.0	0–24	W	5.0	0–11

collimator: coarse or fine; crystals: PET and LiF (200); detector: flow, scintillation or both in tandem. Counting times and masks (used to minimize spectral interference from the excitation source and escape of high energy radiations) are tabulated for both studies. Also tabulated are: concentration ranges, standard deviations and proportionality constants of calibrations, results on check reference materials and test samples, and Cr values obtained with different sets of influence coefficients (one for short range, the other for longer concentration ranges).

Rousseau and Bouchard (1986) have shown a practical application of the Rousseau algorithm in the context of the determination of 15 elements occurring in a very wide range of concentrations, Table II.9.2. A series of 28 NBS (National Institute for Standards and Technology: NIST) reference materials was analysed and the results are discussed. The analyte lines selected were $K-L_{2,3}$ and L_3-M_5; tube target: Rh; operating conditions 45 kV; 25 mA; collimator: fine; crystals: PET and LiF (200); detectors: flow, scintillation or both in tandem. Backgrounds ($\pm$ degrees 2θ) and all net intensities in c/s (corrected for background, spectral overlap and drift) are tabulated. In order

Table II.9.2 Concentration ranges of reference materials used for experimental verification by Rousseau and Bouchard (1986)

Element	Range (%)	Element	Range (%)
Al	0–5	Ni	0–80
Si	0–1	Cu	0–95
Ti	0–93	Zn	5–95
V	0–2	Mo	0–8
Cr	0–24	Sn	0–39
Mn	0–12	W	0–13
Fe	0–98	Pb	0–60
Co	0–12		

to illustrate the flexibility of the method, the intercept and the calibration slope were determined using only two reference materials (lowest and highest concentration). Analysis of unknowns was a three step process: firstly, an estimate of the composition of the sample was calculated using constant influence coefficients previously computed for the Claisse–Quintin algorithm. Secondly, fundamental absorption and enhancement influence coefficients were calculated for all the analytes using the estimated compositions. Once all the influence coefficients were known, a refined composition was obtained by iteration. The comparison between certified composition and results obtained, the reciprocal of the calibration slope, the intercept and the mean absolute error are also tabulated.

The determination of eight elements in brasses and bronzes using energy dispersion was described by Kovacs and Kis-Varga (1986). The isotope Am-241 was used to excite the characteristic lines of Mn, Fe, Ni, Cu, Zn, Sn, Sb and Pb. Measured intensities were corrected for background and dead time and ratioed to the net intensity emitted by a specimen of pure copper to compensate for drift, and then corrected for peak overlap. The de Jongh model was used for matrix effect corrections using empirical influence coefficients. A comparison of results obtained versus known values and goodness of fit for 16 standards are tabulated.

II.9.2 CEMENTS AND RAW MATERIALS IN THE CEMENT INDUSTRY

X-ray fluorescence methods have been applied to the analysis of Portland cement for many years, and more recently to the raw materials used in its manufacture. Table II.9.3 lists the range of concentrations likely to be encountered in cement analysis and the maximum permissible variations in results (ASTM C114); i.e. the performance requirements expected for rapid methods. Two suggested methods (having no official status and published as information only in *Methods for Analytical Atomic Spectroscopy*, 8th ed., American Society for Testing and Materials) are examined.

Moore (1987) proposed a method for the determination of eight elements in the context of wavelength dispersion and fused disc specimens. The following is an outline of the procedure recommended.

(a) The analyst should ascertain the loss on ignition of the sample and of the lithium tetraborate flux (1000°C) to the nearest 0.0001 g.

(b) Weigh a quantity of sample that will yield 0.7000 g upon fusion.

(c) Weigh a quantity of lithium tetraborate that will yield 2.1000 g upon fusion.

(d) Mix cement and flux in a glass bottle by shaking and rolling (a mixer mill may be used but is not essential).

(e) Transfer the mixture to the fusion crucible (graphite, 32 mm diameter,

Table II.9.3 Concentration ranges and maximum permissible variations in results (ASTM C114) for Portland cement

Oxide	Range, %	Maximum difference between duplicates	Maximum difference of the average of duplicates from SRM certificate values
Na$_2$O	0.04–1	0.03	± 0.05
MgO	0.4–6	0.16	± 0.2
Al$_2$O$_3$	3–7	0.20	± 0.2
SiO$_2$	18–25	0.16	± 0.2
P$_2$O$_5$	0.05–0.5	0.03	± 0.03
SO$_3$	1–7	0.10	± 0.1
K$_2$O	0.02–1	0.03	± 0.05
CaO	58–68	0.20	± 0.3
TiO$_2$	0.1–0.6	0.02	± 0.03
Mn$_2$O$_3$	0.04–0.2	0.03	± 0.03
Fe$_2$O$_3$	1–6	0.10	± 0.10
SrO	0.05–0.4		

29 mm high over-all with a cavity of 9 ml capacity) and heat in a muffle furnace at 1000°C for 10 min.

(f) Transfer the cooled bead to a rotary swing mill grinder and grind for 1 min (for other grinders, grinding time is the minimum time required to give stable intensities).

(g) Place a quantity of inert material (boric acid or cellulose makes a good backing material) in the press mould cavity, spread the ground fused powder on top of the backing material and press to the selected pressure (in the range 20 000–50 000 psi) for 30 s.

(h) Prepare a suitable permanent counting monitor by fusing a selected Portland reference material (SRM No. 1014 is suitable) in the 3:1 ratio, permit the melt to cool slowly on a flat graphite surface in a mould of suitable size, grind and polish a flat surface.

(i) Prepare synthetic samples containing only the flux and the oxide of the element to be determined in amounts to give count rates near the maximum encountered in Portland cement samples.

(j) Excitation source: Cr target X-ray tube (or other material of equal or better efficiency for elements with $Z = 12–14$) capable of operation at 50 kV and 50 mA constant potential with input voltage regulated to ± 1%.

(k) Steps to determine pulse height selection are given textually.

(l) Analyte lines and suggestions for analysing crystals, detectors and minimum counts for the counting monitor are tabulated.

(m) Select an exposure time and power setting that will give at least the number of counts for the counting monitor, X-ray path at 250 μm pressure or less controlled to ± 50 μm.

(n) A detailed procedure to be followed for calibration includes equations for the analytical curves and for matrix effect corrections.

Data on precision and accuracy for single determinations are also tabulated in respect of calibrations carried out as prescribed using NIST Portland cement samples identified as (SRM) Nos. 1011, 1013, 1014, 1015 and 1016, or their replacements.

Wheeler (1987) proposed a method for the determination of 12 elements in the context of energy dispersion and pressed pellet specimens. The following is an outline of the procedure recommended.

(a) Weigh 5.000 ± 0.005 g of cement and 0.100 ± 0.005 of grinding aid (a commercially available detergent) and grind in a rotary swing mill with tungsten carbide vial for a pre-selected time (one minute more than the grinding time observed for emitted intensities to stabilize as a function of grinding time on test samples).

(b) Transfer the ground material into the pellet die, level the powder with a spatula, cover with ≈ 11 g of boric acid and press at 30 000 psi for 30 s.

(c) Prepare the standard reference materials NBS Nos. 633, 635, 636, 637, 639 and 1016 and the well homogenized sample used in the grinding time study (for use as a counting monitor) as in (a) and (b) above.

(d) Analytical lines $K-L_{2,3}$ (keV), tube power and current and filter used for strontium are tabulated.

(e) Excitation source is a rhodium target tube with a 0.25 mm beryllium window; a solid-state silicon detector (165 eV FWHM at 10 μs shaping time and 1000 c/s at 5.9 keV Mn $K-L_{2,3}$ peak) and a multichannel analyser are used to integrate pulses.

No details are given for intensity data reduction except that intensities are corrected for spectral interference and that matrix effects are corrected using an exponential function. Data for precision (determined by preparing replicate specimens and analysing each several times over a period of days) and accuracy (three unknowns analysed and reported with known concentrations subsequently supplied by the submitter) data are tabulated for all 12 analytes.

Babikian (1993) proposed a method for the analysis of raw materials in the cement industry, where the output of a high quality product requires the rapid and precise knowledge of the concentrations (Na, Mg, Al, Si, P, S, K, Ca, Ti and Fe) of the input materials. These are likely to include pure to argillaceous limestones, shale and clay, sand, iron ore, fly ash, coal ash and, potentially, suitable industrial waste. The context is somewhat similar to the determination of the major constituents in rocks except that advantage may be taken of the fact that the unknowns are likely to be well categorized. The following is an outline of the step by step procedure recommended, with the proviso that it is most important that the reference materials used for calibration and the unknowns be prepared in an identical manner.

(a) A wavelength dispersive spectrometer with a Rh end-window X-ray tube operated at 30 kV and 80 mA provides the excitation source, the detector is a flow proportional counter, and a coarse collimator is used for all the analytes except calcium.

(b) The samples 'as received' are dried to eliminate surface water and ground to 150 mesh.

(c) Loss on ignition is determined by heating in a muffle furnace at $1050°C$ for 30 min.

(d) To 1.2000 ± 0.0003 g of the ignited sample, pulverized until 100% passes through a 150 mesh sieve, are added 6.000 ± 0.0003 g of lithium tetraborate and 0.1500 ± 0.0003 g of lithium fluoride.

(e) Two drops of ammonium nitrate solution (10% weight by volume) and two drops of lithium bromide solution (50% weight by volume) are added to the mixtures.

(f) The fusion process (commercially available gas burner, Pt–Au crucibles and moulds) consists in preliminary heating at $\approx 850°C$ without agitation for 2 min followed by a 6 min cycle at $1050°C$ with agitation; the melt is poured into a 30 mm mould.

(g) To compensate for drift, a two point normalization process is applied using the expression

$$I_{tc} = I_{lo,tc} + \frac{(I_{ta} - I_{lo,ta}) \times (I_{hi,tc} - I_{lo,tc})}{(I_{hi,ta} - I_{lo,ta})} \tag{II.9.1}$$

where I_{tc} = intensity corrected for drift, i.e., normalized to 'time of calibration'; I_{ta} = intensity at the time of analysis; $I_{lo,ta}$ = intensity of low monitor at the time of analysis; $I_{hi,ta}$ = intensity of high monitor at the time of analysis; $I_{lo,tc}$ = intensity of low monitor at the time of calibration; $I_{hi,tc}$ = intensity of high monitor at the time of calibration.

(h) A truncated Claisse–Quintin model (cross-coefficient term omitted) is used to correct for matrix effects, namely

$$C_i = K_i I_i \left[1 + \sum_j (\alpha_j + \alpha_{jj} C_j) C_j \right] \tag{II.9.2}$$

where the product $K_i I_i$ is the apparent or uncorrected concentration.

The drift monitors (five monitors for 11 elements) were selected from the 19 international reference materials (plus an additional 19 made up of mixtures of equal weights of two reference materials) used for calibration. The wide and short ranges covered by the calibration curves are shown in Table II.9.4. The results obtained on four reference materials (obsidian rock, coal fly ash, blast furnace slag and argillaceous limestone) analyzed as unknowns are compared with their certified values.

Table II.9.4 Calibration ranges and standard error of estimate obtained on raw materials used in the cement industry (Babikian, 1993). Some data are rounded-off

(a) Broad ranges

Oxide	Range (%)	$(1)^a$	Oxide	Range (%)	$(1)^a$
Na_2O	0–9	0.046	SO_3	0–7	0.314
MgO	0.2–39	0.119	K_2O	0–11	0.073
Al_2O_3	0.2–61	0.202	CaO	0.2–98	0.418
SiO_2	0–100	0.308	TiO_2	0–2	0.022
P_2O_5	0–0.5	0.012	Fe_2O_3	0–13	0.083

(b) Restricted ranges

Oxide	Range (%)	$(1)^a$	Oxide	Range (%)	$(1)^a$
SiO2	60–100	0.384	CaO	57–98	0.341
	35–65	0.220		30–61	0.313
	1–38	0.072		0.2–35	0.079
Al_2O_3	0.2–40	0.188			

^{a} (1) = Standard error of estimate.

II.9.3 GEOLOGICAL CONTEXTS

Harvey (1989) presents a detailed procedure for the determination of 48 trace elements (F (9)–U (92)) using a Rh target tube (except for Ag, Cd and In) and wavelength dispersion. Specimens were pressed pellets: a few drops of a mixture of polyvinylpyrrolidine and methylcellulose in the ratio 7:4 dissolved in ethanol and water were added to 8–10 g of sample, mixed and pressed at 10 tons. The resulting damp pellet was dried at $110°C$ for 15 min. The selected characteristic lines were $K-L_{2,3}$, $K-M_3$, L_3-M_5 and L_2-M_4; operating conditions: 40, 60 and 75 kV; 40, 50 and 70 mA; collimator: coarse or fine; crystals: LiF (220), LiF (200), Ge, PET, PX-1, TAP; detectors: flow, scintillation or both in tandem, are listed for all 48 elements. Also tabulated are total counting times, a representative (2σ) detection limit and potential overlapping lines for 20 elements (including the presence of broad 'ghost' peaks occurring at angles corresponding to the theoretically impossible value of $n = 1.5$ in the Bragg equation, in particular with the LiF (220) crystal). The correction of intensities for absorption effects using Compton scatter with provision for crossing absorption edges is described in detail, as are five areas which need to be monitored for quality control. Analytical data are processed in batches of samples so that drift and quality control checks may be made on the batch as a whole. The author notes that the selections are not necessarily the optimum choices but are usually adequate when turn-around time and high productivity are a concern.

Chappell (1991) concludes that X-ray fluorescence sets the standard for the analysis of most elements in rocks above the ppm level, and presents detailed instrumental conditions for the determination of 29 elements (S (16)–U (92) at trace levels), based on some 30 years' experience. Tabulated for each element are selected characteristic lines: $K-L_{2,3}$, L_3-M_5, L_2-M_4, L_3-N_5; tube targets: Sc, Cr, Mo, Rb, W and Au (listed in order of merit); kV: 40, 60 or 100; crystals: Ge, LiF (200), LiF (220); collimator: coarse or fine; detector: flow or scintillation; and principal line overlap interferences. It is considered essential to subtract backgrounds, preferably based on two measurements in most cases, and to compensate for X-ray tube contamination and for absorption effects. The necessity for very fine grinding and the unavoidable introduction of contamination from the grinding vessel are stressed. It is recommended that, for elements lighter than Sc (21), samples be prepared as fused discs to avoid particle size effects. Evidence is presented of the very good agreement (X-ray fluorescence versus isotope dilution) of Rb and Sr determinations on Apollo 11 and 12 samples (Rb content 0.5–10 and 0.5–1.5 ppm respectively).

Eastell and Willis (1990, 1993) investigated the determination of 21 trace elements (S (16)–U (92)) and 11 major and minor elements in the context of a low dilution fusion technique using a 2:1 flux-to-sample ratio. Specimen preparation comprised: adding 2 ± 0.001 g of sample dried at 110°C to 4 ± 0.001 g of flux (80% lithium metaborate, 20% lithium tetraborate, 0.6 g of lithium nitrate as oxidant) dried at 450°C; fusions carried out at 1100°C were successfully used to produce blanks and interference and calibration disc specimens. Even though 6 g discs were found not to be infinitely thick, all 21 elements yielded linear relationships between concentration and the product $I \times \mu$. The net loss on fusion was determined by weighing the mixture before and after fusion, and it was observed that sulphur is only semi-quantitatively retained, depending on whether the sulphide minerals are oxidized completely and rapidly to sulphate. Pulse height selection was used to reduce interference from higher order spectral lines and to reduce background. Detailed examination of wavelength scans showed unusual relative intensities, complex spectral interferences and mutual spectral interferences when certain elements occurred in higher than normal concentrations. Analytical conditions are tabulated for the selected characteristic lines: $K-L_{2,3}$, $K-M_3$, L_3-M_5 and L_2-M_4 (intensities of six additional lines are measured for spectral interference corrections); tube target: Cr, Mo, W or Au; kV: 50 or 60; mA: 45, 55 or 60; crystal: LiF (200) or LiF (220), collimator: coarse or fine; pulse height selection: various settings; detectors: flow, scintillation or both in tandem; and background degrees 2θ. Possible spectral interferences are tabulated for 19 elements. Differences in sample absorption were corrected using either incoherent scatter or mass absorption coefficients calculated from the major element compositions and the flux components. The conditions for the determination of μ using the Mo $K-L_{2,3}$, Compton peak, a comparison of results versus recommended values for 14 international rock reference materials, and limits of detection and counting errors are tabulated for 20 elements, i.e. excluding S. One of the major advantages of the low dilution fusion technique is that the same specimens are used for the determination of Na, Mg, Al, Si, P, K, Ca, Ti, Mn, Fe and Ba (barium is treated as both a trace and a minor constituent). Three methods of calculating influence coefficients (empirical and theoretical) were investigated; because of the low dilution, it was found necessary to use the oxide α coefficient option (based on a modified Claisse–Quintin algorithm in which the coefficients vary as a function of specimen composition) in the NBSGSC program (Tao *et al.*, 1985).

Faced with the need for a simple, rapid, reproducible and low cost method for the analysis of large numbers of samples in the context of regional geological surveys, Ma and Li (1989) developed a method (total counting time 25 min per sample) for the determination of 30 elements (Na (11)–U (92)) in pressed powder pellets. Samples were ground to 200 mesh and mixed with low-pressure polyethylene to produce stable specimens. Analytical conditions are tabulated, listing spectral lines: $K-L_{2,3}$, L_3-M_5 and L_2-M_4; tube target: Rh; kV and mA not specified; collimator: coarse or fine; detector: flow, scintillation or both in tandem; pulse height selection: various; background degrees 2θ; and counting times for peaks and backgrounds. Measured intensities were ratioed to a monitor to compensate for drift and corrected for background and spectral interference (listed for 12 elements). The elements were sorted into two groups for matrix effect correction. Major and minor components (14 elements) formed one group and correction was made using a second order calibration and empirical influence coefficients (function of concentration or intensity) based on regression analysis on 50 reference materials. Results for the remaining 16 trace elements were corrected for matrix effects using either coherent or incoherent scatter of the Rh $K-L_{2,3}$ line. Detection limits, reproducibilities of determinations and comparison of results with three reference materials are also tabulated. The authors concluded that the detection limits for Co, Cu, As and U did not meet the limits required for their regional geochemical prospecting.

Based on the premise that the effectiveness of empirical influence coefficients depends on the set of reference materials chosen, i.e. concentrations accurately known and suitably distributed over the foreseeable concentration range of all the elements, Ward and Valdes (1987) concluded that these conditions can be best fulfilled by preparing synthetic reference materials. These are prepared from calculated mixtures of high purity ignited oxides or carbonates in the form of fused disc specimens. The analyst has complete control over the composition of the discs and all physical properties of the compounds used are erased by the fusion process. The de Jongh model, with empirical coefficients, was chosen for matrix effect corrections. Influence coefficients were obtained by regression based on an average composition for the oxides of Na, Mg, Al, Si, K, Ca, Ti and Fe (SiO_2, being the major constituent, was selected for elimination). A 1 g portion of a pre-ignited sample was added to 4.5 g of lithium tetraborate, mixed and fused in a Pt–Au crucible in an electric furnace at 1150°C. During cooling, the glass was slightly punctured with a Pt–Au needle to allow for rapid cooling and prevent a build-up of internal stresses. A Cr target tube operated at 40 kV was used to generate intensities, which were corrected for background. Sodium was determined separately in pressed pellet specimens. The empirical influence coefficients were compared with the corresponding coefficients calculated from theory, and showed agreement in some cases and significant differences in others. The authors recommend that analysts should combine both sets of coefficients in order to retain the best of both approaches.

To circumvent particle size and mineralogical effects and the lower intensities associated with fused discs containing a heavy absorber, Austen and Russell (1974) proposed using a 1:5 sample-to-flux ratio and application of a mathematical correction to compensate for matrix effects in the context of the determination of Na, Mg, Al, Si, P, K, Ca, Ti, Cr, Mn and Fe in rocks and metallurgical slags. The coefficients were determined using binary and ternary synthetic standard fused discs. Subsequently, Austen and Steele (1975) found it advantageous to use a calibration procedure based on theoretical influence coefficients computed from fundamental parameters. It was noted that the theoretical and experimental coefficients were in good agreement and yielded excellent results (assessed by the analysis of 12 international silicate rock standards, as well as standard samples of iron, a chrome refractory, a manganese ore, ilmenite and apatite), considering that the concentration ranges were very wide, namely:

Na_2O	0.0–8.3%	MgO	0.1–50%	Al_2O_3	0.1–30%
SiO_2	2.5–76%	K_2O	0.0–15%	CaO	0.1–12%
TiO_2	0.0–51%	Cr_2O_3	0.0–32%	Mn_3O_4	0.0–68%
		Fe_2O_3	1.3–94%		

Potts *et al.* (1984) investigated the analysis of silicate rocks for 24 elements using energy dispersion. The elements were divided into two groups; 14 were labelled major constituents and their concentrations were determined following fusion, while the remaining 10 were determined following pelletization. The excitation source was a silver side-window tube operated at various power settings to obtain maximum efficiency. A 1 g amount of sample (pre-dried overnight) was added to 6 g of flux (80% metaborate, 20% lithium tetraborate), mixed and fused in Pt–Au crucibles in a muffle furnace at 1100°C for 20 min, with repeated agitation to ensure homogeneity. A separate ignition loss test was carried out on each rock powder to account for loss of volatile constituents, and a daily trial fusion of the flux was made in order to correct for the small weight loss by adding a calculated small excess of flux to each fusion mixture. Pellets were prepared by mixing to an even consistency a minimum of 6 g of finely ground rock powder with a saturated aqueous solution (5 wt%) of polyvinyl alcohol and compressing at 10 ton in^{-2}. The resulting pellets were then baked at 110°C to remove excess moisture and improve durability. Optimum instrumental operating conditions were investigated and are stated in the text. The significant spectral interferences between the $K-L_{2,3}$ lines of Na–Mg–Al–Si–P; between the $K-L_{2,3}$ and $K-M_3$ lines of K–Ca, Cr–Mn–Fe and Rb–Sr–Y–Zr–Nb; and between the K and L lines of Ti–Ba and As–Pb require accurate peak area quantification to ensure successful analysis. Extensive tests were carried out to establish a library of individual element X-ray profiles, essential to this form of spectral analysis. Escape peaks were removed and pile-up peaks were minimized by restricting count rates (maximum 50% dead time). Calibrations were carried out based on 24 international reference materials. Intensities were measured on duplicate specimens and corrected for drift by frequent intensity measurements of monitors. In the case of the major elements, calibration was based on calculating apparent concentrations using influence coefficients. However, intensity data for trace elements were first corrected using a proprietary correction program. Calibration data (composition range, regression coefficients, counting sensitivity, fitting error and limit of determination) for both fused discs and pellets are tabulated. The quality of the calibrations is discussed separately for each element, as is the accuracy of the method, based on three independent reference materials. The few larger discrepancies are examined in relation to the analytical limitations observed in the calibrations.

Schroeder *et al.* (1980) addressed the analysis of geological materials with special emphasis on rocks in the context of deep sea drilling of the sea floor, i.e. on materials with high concentrations of volatile due to hydration and of which the quantity for analysis is limited. The basis of the proposed procedures follows well proven conventional methods. In total, 23 elements were determined; 10 are classified as occurring in major concentrations (greater than 0.1 wt%) and determined on fused disc specimens, and the remaining trace elements are determined on pressed pellets (Fe and Ti are common to both sets). Because of the variable volatile content of oceanic rocks, it was critical that ignition losses and rock-to-flux ratio be measured as accurately as possible. About 0.5 g of 200 mesh rock powder was transferred into pre-weighed Pt–Au crucibles and the weight loss at 110°C was determined, followed by the determination of ignition loss at 1000°C. The ignition loss of the flux was established daily and an amount was added to the ignited residue to give a flux-to-rock ratio of 5:1. The mixture was fused for 20 min and, after cooling, re-weighed to obtain the actual flux-to-rock ratio. The fused glass was re-melted and cast and annealed under strictly controlled conditions. Pressed pellets, rimmed and backed with boric acid, were prepared using 3 g of sample and pelletized in vacuum at 25 000 psi. A Cr target X-ray tube (60 kV) was used for the major elements except for Mn; the trace elements and Mn were determined separately using a W tube at 60 kV. Tube current was 10, 30 or

40 mA; crystals: TAP, PET and LiF (200), background degrees 2θ, and collimation and counting times are tabulated for both fused disc and pellet specimens. Also tabulated are the reference materials used for calibration, the mean and standard deviation of 10 analyses (majors) of a reference rock, and of 40 analyses (traces) of a reference rock run over a 10 day period. Results of analyses of three reference rocks (majors) and six reference rocks (traces) are listed along with a comparison of analyses for Rb and Sr versus isotope dilution data for 20 specimens. Measured intensities were corrected for background, spectral overlap and drift. The Rasberry–Heinrich model was used for matrix effect correction for the major elements, and the mass absorption coefficient of the specimen (computed using the previously determined major element composition) was used for matrix effect corrections for the trace elements.

II.9.4   DISCUSSION AND SUMMARY

Faced with the prospects of developing a comprehensive multielement analytical scheme for similar materials, analysts are likely to agree on strategy

Table II.9.5 Instrumental parameters relating to Griffiths and Webster (1986) and to Rousseau and Bouchard (1986)[a]

Element	Lab.	Line	kV	mA	Crystal	Collimator	Detector	time (s)
Al	1	$K-L_{2,3}$	50	50	PET	c	F	20
	2	$K-L_{2,3}$	60	45	PET	c	F	20
	3	$K-L_{2,3}$	45	25	PET	f	F	40
Ti	1	$K-L_{2,3}$	40	10	LiF	f	F	10
	2	$K-L_{2,3}$	60	20	LiF	f	F	4
	3	$K-L_{2,3}$	45	25	LiF	f	F	40
Cr	1	$K-L_{2,3}$	50	50	LiF	f	F	10
	2	$K-L_{2,3}$	60	45	LiF	f	F	40
	3	$K-L_{2,3}$	45	25	LiF	f	F	40
Ni	1	$K-L_{2,3}$	40	10	LiF	f	FS	10
	2	$K-L_{2,3}$	60	30	LiF	f	S	20
	3	$K-L_{2,3}$	45	25	LiF	f	F	40
Mo	1	$K-L_{2,3}$	60	40	LiF	f	S	10
	2	$K-L_{2,3}$	60	30	LiF	f	S	10
	3	$K-L_{2,3}$	45	45	LiF	f	S	40
W	1	L_2-M_4	60	40	LiF	f	FS	20
	2	L_3-N_5	60	45	LiF	f	FS	20
	3	L_3-M_5	45	45	LiF	f	F	40

[a] Labs. 1 and 2 relate to Griffiths and Webster (1986); Lab. 3 relates to Rousseau and Bouchard (1986); LiF crystal is LiF (200). Collimators: c = coarse, f = fine; detector: F = flow, S = scintillation, FS = F + S in tandem.

Table II.9.6 Examples of instrumental parameters selected for the analysis of geological materials using wavelength dispersion

Element (and Z)	Ref[a]	Spc[b]	Line[c]	Tube	kV	mA	Xtl[d]	Coll[e]	Det[f]	Bkg[g]	Pk[h]	Bk[h]
F (9)	1	pp	1	Rh	40	70	PX-1	c	F	±		
Na (11)	3	fd	1	Cr	50	55	TAP	c	F	±	160	80
	4	pp	1	Rh			TAP	c	F	+	40	20
	5	fd	1	Cr	60	40	TAP	c	F	+	95	85
Mg (12)	3	fd	1	Cr	50	55	TAP	f	F	±	320	160
	4	pp	1	Rh			TAP	c	F	+	20	10
	5	fd	1	Cr	60	40	TAP	c	F	−	90	60
P (15)	3	fd	1	Cr	50	55	Ge	c	F	±	80	40
	4	pp	1	Rh			Ge	c	F	+	20	10
	5	fd	1	Cr	60	40	PET	c	F	+	100	50
S (16)	1	pp	1	Rh	40	70	PET	c	F	±		
	2	fd	1	Sc	40		Ge	f	F			
	3	fd	1	Cr	50	55	Ge	c	F	±	80	40
Ca (20)	3	fd	1	Cr	50	55	200	f	F	±	16	16
	4	pp	1	Rh			200	f	F			
	5	fd	1	Cr	60	10	200	f	F	+	35	10
V (23)	1	pp	1	Rh	40	70	220	f	F	−		
	2		1	W	60		220	f	F			
	3	fd	1	W	50	55	220	f	F	−	80	40
	4	pp	1	Rh			200	c	F	+		
	5	pp	1	W	60	30	200	f	FS	−	130	50
Mn (25)	1	pp	1	Rh	60	50	220	f	F	±		
	2		1	W	60		200	f	F			
	3	fd	1	Cr	50	55	220	f	F	±	160	80
	4	pp	1	Rh			200	c	S	+		
	4	pp	1	Rh			220	c	S	+		
	5	fd	1	Cr	60	30	200	c	F	−	25	25
Ni (28)	1	pp	1	Rh	70	40	200	f	FS	−		
	2		1	Au	60		200	f	F			
	2		1	Au	60		200	c	S			
	3	fd	1	Au	60	45	220	f	FS	−	80	40
	4	pp	1	Rh			200	c	S	+		
	5	pp	1	W	60	30	200	f	FS	−	80	40
Ga (31)	1	pp	1	Rh	75	40	200	f	FS	±		
	2		1	Mo	100		200	c	S			
	3	fd	1	Mo	50	55	200	f	FS	±	80	40
	4	pp	1	Rh			200	c	S	+		

Table II.9.6 (*continued*)

Element (and Z)	Ref[a]	Spc[b]	Line[c]	Tube	kV	mA	Xtl[d]	Coll[e]	Det[f]	Bkg[g]	Pk[h]	Bk[h]
Rb (37)	1	pp	1	Rh	75	40	220	f	S	±		
	2		1	Mo	100		200	f	S			
	3	fd	1	W	50	55	200	f	S		80	40
	4	pp	1	Rh			200	c	S	+		
	5	pp	1	W	60	30	200	c	FS	+	80	80
Nb (41)	1	pp	1	Rh	75	40	220	f	S	±		
	2		1	Rh	100		200	f	S			
	3	fd	1	W	50	55	200	f	S		120	40
	4	pp	1	Rh			200	c	S	+		
	5	pp	1	W	60	30	200	c	FS	−	100	80
Sn (50)	1	pp	1	Rh	75	40	220	f	S	±		
	2		1	Au	100		220	f	S			
	4	pp	1	Rh			200	c	S	−		
Ba (56)	1	pp	3	Rh	50	60	200	f	F	+		
	2		4	W	60		220	f	F			
	2		5	Cr	60		220	f	F			
	3	fd	3	Cr	50	55	200	f	F	±	80	40
	4	pp	1	Rh			200	c	S	−		
	4	pp	3	Rh			200	c	F	+		
	4	pp	4	Rh			200	c	F	+		
	5	pp	4	W	60	30	200	c	FS	+	100	100
Nd (60)	1	pp	3	Rh	50	60	220	f	F	±		
	2		4	W	60		220	f	F			
	3	fd	3	W	50	55	220	f	F	+	200	40
Ta (73)	1	pp	3	Rh	75	40	220	f	FS	−		
Pb (82)	1	pp	4	Rh	75	40	220	f	S	±		
	2		4	Mo	100		200	f	S			
	3	fd	4	W	50	55	200	f	S	+	80	40
	4	pp	4	Rh			200	c	S	+		
Th (90)	1	pp	3	Rh	75	40	220	f	S	±		
	2		3	Mo	100		200	f	S			
	3	fd	3	W	50	55	200	f	S		80	40
	4	pp	3	Rh			200	c	S	+		

[a] Ref: 1, Harvey (1989); 2, Chappell (1991); 3, Eastell and Willis (1990, 1993); 4, Ma and Li (1989); 5, Schroeder *et al.* (1980).

[b] Spc = specimen: pp = pressed pellet; fd = fused disc.

[c] Line: $1 = K-L_{2,3}$; $2 = K-M_3$; $3 = L_3-M_5$; $4 = L_2-M_4$; $5 = L_3-N_5$.

[d] Xtl = crystal: 200 = LiF (200); 220 = LiF (220).

[e] Coll = collimator: c = coarse; f = fine.

[f] Det = detector: F = flow; S = scintillation; FS = F + S.

[g] Bkg = background: ±, measured at higher and lower 2θ; +, measured at higher 2θ only; −, measured at lower 2θ only.

[h] Pk, Bk are counting times (seconds) at peak and background positions, respectively.

in many cases, but will no doubt diverge in specific cases in order to eliminate or at least minimize a particular problem. This is illustrated in Tables II.9.5 and II.9.6 in the context of the determination of major and trace elements in alloys and geological materials.

Thus, depending on the available laboratory resources and the priorities, analysts have at some time or another been obliged to consider the following trade-offs.

(a) One analyst will dissolve alloys and measure intensities in aqueous specimens in order to solve the problem of samples submitted in a variety of forms, whereas another will absorb aqueous solutions in particular media in order not to have to deal with liquid specimens.

(b) Analysts will accept a longer specimen preparation time, e.g. two grinding procedures (one in a metallic and one in a ceramic material) as a better means of dealing with contamination.

(c) Analysts will accept longer counting times to compensate for diluted specimens.

(d) Analysts will accept measuring the intensity of more than one line; for example: two K lines, two L lines or a K and an L line for one or more analytes, in order to compare concentrations and hence detect an unexpected anomaly.

(e) Analysts may prefer to develop in-house software to allow better handling of specific aspects of a given context, a feature which may not be readily available commercially.

(f) Analysts may accept a simpler empirical correction approach, even though it will be of little help if 'something goes astray'.

(g) Analysts may decide to maximize automation, which involves an immediate expense but which is likely to be cost effective in the long run.

The list is almost endless. When all is said and done, the main concern of the analyst is *not* to find the *one and only combination* of K, I, M and S factors but to develop one or more procedures that produce valid results, i.e. that do not mislead the user in any one given context. The next concern is that the necessary steps should be taken to maintain this process over a more or less lengthy time span. This either requires close visual monitoring of the analytical process by the analyst or, preferably, if the samples can be well categorized, a sophisticated software package that will control the process and flag anomalies. Finally, the analyst must be ever ready to incorporate improvements (better precision, accuracy, limits of detection and productivity, preferably at lower cost), which seems to be an ongoing expectation in most analytical laboratories.

REFERENCES AND BIBLIOGRAPHY

Ahlgren, L., Grönberg, T. and Mattson, S. (1980) In Vivo X-Ray Fluorescence Analysis for Medical Diagnosis. *Adv. X-Ray Anal.* **23**, 185–191.

Andermann, G. and Kemp, J. W. (1958) Scattered X-Rays as Internal Standards in X-Ray Emission Spectroscopy. *Anal. Chem.* **30**, 1306–1309.

Arai, T., Shoji, T. and Omote, K (1986) Measurement of the Spectral Distribution Emitted from X-Ray Spectrographic Tubes. *Adv. X-Ray Anal.* **29**, 413–426.

Austin, C. E. and Russell, B. G. (1974) *The Analysis of Silicate Rocks and Metallurgical Slags by X-Ray Fluorescence Spectrometry.* Report No. 1599, National Institute for Metallurgy, South Africa. 15 pp.

Austen, C. E. and Steele, T. W. (1975) The Computer Calculation, from Fundamental Parameters, of Influence Coefficients for X-Ray Spectrometry. *Adv. X-Ray Anal.* **18**, 362–371.

Babikian, S. H. (1993) Raw Materials Analysis by XRF in the Cement Industry. *Adv. X-Ray Anal.* **36**, 139–144.

Bambynek, W., Crassemann, B., Fink, R. W. *et al.* (1972) *Rev. Mod. Phys.* **44**, 716.

Bearden, J. A. and Burr, A. F. (1967) Reevaluation of X-Ray Atomic Energy Levels. *Rev. Mod. Phys.* **39**, 125–142.

Beattie, H. J. and Brissey, R. M. (1954) Calibration Method for X-Ray Fluorescence Spectrometry. *Anal. Chem.* **26**, 980–983.

Beatty, R. T. (1912) *Proc. R. Soc. (London)* **A87**, 511.

Bertin, E. P. (1975) *Principles and Practice of X-Ray Spectrometric Analysis*, 2nd ed., Plenum Press, New York.

Birks, L. S. (1959) *X-Ray Spectrochemical Analysis.* Interscience Publishers, Inc., New York.

Birks, L. S. (1969) *X-Ray Spectrochemical Analysis*, 2nd ed. pp. 121–128, Interscience Publishers, Inc., New York.

Blanquet, P. (1966) PhD Thesis, Université de Paris.

Broll, N. (1986) Quantitative X-Ray Fluorescence Analysis. Theory and Practice of the Fundamental Coefficient Method. *X-Ray Spectrom.* **15**, 271–285.

Broll, N. (1990) Fundamental Coefficient Method Applied to a Quasi-Monochromatic Excitation. *X-Ray Spectrom.* **19**, 193–195.

Broll, N. and Tertian, R. (1983) Quantitative X-Ray Fluorescence Analysis by Use of Fundamental Influence Coefficients. *X-Ray Spectrom.* **12**, 30–37.

Broll, N., Caussin, P. and Peter, M. (1992) Matrix Correction in X-Ray Fluorescence Analysis by the Effective Coefficient Method. *X-Ray Spectrom.* **21**, 41–49.

Brown, D. B., Gilfrich, J. V. and Peckerar, M. C. (1975) Measurement and Calculation of Absolute Intensities of X-Ray Spectra. *J. Appl. Phys.* **46**, 4537.

Budesinsky, B. W. (1979) Determination of Correction Constants in X-Ray Fluorescence Spectrometry by a Multivariate Least-Squares Method. *Anal. Chim. Acta* **104**, 1–9.

Budesinsky, B. W. (1980) Multielement X-Ray Fluorescence Spectrometry of Solutions. *X-Ray Spectrom.* **9**, 13–18.

Campbell, J. L. and De Forge, D. (1989) Semi-Empirical Schemes for X-Ray Attenuation Coefficients in the Context of PIXE. *X-Ray Spectrom.* **18**, 235–242.

Chappell, B. W. (1991) Trace Element Analysis of Rocks by X-Ray Spectrometry. *Adv. X-Ray Anal.* **34**, 263–276.

Claisse, F. (1957) *Accurate X-ray Fluorescence Analysis without Internal Standards.* Norelco Rep. 4, 3–17, 17, 19, 95–6. (See also *Que. Prov. Dept. Mines Prelim. Rept.* 327 (1956), 24 pp.

Claisse, F. and Quintin, M. (1967) Generalization of the Lachance–Traill Method for the Correction of the Matrix Effect in X-Ray Fluorescence Analysis. *Can. Spectrosc.* **12**, 129–134.

Claisse, F. and Samson, C. (1962) Heterogeneity Effects in X-Ray Analysis. *Adv. X-Ray Anal.* **5**, 335–354.

Criss, J. W. and Birks, L. S. (1968) Calculation Methods for Fluorescent X-Ray Spectrometry. *Anal. Chem.* **40**, 1080–1086.

Criss, J. W., Birks, L. S. and Gilfrich, J. V. (1978) Versatile X-Ray Analysis Program Combining Fundamental Parameters and Empirical Coefficients. *Anal. Chem.* **50**, 33–37.

Dalton, J. L. (1969) PhD Thesis, Carleton University, Ottawa, Canada.

Dalton, J. L. (1969) The Determination of Mass Attenuation Coefficients for Titanium, Vanadium, Iron, Nickel, Copper, Praseodymium, Gadolinium and Erbium. Research Report R211, Energy, Mines and Resources, Ottawa, Canada, 37 pp.

Dalton, J. L. and Goldak, J. (1969) The Determination of Some X-Ray Attenuation Coefficients in the Range 26.3 to 2.3 keV. *Can. Spectrosc.* **14**, 171–173.

de Jongh, W. K. (1973) X-Ray Fluorescence Analysis Applying Theoretical Matrix Corrections. Stainless Steels. *X-Ray Spectrom.* **2**, 151–158.

de Jongh, W. K. (1979) The Atomic Number $Z = 0$: Loss and Gain on Ignition in XRF Analysis Treated by the JN-Equation. *X-Ray Spectrom.* **8**, 52–56.

Doster, J. M. and Gardner, R. P. (1982) The Complete Spectral Response for EDXRF Systems—Calculation by Monte Carlo and Analysis Application. I. Homogeneous Samples. *X-Ray Spectrom.* **11**, 173–180.

Duane, W. and Hunt, F. L. (1915) X-Ray Wavelengths. *Phys. Rev.* 6, 166–171.

Eastell, J. and Willis, J. P. (1990) A Low Dilution Fusion Technique for the Analysis of Geological Samples. 1—Method and Trace Element Analysis. *X-Ray Spectrom.* **19**, 3–14.

Eastell, J. and Willis, J. P. (1993) A Low Dilution Fusion Technique for the Analysis of Geological Samples. 2—Major and Minor Element Analysis and the Use of Influence/Alpha Coefficients. *X-Ray Spectrom.* **22**, 71–79.

Ebel, H. and Poehn, Ch. (1989) Quantitative X-Ray Fluorescence Analysis of Specimens with Rough Surfaces: Monte Carlo Approach. *X-Ray Spectrom.* **18**, 101–104.

Eich, J. D., Caul, H. J., Smith, D. L. and Rasberry, S. D. (1967) Analysis of Gold and Platinum Group Alloys by X-Ray Emission with Correction for Interelement Effects. *Appl. Spectrosc.* **21**, 324–328.

Feather, C. E. and Willis, J. P. (1976) A Simple Method for Background and Matrix Correction of Spectral Peaks in Trace Element Determination by X-Ray Fluorescence Spectrometry. *X-Ray Spectrom.* **5**, 41–48.

Feng, L. (1993) A Simple Approach to Multilayer Thin Film Analysis Based on Theoretical Calculations Using Fundamental Parameters Method. *Adv. X-Ray Anal.* **36**, 279–286.

Friedman, H. and Birks, L. S. (1948) Geiger-Counter Spectrometer for X-Ray Fluorescence Analysis. *Rev. Sci. Instrum.* **19**, 323–330.

Freund, H. U. (1975) Recent Experimental Values for K Shell X-ray Fluorescence Yields. *X-Ray Spectrom.* **4**, 90–1.

Gardner, R. P. and Doster, J. M. (1979) The Reduction of Matrix Effects in X-Ray Fluorescence Analysis by the Monte Carlo, Fundamental Parameters Method. *Adv. X-Ray Anal.* **22**, 343–356.

Gilfrich, J. V. and Birks, L. S. (1968) Spectral Distribution of X-Ray Tubes for Quantitative X-Ray Fluorescence Analysis. *Anal. Chem.* **40**, 1077–1080.

Gilfrich, J. V., Burkhalter, P. G., Whitlock, R. R., Warden, E. S. and Birks, L. S. (1971) Spectral Distribution of a Thin Window Rhodium Target X-Ray Spectrographic Tube. *Anal. Chem.* **43**, 934–936.

Gillam, E. and Heal, H. T. (1952) Some Problems in the Analysis of Steels by X-Ray Fluorescence. *Br. J. Appl. Phys.* **3**, 353–358.

Gillieson, A. H., Reed, D. J., Milliken, K. S. and Young, M. J. (1965) X-Ray Spectrochemical Analysis of High-Temperature Alloys. *ASTM Spec. Tech. Publ.* No. 376, 3–22.

Green, M. (1962) The Target Absorption Correction in X-Ray Microanalysis. *X-Ray Optics and Microanalysis*, p. 361. Academic Press, New York.

Green, M. and Cosslett, V. E. (1968) *Br. J. Appl. Phys. Ser.* **21**, 425.

Griffith, J. M. and Webster, H. W. M. (1986) X-Ray Analysis of Nickel-Base Alloys with Theoretically Derived Inter-Element Correction Coefficients. *X-Ray Spectrom.* **15**, 61–72.

Gunicheva, T. N., Finkel'shtein, A. L. and Afonin, V. P. (1990) A Matrix Effect Correction Algorithm for X-Ray Fluorescence Analysis of Steels. *X-Ray Spectrom.* **19**, 237–242.

Harvey, P. K. (1989) Automated X-Ray Fluorescence in Geochemical Exploration. In *X-Ray Fluorescence Analysis in the Geological Sciences*, Ahmedali, S. T., ed., Geological Association of Canada Short Course 7, 221–258.

Harvey, P. K. (1992) X-Ray Fluorescence Determination of Trace Elements in Geological Materials: An Iterative Approach to Compton Scatter Corrections for Matrix Absorption. *X-Ray Spectrom.* **21**, 3–9.

Heinrich, K. F. T. (1966) X-Ray Absorption Uncertainty, in *The Electron Microprobe*, pp. 296–377, E. D. McKinley, K. F. T. Heinrich and D. B. Wittry, eds. John Wiley & Sons, Inc., New York.

Heinrich, K. F. T. (1986) Mass Absorption Coefficients for Electron Probe Micro-analysis, in *Proc. 11th Int. Congress on X-Ray Optics and Microanalysis*. University of Western Ontario, London, Canada, p. 72.

Heinrich, K. F. T., Fiori, C. E. and Myklebust, R. L. (1979) *J. Appl. Phys.* **50**, 5589.

Helsen, J. A. and Vrebos, B. A. R. (1984) Monte Carlo Simulations of XRF Intensities in Non-Homogeneous Matrices. *Spectrochim. Acta, Part B,* **39B**, 751–759.

Holmes, G. S. (1981) The Limitations of Accurate "Thin Films" X-Ray Fluorescence Analysis of Natural Particulate Matter: Problems and Solution. *Chem. Geol.* **33**, 333–353.

Howarth, R. J. and Thompson, M. (1976) Duplicate Analysis in Geochemical Practice. Part II. Examination of Proposed Method and Examples of Its Use. *Analyst* **101**, 699–709. (For Part I see Thompson and Howarth, 1976).

Hower, J. (1959) Matrix Correction in the X-Ray Spectrographic Trace-Element Analysis of Rocks and Minerals. *Am. Mineral.* **44**, 19–32.

Ito, M., Sato, S. and Narita, M. (1981) Comparison of the Japanese Industrial Standard and α-Correction Methods for X-Ray Fluorescence Analysis of Steels. *X-Ray Spectrum.* **10**, 103–108.

Jenkins, R. (1979) Current Status of Quantitative Methods in Energy and Wavelength Dispersive X-Ray Spectrometry, in *Keynote Lectures, XXI Colloquim Spectroscopicum Internationale*, Heyden & Son Ltd., London.

Jenkins, R., Gould, R. W. and Gedcke, D. (1981) *Quantitative X-Ray Spectrometry.* Marcel Dekker, Inc., New York.

Ji, A., Yuan, N. and Tao, G. Y. (1980) Application of Rasberry-Heinrich Equation for Correction of Inter-element Effects—Analysis of Cerium Doped Strontium-Barium Niobate Single Crystals by X-Ray Fluorescence Spectrometry. *Res. Inorg. Mater.* 179–182.

Johnson, R. G. and Fleming, S. L., II (1987) Energy-Dispersive X-Ray Fluorescence Analysis of Massive Sulphides Using Fundamental Influence Coefficients. *X-Ray Spectrom.* **16**, 167–170.

Klimasara, A. J. (1992) Automated Quantitative XRF Analysis Software in Quality Control Applications. *Adv. X-Ray Anal.* **35**, 111–116.

Kovacs, P. and Kis-Varga, M. (1986) Application of an Empirical Correction Method for the X-Ray Fluorescence Analysis of Copper Alloys. *X-Ray Spectrum.* **15**, 221–225.

Kramers, H. A. (1923) Theory of X-Ray Absorption and of the Continuous X-Ray Spectrum. *Philos. Mag.* **46**, 836–871.

Kulenkampff, H. (1922) Continuous X-Ray Spectrum. *Ann. Phys.* **69**, 548–596.

Kunz, F. and Belitz, R. (1993) X-Ray Fluorescence Critical Sample Thickness and Volume of Materials Excited in Catalysts. *Adv. X-Ray Anal.* **36**, 145–154.

Lachance, G. R. (1970) Fundamental Coefficients for X-Ray Spectrochemical Analysis. *Can. Spectrosc.* **15**, 64–72.

Lachance, G. R. (1979) The Family of Alpha Coefficients in X-Ray Fluorescence Analysis. *X-Ray Spectrum.* **8**, 190–195.

Lachance, G. R. (1980) A Practical Relation Between Atomic Numbers and Alpha Coefficients. *X-Ray Spectrom.* **9**, 195–197.

Lachance G. R. (1981) The Role of Alpha Coefficients in X-Ray Spectrometry. *International Conference on Industrial Inorganic Elemental Analysis.* Metz. France, June 3.

Lachance, G. R. (1988) Defining and Deriving Theoretical Influence Coefficients in X-Ray Spectrometry. *Adv. X-Ray Anal.* **31**, 471–478.

Lachance, G. R. and Claisse, F. (1980) A Comprehensive Alpha Coefficient Algorithm. *Adv. X-Ray Anal.* **23**, 87–92.

Lachance, G. R. and Traill, R. J. (1966) A Practical Solution to the Matrix Problem in X-Ray Analysis. *Can. Spectrosc.* **11**, Part I, 43–48; Part II, 63–71.

Lankosz, M., Holynska, B., Pella, P. A. and Blackburn, D. H. (1993) Research in the Quantitative Analysis of Individual Particles by X-Ray Fluorescence Spectrometry. *Adv. X-Ray Anal.* **36**, 11–16.

Leroux, J. (1961) Method for Finding Mass Absorption Coefficients by Empirical Equations and Graphs. *Adv. X-Ray Anal.* **5**, 153–160.

Leroux, J. and Thinh, T. P. (1977) Revised Tables of X-Ray Attenuation Coefficients.

Claisse Scientific Corporation, ed., Quebec, Canada; see also Thinh, T. P. and Leroux (1979) *X-Ray Spectrom.* **8**, 85 and corrigendum in *X-Ray Spectrom.* **10**, v (1981).

Levine, H. S. and Higgins, K. L. (1991) Phosphorus Determination in Borophosphosilicate or Phosphosilicate Glass Films on a Si Wafer by Wavelength Dispersive X-Ray Spectroscopy. *Adv. X-Ray Anal.* **34**, 299–305.

Liebhafsky, H. A. and Zemany, P. D. (1956) Film Thickness by X-Ray Emission Spectrography. *Anal. Chem.* **28**, 455–459.

Loomis, T. C. and Keith, H. D. (1976) Spectral Distributions of X-Rays Produced by a General Electric EA-75 Cr/W Tube at Various Applied Constant Voltages. *X-Ray Spectrom.* **5**, 104–114.

Lucas-Tooth, H. J. and Price, B. J. (1961) A Mathematical Method for the Investigation of Interelement Effects in X-Ray Fluorescent Analyses. *Metallurgia* **64**, 149–152.

Lucas-Tooth, H. J. and Pyne, C. (1964) The Accurate Determination of Major Constituents by X-Ray Fluorescent Analysis in the Presence of Large Interelement Effects. *Adv. X-Ray Anal.* **7**, 523–541.

Ma, G. Z. and Li, G. H. (1989) Application of X-Ray Fluorescence Spectrometry to the Analysis of Geochemical Prospecting Samples in China. *X-Ray Spectrom.* **18**, 199–205.

Mainardi, R. T., Fernandez, J., Bonetto, R. and Riveros, J. A. (1981) A Theoretical Procedure to Determine Coefficients from Rasberry–Heinrich Calibration Curve in XRF Spectroscopy. *X-Ray Spectrom.* **10**, 74–77.

Mantler, M. (1984) Lama III—A Computer Program for Quantitative XRFA of Bulk Specimens and Thin Film Layers. *Adv. X-Ray Anal.* **27**, 433–440.

Mantler, M. (1993) Quantitative XRFA of Light Elements by the Fundamental Parameters Method. *Adv. X-Ray Anal.* **36**, 27–33.

McKee, C. and Renault, J. (1993) XRF Analysis of Pb in Occasional Paint Chips. *Adv. X-Ray Anal.* **36**, 167–169.

McMaster, W. H., Del Grande, N. K., Mallett, J. H. and Hubbell, J. H. (1969) Compilation of X-ray Cross Sections. *U.S. Atomic Energy Commission Report* UCRL-50174, Section II, Rev. 1. Lawrence Radiation Laboratory, University of California.

Moeller, T. (1952) *Inorganic Chemistry, An Advanced Textbook.* John Wiley & Sons, Inc. New York.

Molchanova, E. I., Smagunova, A. N., Gunicheva, T. N. and Smagunov A. V. (1992) Dependence of the Accuracy of the Results of Steel X-Ray Fluorescence Analysis on the Method of Considering the Variation of α Coefficients with Sample Chemical Composition. *X-Ray Spectrom.* **21**, 149–153.

Moore, C. W. (1987) Spectrochemical Analysis of Portland Cement by Fusion with Lithium Tetraborate Using an X-Ray Spectrometer, in *Methods for Analytical Atomic Spectroscopy*, 8th ed., E2 SM 10-26 ASTM, 949–955.

Mosichev, V. I., Pershin, N. V. and Nikolaev, G. I. (1984) A New General Equation For Correction of Interelement Effects in X-Ray Fluorescence Analysis. *Can. Spectrosc.* **28**, 39–62.

Nielson, K. K. (1979) Progress in X-Ray Fluorescence Correction Methods Using Scattered Radiation. *Adv. X-Ray Anal.* **22**, 303–315.

Noakes, G. E. (1954) An Absolute Method of X-Ray Fluorescence Analysis Applied to Stainless Steels. *ASTM Spec. Tech. Publ.* No. 157, 57–62.

Pella, P. A. and Soares, C. G. (1991) Secondary Target X-Ray Excitation for *in vivo* Measurement of Lead in Bone. *Adv. X-Ray Anal.* **34**, 293–298.

Pella, P. A., Feng, L. and Small, J. A. (1985) An Analytical Algorithm for Calculation of Spectral Distributions of X-Ray Tubes for Quantitative X-Ray Fluorescence Analysis. *X-Ray Spectrom.* **14**, 125–135.

Pella, P. A., Feng, L. and Small, J. A. (1991) Addition of *M*- and *L*-Series Lines to NIST Algorithm for Calculation of X-Ray Tube Output Spectral Distributions. *X-Ray Spectrom.* **20**, 109–110.

Poehn, C., Wernisch, J, and Hanke. W. (1985) Least-Squares Fits of Fundamental Parameters for Quantitative X-Ray Analysis as a Function of Z ($11 \leqslant Z \leqslant 83$) and E (1 keV $\leqslant E \leqslant 50$ keV). *X-Ray Spectrom.* **14**, 120–124.

Potts, P. J., Webb, P. C. and Watson, J. S. (1984) Energy-Dispersive X-Ray Fluorescence Analysis of Silicate Rocks for Major and Trace Elements. *X-Ray Spectrom.* **13**, 2–15.

Prange, A. and Schwenke, H. (1992) Trace Element Analysis Using Total-Reflection X-Ray Fluorescence Spectrometry. *Adv. X-Ray Anal.* **35**, 899–923.

Rasberry, S. D. and Heinrich, K. F. J. (1974) Calibration for Interelement Effects in X-Ray Fluorescence Analysis. *Anal. Chem.* **46**, 81–89.

Rosa, R. and Armigliato, A. (1989) Monte Carlo Simulation of Thin-Film X-Ray Microanalysis at High Energies. *X-Ray Spectrom.* **18**, 19–24.

Rousseau, R. M. (1984a) Fundamental Algorithm Between Concentration and Intensity in XRF Analysis 1—Theory. *X-Ray Spectrom.* **13**, 115–120.

Rousseau, R. M. (1984b) Fundamental Algorithm Between Concentration and Intensity in XRF Analysis 2—Practical Application. *X-Ray Spectrom.* 121–125.

Rousseau, R. M. (1987) A Comprehensive Alpha Coefficient Algorithm (a Second Version). *X-Ray Spectrom.* **16**, 103–108.

Rousseau, R. M. and Bouchard, M. (1986) Fundamental Algorithm Between Concentration and Intensity in XRF Analysis 3—Experimental Verification. *X-Ray Spectrom.* **15**, 207–215.

Rousseau, R. and Claisse, F. (1974) Theoretical Alpha Coefficients for the Claisse–Quintin Relation for X-Ray Spectrochemical Analysis. *X-Ray Spectrom.* **3**, 31–36.

Salem, S. I., Falconer, T. H. and Windrell, R. W. (1972) *Phys. Rev., A* **A6**, 2147.

Schreiber, T. P. and Wims, A. M. (1982) Relative Intensity Factors for *K*, *L* and *M* Shell X-Ray Lines. *X-Ray Spectrom.* **11**, 42–45.

Schroeder, B., Thompson, G., Sulanowska, M. and Ludden, J. N. (1980) Analysis of Geologic Materials Using an Automated X-Ray Fluorescence System. *X-Ray Spectrom.* **9**, 198–205.

Schofield, J. H. (1974) *Phys. Rev. A* **A9**, 1041.

Sherman, J. (1954) The Correlation Between Fluorescent X-Ray Intensity and Chemical Composition. *ASTM Spec. Tech. Publ.* No. 1157, 27–33.

Sherman, J. (1955) The Theoretical Derivation of Fluorescent X-Ray Intensities from Mixtures. *Spectrochim. Acta* **7**, 283–306.

Sherman, J. (1958) A Theoretical Derivation of the Composition of Mixable Specimens from Fluorescent X-Ray Intensities. *Adv. X-Ray Anal.* **1**, 231–251.

Shiraiwa, T. and Fujino, N. (1966) Theoretical Calculation of Fluorescent X-Ray Intensities in Fluorescent X-Ray Spectrochemical Analysis. *Jpn. J. Appl. Phys.* **5**, 886–899.

Shiraiwa, T. and Fujino, N. (1967) Theoretical Calculation of Fluorescent X-Ray Intensities of Nickel–Iron–Chromium Ternary Alloys. *Bull. Chem. Soc. Jpn.* **40**, 2289–2296.

Sieber, J. R. (1993) Quantitative XRF Determinations of Additive Elements in Greases for Manufacturing Specifications. *Adv. X-Ray Anal.* **36**, 155–166.

Sproull, W. T. (1946) *X-Rays in Practice*. McGraw-Hill Book Co. Inc., New York.

Spruch, G. M. and Spruch, L. (1974) *The Ubiquitous Atom*. Charles Scribner's Sons, New York.

Tao, G. Y., Pella, P. A. and Rousseau, R. M. (1985) NBSGSC—A FORTRAN Program for Quantitative X-Ray Fluorescence Analysis. *NBS Tech. Note* 1213, National Institute for Standards and Technology, Gaithersburg, USA.

Tertian, R. (1969a) A Rapid and Accurate X-Ray Determination of the Rare Earth Elements in Solid or Liquid Materials Using the Double Dilution Method. *Adv. X-Ray Anal.* **12**, 546.

Tertian, R. (1969b) Quantitative Chemical Analysis With X-Ray Fluorescence Spectrometry—An Accurate and General Mathematical Correction Method for the Interelement Effects. *Spectrochim. Acta, Part B* **24B**, 447–471.

Tertian, R. (1972a) PhD Thesis, Université de Paris.

Tertian, R. (1972b) X-Ray Fluorescence Analysis of Liquid and Solid Solution Specimens—a Theoretical Account of the Double Concentration Method and Its Performances. *Spectrochim. Acta, Part B*, **27B**, 159–183.

Tertian, R. (1973) A New Approach to the Study and Control of Interelement Effects in the X-Ray Fluorescence Analysis of Metal Alloys and Other Multi-Component Systems. *X-Ray Spectrom.* **2**, 95–108.

Tertian, R. (1974) Concerning Interelement Crossed Effects in X-Ray Fluorescence Analysis. *X-Ray Spectrom.* **3**, 102–108.

Tertian, R. (1975) A Self-Consistent Calibration Method for Industrial X-Ray Spectrometric Analyses. *X-Ray Spectrom.* **4**, 52–61.

Tertian, R. (1976) An Accurate Coefficient Method for X-Ray Fluorescence Analysis. *Adv. X-Ray Anal.* **19**, 85–109.

Tertian, R. (1986) Mathematical Matrix Correction Procedures for X-Ray Fluorescence Analysis. A Critical Survey. *X-Ray Spectrom.* **15**, 177–190.

Tertian, R. (1987) The Claisse–Quintin and Lachance–Claisse Alpha Correction Algorithms and Their Modifications. A Critical Examination. *X-Ray Spectrom.* **16**, 261–266.

Tertian, R . (1988) Unification of Fundamental Matrix Correction Methods in X-Ray Fluorescence Analysis. Arguments for a New Binary Coefficient Approach. *X-Ray Spectrom.* **17**, 89–98.

Tertian, R. and Broll, N. (1984) Spectral Intensity Distribution from X-Ray Tubes—Calculated Versus Experimental Evaluations. *X-Ray Spectrom.* **13**, 134–141.

Tertian, R. and Claisse, F. (1982) *Principles of Quantitative X-Ray Fluorescence Analysis*. Heyden, London.

Tertian, R. and Vié le Sage, R. (1977) Crossed Influence Coefficients for Accurate X-Ray Fluorescence Analysis of Multicomponent Systems. *X-Ray Spectrom.* **6**, 123–131.

Thompson, M., Chairman, (1987) Recommendations for the Definition, Estimation and Use of the Detection Limit. Report from the Statistical Sub-Committee of the Analytical Methods Committee, Analytical Division, Royal Society of Chemistry *Analyst* **112**, 199–204.

Thompson, M. and Howarth, R. J. (1976) Duplicate Analysis in Geochemical Practice. Part I. Theoretical Approach and Estimation of Analytical Reproducibility. *Analyst* **101**, 690–698.

Thompson, M. and Howarth, R. J. (1978) A New Approach to the Estimation of Analytical Precision. *J. Geochem. Explor.* **9**, 23–30.

Traill, R. J. and Lachance, G. R. (1965) A New Approach to X-ray Spectrochemical Analysis. *Geol. Surv. Can. Paper* 64–87, 22 pp.

Verghese, K., Mickael, M., He, T. and Gardner, R. P. (1988) A New Analysis Principle for EDXRF: The Monte Carlo-Library Least-Squares Analysis Principle. *Adv. X-Ray Anal.* **31**, 461–469.

von Hamos, L. (1945) On the Determination of Very Small Quantities of Substances by the Arkiv. X-Ray Micro-Analyser. *Ark. Mat. Astron. Fys.* 31a, 1–11.

Vrebos, B. A. R. and Pella, P. A. (1988) Uncertainties in Mass Absorption Coefficients in Fundamental Parameter X-Ray Fluorescence Analysis. *X-Ray Spectrom.* **17**, 3–12.

Ward, J. and Valdes, J. (1987) Analysis of Fused Samples Using Influence Coefficients Obtained from Synthetic Standards. *X-Ray Spectrom.* **16**, 61–66.

Weber, F., Mantler, M. and Wariwoda, L. (1993) XRFA of Carbon in Steels. *Adv. X-Ray Anal.* **36**, 41–46.

Wheeler, B. D. (1987) X-Ray Emission Spectrometric Analysis of Portland Cement by the Energy-Dispersive Technique. in *Methods for Analytical Atomic Spectroscopy*, 8th ed. E2 SM 10-34, ASTM 992–996.

White, R. L. and Huang, T. C. (1989) Determination of Ultra-Thin Carbon Coating Thickness by X-Ray Fluorescence Technique. *Adv. X-Ray Anal.* **32**, 331–339.

Willis, J. E. (1988) Simultaneous Determination of the Thickness and Composition of Thin Film Samples Using Fundamental Parameters. *Adv. X-Ray Anal.* **31**, 175–180.

Willis, J. P. (1989) Compton Scatter and Matrix Correction for Trace Element Analysis of Geological Materials, in *X-Ray Fluorescence Analysis in the Geological Sciences*, Ahmedali, S. T., ed., Geological Association of Canada Short Course 7, 221–258.

Willis, J. P. (1991) Mass Absorption Coefficient Determination Using Compton Scattered Tube Radiation: Applications, Limitations and Pitfalls. *Adv. X-Ray Anal.* **34**, 243–261.

Yager, D. B. and Quick, J. E. (1993) SUPERXAP—A Personal-Computer-Based Program for Energy-Dispersive X-Ray Spectra Analysis. *Adv. X-Ray Anal.* **36**, 17–25.

Zemany, P. D. and Liebhafsky, H. A. (1956) Plating Thickness by the Attenuation of Characteristic X-Rays. *J. Electrochem. Soc.* **103**, 157–159.

Zimmerman, J. B. and Ingles, J. C. (1972) The Determination of High Concentrations of Elements in Multi-Element Solutions by X-Ray Fluorescence Spectrometry Using a Spiking–Mathematical Correction Technique. *Can. Spectrosc.* **17**, 156–163.

AUTHOR INDEX

Afonin, Y.P., 372
Ahlgren, L., 367
Arai, T., 14, 15
Armigliato, A., 87
Austen, C.E., 381
Andermann, G., 183

Babikian, S.H., 376
Bambynek, W., 30, 296, 297
Bearden, J.A., 7, 8, 11, 282
Beattie, H.J., 64, 96, 129, 152
Beatty, R.T., 13
Belitz, R., 369
Bertin, E.P., 40, 194, 202, 236, 237
Birks, L.S., 15, 63, 64, 65, 70, 75, 76,
    93, 98, 99, 107, 116, 123, 125, 129,
    131, 152, 161, 162, 284, 285, 308,
    311, 312, 322, 336, 340
Blackburn, D.H., 367
Blanquet, P., 246
Bonetto, R., 357
Bouchard, M., 319, 324, 373, 383
Brissey, R.M., 64, 96, 129, 152
Broll, N., 16, 29, 50, 74, 81, 100, 120,
    140, 205, 265, 284, 294, 295, 296,
    297, 300, 324
Brown, D.B., 16
Budesinsky, B.W., 351, 356
Burkhalter, P.G., 285
Burr, A.F., 7, 8, 11, 282

Campbell, J.L., 205
Caul, H.J., 357
Caussin, P., 205

Chappell, B.W., 378, 385
Claisse, F., xx, 48, 57, 97, 130, 143,
    135, 139, 140, 145, 146, 153, 177,
    181, 183, 184, 216, 224, 246, 252,
    361
Cosslett, V.E., 17
Crasemann, B., 30, 296, 297
Criss, J.W., 63, 64, 65, 70, 75, 93, 99,
    107, 116, 123, 125, 131, 152, 161,
    162, 308, 311, 312, 322, 336, 340
Dalton, J.L., 20, 21, 22, 50, 285, 288,
    294
De Forge, D., 205
de Jongh, W.K., 105, 225, 327, 329,
    330
Del Grande, N.K., 23, 26, 29, 50, 290,
    291, 293, 295
Doster, J.M., 86, 87
Duane, W., 13

Eastell, J., 379, 385
Ebel, H., 87
Eick, J.D., 357

Falconer, T.H., 295
Feather, C.E., 184, 186, 189, 362
Feng, L., 16, 17, 284, 369
Fernandez, J., 357
Fink, R.W., 30, 296, 297
Finkelshtein, A.L., 372
Fiori, C.E., 295
Fleming, S.L. II, 325
Freidman, H., 129
Freund, H.U., 297

Fujino, N., 48, 56, 58, 59, 63, 76, 106

Gardner, R.P., 86, 87
Gedcke, D., 43, 87, 252
Gilfrich, J.V., 15, 16, 98, 284, 285, 312, 336, 340
Gillam, E., 59, 129, 160
Gillieson, A.H., 357
Goldak, J., 21, 50, 288, 294
Gould, R.W., 43, 87, 252
Green, M., 17
Griffiths, J.M., 372, 373, 383
Grönberg, T., 367
Gunicheva, T.N., 359, 372

Hanke, W., 281, 282, 284, 291, 292, 293, 294, 295, 296, 297
Harvey, P.K., 362, 364, 378, 385
He, T., 87
Heal, H.T., 59, 129, 160
Heinrich, K.F.J., 21, 74, 75, 137, 163, 165, 206, 289, 290, 292, 293, 294, 295, 301, 307, 309, 347, 355, 356, 357
Helsen, J.A., 87
Higgins, K.L., 369
Holmes, G.S., 368
Holynska, B., 367
Howarth, R.J., 371
Hower, J., 185
Huang, T.C., 370
Hubbell, J.H., 23, 26, 29, 50, 290, 291, 293, 295
Hunt, F.L., 13

Ingles, J.C., 353
Ito, M., 170

Jenkins, R., 43, 87, 252
Ji, A., 356
Johnson, R.G., 325

Keith, H.D., 16
Kemp, J.W., 183
Kis-Varga, M., 374
Klimasara, A.J., 351
Kovacs, P., 374
Kramers, H.A., 14, 284
Kulenkampff, H., 14
Kunz, F., 369

Lachance, G.R., 76, 97, 99, 104, 125, 137, 139, 153, 155, 225, 338, 355
Lankosz, M., 367
Leroux, J., 68, 206, 289, 290, 293
Levine, H.S., 369
Li, G.H., 380, 385
Liebhafsky, H.A., 210, 212
Loomis, T.C., 16
Lucas-Tooth, H.J., 166, 167, 168, 169
Ludden, J.N., 356, 382, 385

Ma, G.Z., 380, 385
Mainardi, R.T., 357
Mallett, J.H., 23, 26, 29, 50, 290, 291, 293, 295
Mantler, M., 215, 367, 368
Mattson, S., 367
McKee, C., 367
McMaster, W.H., 23, 26, 29, 50, 290, 291, 293, 295
Mickael, M., 87
Milliken, K.S., 357
Molchanova, E.I., 359
Moeller, T., 3
Moore, C.W., 374
Mosichev, V.I., 358
Myklebust, R.L., 295

Narita, M., 170
Nielson, K.K., 202
Nikolaev, G.I., 358
Noakes, G.E., 352, 354

Omote, K., 14, 15 $

Peckerar, M.C., 16
Pella, P.A., 16, 17, 145, 284, 294, 367, 380
Pershin, N.V., 358
Peter, M., 205
Poehn, C., 87, 281, 282, 284, 291, 292, 293, 294, 295, 296, 297
Potts, P.J., 381
Prange, A., 208
Price, B.J., 166
Pyne, C., 167, 168, 169

Quick, J.E., 365
Quintin, M., 97, 130, 134, 145, 146, 153

Rasberry, S.D., 75, 137, 163, 165, 301, 307, 309, 347, 355, 356, 357
Reed, D.J., 357
Renault, J., 367
Riveros, J.A., 357
Rosa, R., 87
Rousseau, R., 135, 139, 146
Rousseau, R.M., 102, 145, 155, 265, 319, 324, 373, 380, 383
Russell, B.G., 381

Salem, S.I., 295
Samson, C., 246
Sato, S., 170
Schreiber, T.P., 295, 296
Schroeder, B., 356, 382, 385
Schwenke, H., 208
Scofield, J.H., 295
Sherman, J., 59, 63, 64, 76, 86, 102, 160, 177
Shiraiwa, T., 48, 56, 58, 59, 63, 76, 106
Shoji, T., 14, 15
Sieber, J.R., 368
Smagunov, A.V., 359
Smagunova, A.N., 359
Small, J.A., 16, 17, 284
Smith, D.L., 357
Soares, C.G., 367
Sproull, W.T., 24
Spruch, G.M., 3
Spruch, L., 3
Steele, T.W., 381
Sulanowska, M., 356, 382, 385

Tao, G.Y., 145, 356, 380
Tertian, R., 16, 48, 57, 76, 81, 94, 95, 97, 100, 125, 126, 137, 139, 140, 144, 145, 146, 153, 155, 177, 181,
183, 184, 216, 224, 252, 265, 284, 325, 348, 349, 361
Thinh, T.P., 68, 206, 290, 293
Thompson, G., 356, 383, 385
Thompson, M., 271, 272, 371
Traill, R.J., 97, 125, 338, 355

Valdes, T., 380
Verghese, K., 87
Vié le Sage, R., 100, 153
von Hamos, L., 59, 96, 160
Vrebos, B.A.R., 87, 294

Ward, J., 380
Warden, E.S., 285
Wariwoda, L., 368
Watson, J.S., 381
Webb, P.C., 381
Weber, F., 368
Webster, H.W.M., 372, 373, 383
Wernisch, J., 281, 282, 284, 291, 292, 293, 294, 295, 296, 297
Wheeler, B.D., 376
White, R.L., 370
Whitlock R.R., 285
Willis, J.E., 215
Willis, J.P., 184, 186, 189, 208, 362, 379, 385
Wims, A.M., 295, 296
Windrell, R.W., 295

Yager, D.B., 365
Young, N.J., 357
Yuan, N., 356

Zemany, P.D., 210, 212
Zimmerman, J.B., 353

SUBJECT INDEX

Absorption
 coefficients, 18–20
 edge, 21–4, 37
 jump factor, 41, 294
 jump ratio, 27–9, 41, 294
 of X-rays, 18
 path length, 40
Accuracy, 241, 243–4, 371
Addition method, *see* Standard addition
 method
Algorithms, correction for matrix effects
 based on
 alpha influence coefficients (semi-
 empirical), 126–8
 Claisse–Quintin, 97–8, 134–8, 265,
 340–4
 COLA, 145, 265, 344–7
 Lachance–Traill, 125, 325
 numerical examples, 138–9, 141–4,
 147–51, 339–49
 Tertian, 137–8, 146, 265, 348–50
 empirical influence coefficients
 Beattie–Brissey, 96, 129
 JIS, 170, 265
 Lachance–Traill, 163, 264, 351, 355
 Lucas-Tooth and Price, 166, 357
 Lucas-Tooth and Pyne, 168, 358
 Rasberry–Heinrich, 163–6, 355–7
 fundamental influence coefficients
 Broll–Tertian, 102, 265, 323–4
 de Jongh, 105, 266, 327–30
 Lachance–Traill, 106, 325–7
 Rousseau, 102, 265, 324–5

Algorithms for the calculation of
 fundamental parameters
 absorption jump factors, 294–5
 absorption jump ratios, 294
 characteristic lines, 284
 critical excitation energies, 281–2
 fluorescence yield, 296–7
 mass attenuation coefficients, 286,
 290–3
 relative line intensities within series,
 295–6
Alpha influence coefficients, 312,
 337–50
Analytical context, 240–45
Analytical precision, 271
Analytical procedures
 brasses and bronzes, 374
 cement and raw materials, 374–8
 geological materials and metallurgical
 slags, 378–85
 steels and alloys, 372–3, 383
Analysing crystals selection, 254
Angles
 incident excitation source, 40
 emergent characteristic X-radiations
 (take-off), 40
Apparent concentration, 168, 170
 see also Relative intensity to pure
 element
Atomic structure, 4–6
 Bohr model, 4
 energy levels, 5
 quantum numbers, 6

Attenuation coefficient
  effective, 42–3, 51, 65, 176
  linear, 18
  mass, 19
  see also Absorption coefficient
Auger effect, 19, 30, 42

Background, 252
Beattie–Brissey algorithm, 96, 129
Binary influence coefficient arrays, 115,
  119, 131–7, 141–2
Binding (Critical excitation) energy, 6–8
Bragg equation, 35
Bremsstrahlung, see Continuum
Broll–Dertian algorithm, 102, 265,
  323–4

Calibration
  constant (factor), 87–9, 121, 260
  definitions of, 260–1
  empirical methods, 163
  models, 261–8
  validity of, 266–77
Characteristic X-radiations
  algorithms for calculating keV of, 284
  notations, 7, 10
  probability of excitation, 41–2, 300
  selection of, 253
Claisse–Quintin algorithm, 97–8,
  134–8, 340–4
Coefficient of variation (relative
  standard deviation), 251
Coherent (Rayleigh) scatter, 24
COLA algorithm, 145, 221, 265, 344–6
Collimator selection, 253
Compton (incoherent) scatter, 24–26
Confidence limits, 244
Continuum
  short wavelength limit, 13
  spectral distribution, 12, 14
  see also Spectral distribution
Correction for
  background, 255–6
  dead time, 258–9
  instrumental drift, 266–70
  spectral overlap, 256–8
Counting (drift) monitors, 267–70, 365
Counting statistics, 243–4, 251–2
Criss–Birks (Fundamental parameters)
  algorithm, 67, 70
Critical excitation energy, 8, 281–2

Critical excitation wavelength, 9, 282
Critical (infinite)
  thickness, 208–9, 369
  width, 208
Cross-coefficients, see Third element
  effect

Dead time, 258–9
de Jongh algorithm, 105–7, 266,
  327–30
Detection limit, 271–3
Dilution methods, 177, 195
Detector selection, 254–5
Double dilution method
  numerical example, 179–82
  practical considerations, 183, 361–2
  theoretical aspects, 177–9

Effective
  absorption effect, 78
  mass attenuation coefficients, 43, 51
  path length, 40
  wavelength range, 45, 64, 152
Electromagnetic spectrum, 3
Electronic states (levels), 5–8
Emission from X-ray tubes
  continuum, 12, 14
  characteristic lines, 15–17
  total emitted intensity, 13
  see also Spectral distribution from
    X-ray tubes
Empirical influence coefficients
  Algorithms
    Beattie–Brissey, 96, 129
    Criss–Birks, 161–3
    JIS, 170–1
    Lachance–Traill, 163
    Lucas-Tooth and Price, 166–7
    Lucas-Tooth and Pyne, 167–70
    Rasberry–Heinrich, 163–6
  concept of, 160–1
Energy dispersive spectrometer,
  configuration of, 36
Energy levels, 5–8
Enhancement
  schematic representation, 46–7, 86
  see also Secondary fluorescence
    emissions
Equivalent wavelength, 152–3
Excitation probability (excitation
  factor), 41, 299–300

Excitation sources
  X-ray tubes, 12–18
  radioactive materials, 11, 374
  selection, 252–3
External standard method, 190–2

Fluorescence emissions (definitions), 34
Fluorescence yield, 30–1, 42, 296–7
Fundamental influence coefficients,
  93–121
  algorithms, 99–107
    Broll–Tertian, 100–2, 114, 117–8
    de Jongh, 105–7, 118–9
    Lachance–Traill, 104–5, 114, 118
    Rousseau, 102–4, 114, 117–8, 319
  binary arrays, 115, 119, 131–7, 141–2
  concept, 95
  interrelation between coefficients,
    333–6
  numerical examples, 107–20
Fundamental parameters approach,
  63–75
  algorithms, 67, 70, 77–81, 83
  application, 310–1
  concept, 64–7
  experimental confirmation, 74–6, 309,
    319
  numerical examples, 68–73, 79–80,
    82–5
Fused disc specimens, 219–36
  correction for gain on fusion
    (oxidation), 227–36
  correction for loss on fusion
    (volatiles), 224–6, 231–6
  flux factor, 221, 222, 235
  influence coefficients modified for
    flux, 219–23
  numerical examples, 222–3, 225–7,
    229–31, 233–6

Gain on fusion (ignition), 227–36
Gaussian (normal) distribution, 243–4,
  251
Global approaches for correction for
  matrix effects, definition of, 175
  methods
    double dilution, 177–83
    high dilution – heavy absorber,
      195–9
    incoherent (Compton) scatter,
      183–9

internal standard, 190–2
matrix matching, 199–201
standard addition, 193–5
'ghost' peaks, 378

Heterogeneity effects, 216
High dilution – heavy absorber method
  theoretical aspects, 195–6, 365
  numerical example, 196–9

Incoherent (Compton) scatter, 25
Incoherent (Compton) scatter methods
  application, 362–4
  theoretical aspects, 183–6
  numerical example, 187–9
Infinite (critical)
  thickness, 208–9, 369
  width, 208
Influence coefficients, concept of, 94–5
Intensity measurement,
  backgrounds,
  counting statistics, 251–2
  line selection,
  valid net, 87, 271
Intensity model, *see* Empirical influence
  coefficients, Lucas–Tooth and
  Price
Interelement effects, *see* Matrix effects
Internal standard method
  theoretical aspects, 190–1
  numerical example, 191–2
Iterative process, 64, 66–7, 311, 339,
  340, 344, 346

JIS (Japanese Industrial Standard)
  algorithm, 170–1, 266
Jump factor, 41, 294

Kramers equation, 14
Kulenkampff equation, 14

Lachance–Traill algorithm
  using alpha (semi-empirical)
    coefficients, 338–40
  using fundamental influence
    coefficients, 104–5
  using empirical influence coefficients,
    264, 351–5
Least-squares technique
  application of, 168, 351, 357
  models, 199–201, 261–4

Limit of detection, 271–273
Line intensities within series, 28–9, 295–6
Line (spectral) overlap, 256–8, 272–3
Liquid specimens, 250, 353
    advantages, 236
    disadvantages, 236
Loss on fusion (ignition), 224–7, 231–4
Low concentrations, 204–5
Low Z elements
    analytes, 204–6
    matrices, 205–8
    critical (infinite) thickness, 208–9, 369
    critical (infinite) width, 208
Lucas–Tooth and Price algorithm, 166, 357
Lucas–Tooth and Pyne algorithm, 168–70, 358

Mathematical (Numerical) correction methods, 125, 175
Mass attenuation coefficients, 19
    algorithms for the calculation of, 21, 286–93
    effective, 43, 51
    experimental determination, 20, 285–8
Mass scattering coefficient, 19
Matrix effects, schematic illustration
    absorption, 85–6
    enhancement, 86
Matrix matching methods, 199–201, 262
Mineralogical effects, 247
Minimum wavelength, 13, 38
Modified scatter see Incoherent scatter
Monte Carlo method, 86–7, 387

Net intensity, 87
Normal (Gaussian) distribution, 243–4, 251

Operating conditions (instrumental parameters), 63, 372–3, 383–5
Optical notation, 8
Oxide systems, 216–9
Oxide influence coefficients, 217–9

Particle size effect, see Heterogeneity effects
Peak stripping, see Correction for spectral overlap

Pelletizing, see Specimen preparation, pressed pellets
Photoelectric absorption coefficients, 19, 26
Powder specimens,
    physical factors, 88
Precision, 241, 243, 251, 271, 371
Pressed pellets, see Specimen preparation
Primary fluorescence emissions, 37–45
    description of, 37–9
    mathematical expression for, 39–43
    experimental verification, 44–5, 301–2
    relative contribution, 58
Probability curve, 244

Quality control, 270–1, 378

Rasberry–Heinrich algorithm, 163–6, 355–7
Rayleigh scatter, see Coherent scatter
Regression analysis
    exponential model, 264
    hyperbolic model, 264
    linear models, 262–3
    second order (quadratic) model, 264
Relative intensities
    of characteristic lines within series, 28–29
    to counting monitors, 267–70, 377
    to pure element, 66
Rousseau algorithm, 102–104, 265, 324–5

Sample, definition of, 242
Scattered radiation methods, 183–9, 362–4, 378
Scattering of X-rays, 24–6
    coherent, 24
    incoherent, 25–6
Secondary fluorescence emissions, 45–54
    description of, 45–6
    mathematical expression for, 47–9
    experimental verification, 49–53, 302–4
    relative contribution, 58
Semi-empirical influence coefficients, see Alpha coefficients

Sensitivity, 260
  see also Calibration constant
Short wavelength limit, 13
  see also Minimum wavelength
Siegbahn notation, 10
Software, 273–5
Specimen, definition of, 242
Specimen preparation
  bulk solids, 245–6
  fused discs, 247–50, 370, 374, 377,
    379–82
  pressed pellets, 246–8, 376, 378,
    380–382
Specimen thickness
  linear region (thin specimens), 210
  exponential region (intermediate
    thickness specimens), 211
  constant intensity region (thick
    specimens), 211
Spectral distribution from X-ray tubes,
  14–18, 65
  algorithms for the calculation of
    Tertian–Broll, 16
    Pella *et al.*, 16, 18
  experimentally measured
    chromium target, 286
    rhodium target, 287
    tungsten target, 15, 285
Spectrometer configurations, 34–37
  energy dispersion, 36
  wavelength dispersion, 35
    sequential, 35
    simultaneous, 35
Standard addition method
  numerical example, 194–5
  theoretical aspects, 193–4

Standard deviation, 244, 251
Synthetic reference materials, 380–1

Tertian algorithm, 146–7, 348–350
Tertiary fluorescence emissions, 54–9
  description of, 54–6
  mathematical expression for, 55–7
  relati ve contribution, 58–9
Thin film methods
  thickness of, 211–15
  thickness and composition, 215–6,
    369
Third element effect, 131, 143–7, 340–3
  numerical examples, 147–51
Total reflection X-ray fluorescence
  spectrometry, 208
Transition levels, 10

Unmodified scatter, *see* Coherent scatter

Wavelength dispersive spectrometer,
  configuration of
  sequential model, 35
  simultaneous model, 35

X-ray spectra
  characteristic lines, 9–10
  continuum, 12–14
X-ray tubes
  emanations from, 12–18
  end window, 15
  side window, 15
X-rays
  interaction with matter, 18–19
  notation, 8